Springer Texts in Statistics

Series Editors:

G. Casella
S. Fienberg
I. Olkin

Springer Texts in Statistics

For other titles published in this series, go to
http://www.springer.com/series/417

V.G. Kulkarni

Introduction to Modeling and Analysis of Stochastic Systems

Second Edition

 Springer

Prof. V.G. Kulkarni
University of North Carolina
Department of Statistics and Operations Research
Chapel Hill, NC 27599-3260
USA
vkulkarn@email.unc.edu

ISBN 978-1-4419-1771-3   e-ISBN 978-1-4419-1772-0
DOI 10.1007/978-1-4419-1772-0
Springer New York Dordrecht Heidelberg London

Library of Congress Control Number: PCN applied for

Printed on acid-free paper

Springer is part of Springer Science+Business Media (www.springer.com)

To my Sons
Milind, Ashwin and Arvind

Preface to the Second Edition

What Is New in the Second Edition?

Students and instructors will notice significant changes in the second edition.

1. The chapters on probability have been removed. Abbreviated versions of these chapters appear as Appendices A, B, C, and D.
2. A new chapter (Chapter 3) on Poisson processes has been added. This is in response to the feedback that the first edition did not give this important class of stochastic processes the attention it deserves.
3. A new chapter (Chapter 7) on Brownian motion has been added. This is to enable instructors who would like the option of covering this important topic. The treatment of this topic is kept at a sufficiently elementary level so that the students do not need background in analysis or measure theory to understand the material.
4. The chapters on design and control of stochastic systems in the first edition have been deleted. Instead, I have added case studies in Chapters 2, 4, 5, and 6. The instructor can use these to talk about the design aspect of stochastic modeling. The control aspect is entirely deleted. This has necessitated a change of title for the new edition.
5. Several typos from the first edition have been corrected. If there are new typos in the second edition, it is entirely my fault. I would appreciate it if the readers would kindly inform me about them. A current list of corrections is available on the Web at www.unc.edu/~vkulkarn/ugcorrections2.pdf.

Who Is This Book For?

This book is meant to be used as a textbook in a junior-or senior-level undergraduate course on stochastic models. The students are expected to be undergraduate students in engineering, operations research, computer science, mathematics, statistics, business administration, public policy, or any other discipline with a mathematical core.

Students are expected to be familiar with elementary matrix operations (addition, multiplication, and solving systems of linear equations, but not eigenvalues or eigenvectors), first-year calculus (derivatives and integrals of simple functions, but not differential equations), and probability. The necessary material on probability is summarized in the appendices as a ready reference.

What Is the Philosophy of This Book?

As the title suggests, this book addresses three aspects of using stochastic methodology to study real systems.

1. **Modeling.** The first step is to understand how a real system operates and the purpose of studying it. This enables us to make assumptions to create a model that is simple yet sufficiently true to the real system that the answers provided by the model will have some credibility. In this book, this step is emphasized repeatedly by using a large number of real-life modeling examples.

2. **Analysis.** The second step is to do a careful analysis of the model and compute the answers. To facilitate this step, the book develops special classes of stochastic processes in Chapters 2, 3, 4, 5, and 7: discrete-time Markov chains, Poisson processes, continuous-time Markov chains, renewal processes, cumulative processes, semi-Markov processes, Brownian motion, etc. For each of these classes, we develop tools to compute the transient distributions, limiting distributions, cost evaluations, first-passage times, etc. These tools generally involve matrix computations and can be done easily in any matrix-oriented language (e.g., MATLAB). Chapter 6 applies these tools to queueing systems.

3. **Design.** In practice, a system is described by a small number of parameters, and we are interested in setting the values of these parameters so as to optimize the performance of the system. This is called "designing" a system. The performance of the system can be computed as a function of the system parameters using the tools developed here. Then the appropriate parameter values can be determined to minimize or maximize this function. This is illustrated by case studies in Chapters 2, 4, 5, and 6.

How Is This Book Intended to Be Used?

Typically, the book will be used in a one-semester course on stochastic models. The students taking this course will be expected to have a background in probability. Hence, Appendices A through D should be used to review the material. Chapters 2, 3, 4, 5, and 6 should be covered completely. Chapter 7 should be covered as time permits.

There are many running examples in this book. Hence the instructor should try to use them in that spirit. Similarly, there are many running problems in the problem section. The instructor may wish to use a running series of problems for homework.

What Is So Different about This Book?

This book requires a new mind-set: a numerical answer to a problem is as valid as an algebraic answer to a problem! Since computational power is now conveniently and cheaply available, the emphasis in this book is on using the computer to obtain numerical answers rather than restricting ourselves to analytically tractable examples.

There are several consequences of this new mind-set: the discussion of the transient analysis of stochastic processes is no longer minimized. Indeed, transient analysis is just as easy as the limiting analysis when done on a computer. Secondly, the problems at the end of each chapter are designed to be fairly easy, but may require use of computers to do numerical experimentation.

Software to Accompany the Book

A software package called MAXIM is available for use with this textbook. It is a collection of over 80 programs written in MATLAB. These programs can be used directly as function files from MATLAB. The user's manual describing these programs is in a file called readme. These programs can be accessed via a graphical user interface (GUI). The GUI is designed to run on PCs with Windows software. Since the software is an evolving organism, I have decided not to include any information about it in the book for fear that it will become outdated very soon. The software and any relevant information can be downloaded from www.unc.edu/~vkulkarn/Maxim/maximgui.zip. The user will need to have MATLAB installed on his or her machine in order to use the software.

Vidyadhar G. Kulkarni
University of North Carolina
Department of Statistics and Operations Research
Chapel Hill, NC 27599-3260
USA
email: vkulkarn@email.unc.edu

Contents

Chapter 1
Introduction

1.1 What Is a Stochastic Process?

A stochastic process is a probability model that describes the evolution of a system evolving randomly in time. If we observe the system at a set of discrete times, say at the end of every day or every hour, we get a discrete-time stochastic process. On the other hand, if we observe the system continuously at all times, we get a continuous-time stochastic process. We begin with examples of the discrete- and continuous-time stochastic processes.

1.2 Discrete-Time Stochastic Processes

Consider a system that evolves randomly in time. Suppose this system is observed at times $n = 0, 1, 2, 3, \ldots$. Let X_n be the (random) state of the system at time n. The sequence of random variables $\{X_0, X_1, X_2, \ldots\}$ is called a (discrete-time) *stochastic process* and is written as $\{X_n, n \geq 0\}$. Let S be the set of values that X_n can take for any n. Then S is called the *state space* of the stochastic process $\{X_n, n \geq 0\}$. We illustrate this with several examples below.

Example 1.1. (Examples of Discrete-Time Stochastic Processes).

(a) Let X_n be the temperature (in degrees Fahrenheit) recorded at Raleigh-Durham Airport at 12:00 noon on the nth day. The state space of the stochastic process $\{X_n, n \geq 0\}$ can be taken to be $S = (-50, 150)$. Note that this implies that the temperature never goes below $-50°$ or over $150°$F. In theory, the temperature could go all the way down to absolute zero and all the way up to infinity!

(b) Let X_n be the outcome of the nth toss of a six-sided die. The state space of $\{X_n, n \geq 0\}$ is $\{1, 2, 3, 4, 5, 6\}$.

(c) Let X_n be the inflation rate in the United States at the end of the nth week. Theoretically, the inflation rate can take any real value, positive or negative. Hence the state space of $\{X_n, n \geq 0\}$ can be taken to be $(-\infty, \infty)$.

V.G. Kulkarni, *Introduction to Modeling and Analysis of Stochastic Systems*,
Springer Texts in Statistics, DOI 10.1007/978-1-4419-1772-0_1,
© Springer Science+Business Media, LLC 2011

(d) Let X_n be the Dow Jones index at the end of the nth trading day. The state space of the stochastic process $\{X_n, n \geq 0\}$ can be taken to be $[0, \infty)$.

(e) Let X_n be the number of claims submitted to an insurance company during the nth week. The state space of $\{X_n, n \geq 0\}$ is $\{0, 1, 2, 3, \ldots\}$.

(f) Let X_n be the inventory of cars at a dealership at the end of the nth working day. The state space of the stochastic process $\{X_n, n \geq 0\}$ is $\{0, 1, 2, 3, \ldots\}$.

(g) Let X_n be the number of PCs sold at an outlet store during the nth week. The state space of the stochastic process $\{X_n, n \geq 0\}$ is the same as that in (f).

(h) Let X_n be the number of reported accidents in Chapel Hill during the nth day. The state space of the stochastic process $\{X_n, n \geq 0\}$ is the same as that in (f). ■

1.3 Continuous-Time Stochastic Processes

Now consider a randomly evolving system that is observed continuously at all times $t \geq 0$, with $X(t)$ being the state of the system at time t. The set of states that the system can be in at any given time is called the *state space* of the system and is denoted by S. The process $\{X(t), t \geq 0\}$ is called a *continuous-time stochastic process* with state space S. We illustrate this with a few examples.

Example 1.2. (Examples of Continuous-Time Stochastic Processes).

(a) Suppose a machine can be in two states, up or down. Let $X(t)$ be the state of the machine at time t. Then $\{X(t), t \geq 0\}$ is a continuous-time stochastic process with state space $\{$up, down$\}$.

(b) A personal computer (PC) can execute many processes (computer programs) simultaneously. You can see them by invoking the task manager on your PC. Let $X(t)$ be the number of processes running on such a PC at time t. Then $\{X(t), t \geq 0\}$ is a continuous-time stochastic process with state space $\{0, 1, 2, \ldots, K\}$, where K is the maximum number of jobs that can be handled by the PC at one time.

(c) Let $X(t)$ be the number of customers that enter a bookstore during time $[0, t]$. Then $\{X(t), t \geq 0\}$ is a continuous-time stochastic process with state space $\{0, 1, 2, \ldots\}$.

(d) Let $X(t)$ be the number of emails that are in your inbox at time t. Then $\{X(t), t \geq 0\}$ is a continuous-time stochastic process with state space $\{0, 1, 2, \ldots\}$.

(e) Let $X(t)$ be the temperature at a given location at time t. Then $\{X(t), t \geq 0\}$ is a continuous-time stochastic process with state space $(-150, 130)$. Note that this implies that the temperature never goes below $-150°$ or above $130°$ Fahrenheit.

(f) Let $X(t)$ be the value of a stock at time t. Then $\{X(t), t \geq 0\}$ is a continuous-time stochastic process with state space $[0, \infty)$. In practice, the stock values are discrete, being integer multiples of .01. ■

1.4   What Do We Do with a Stochastic Process?

In this book, we shall develop tools to study systems evolving randomly in time and described by a stochastic process $\{X_n, n \geq 0\}$ or $\{X(t), t \geq 0\}$. What do we mean by "study" a system? The answer to this question depends on the system that we are interested in. Hence we illustrate the idea with two examples.

Example 1.3. (Studying Stochastic Processes).

(a) Consider the stochastic process in Example 1.1(a). The "study" of this system may involve predicting the temperature on day 10; that is, predicting X_{10}. However, predicting X_{10} is itself ambiguous: it may mean predicting the expected value or the cdf of X_{10}. Another "study" may involve predicting how long the temperature will remain above 90° Fahrenheit assuming the current temperature is above 90°. This involves computing the mean or distribution of the random time T when the temperature first dips below 90°.

(b) Consider the system described by the stochastic process of Example 1.1(e). The study of this system may involve computing the mean of the total number of claims submitted to the company during the first 10 weeks; i.e., $E(X_1 + X_2 + \cdots + X_{10})$. We may be interested in knowing if there is a long-term average weekly rate of claim submissions. This involves checking if $E(X_n)/n$ goes to any limit as n becomes large.

(c) Consider the system described by the stochastic process of Example 1.1(f). Suppose it costs $20 per day to keep a car on the lot. The study of the system may involve computing the total expected cost over the first 3 days; i.e., $E(20X_1 + 20X_2 + 20X_3)$. We may also be interested in long-run average daily inventory cost, quantified by the limit of $E(20(X_1 + X_2 + \cdots + X_n)/n)$ as n becomes large. ∎

Example 1.4. (Studying Stochastic Processes).

(a) Consider the two-state model of a machine as described in Example 1.2(a). The typical quantities of interest are:

1. The probability that the machine is up at time t, which can be mathematically expressed as $P(X(t) = \text{up})$.
2. Let $W(t)$ be the total amount of time the machine is up during the interval $[0, t]$. We are interested in the expected duration of the time the machine is up, up to time t, namely $E(W(t))$.
3. The long-run fraction of the time the machine is up, defined as

$$\lim_{t \to \infty} \frac{E(W(t))}{t}.$$

4. Suppose it costs $c per unit time to repair the machine. What is the expected total repair cost up to time t? What is the long-run repair cost per unit time?

(b) Consider the computer system of Example 1.2(b). The typical quantities of interest are

1. the probability that the computer system is idle at time t;
2. the expected number of jobs in the system at time t, viz. $\mathsf{E}(X(t))$; and
3. the expected time it takes to go from idle to full. ∎

The examples above show a variety of questions that can arise in the study of a system being modeled by a stochastic process $\{X_n, n \geq 0\}$ or $\{X(t), t \geq 0\}$. We urge the reader to think up plausible questions that may arise in studying the other systems mentioned in Examples 1.1 and 1.2.

In order to answer the kinds of questions mentioned above, it is clear that we must be able to compute the joint distribution of $(X_1, X_2, \ldots, X_n)$ for any n in the discrete-time case or $(X(t_1), X(t_2), \cdots, X(t_n))$ for any $0 \leq t_1 \leq t_2 \leq \ldots \leq t_n$ in the continuous-time case. Then we can use the standard probability theory (see Appendix) to answer these questions. How do we compute this joint distribution? We need more information about the stochastic process before it can be done. If the stochastic process has a simple (yet rich enough to capture the reality) structure, then the computation is easy. Over the years, researchers have developed special classes of stochastic processes that can be used to describe a wide variety of useful systems and for which it is easy to do probabilistic computations. The two important classes are discrete-time Markov chains and continuous-time Markov chains. We begin with discrete-time Markov chains in the next chapter.

Chapter 2
Discrete-Time Markov Models

2.1 Discrete-Time Markov Chains

Consider a system that is observed at times $0, 1, 2, \ldots$. Let X_n be the state of the system at time n for $n = 0, 1, 2, \ldots$. Suppose we are currently at time $n = 10$. That is, we have observed $X_0, X_1, \ldots, X_{10}$. The question is: can we predict, in a probabilistic way, the state of the system at time 11? In general, X_{11} depends (in a possibly random fashion) on $X_0, X_1, \ldots, X_{10}$. Considerable simplification occurs if, given the complete history $X_0, X_1, \ldots, X_{10}$, the next state X_{11} depends only upon X_{10}. That is, as far as predicting X_{11} is concerned, the knowledge of $X_0, X_1, \ldots, X_9$ is redundant if X_{10} is known. If the system has this property at all times n (and not just at $n = 10$), it is said to have a *Markov property*. (This is in honor of Andrey Markov, who, in the 1900s, first studied the stochastic processes with this property.) We start with a formal definition below.

Definition 2.1. (Markov Chain). A stochastic process $\{X_n, n \geq 0\}$ on state space S is said to be a discrete-time Markov chain (DTMC) if, for all i and j in S,

$$P(X_{n+1} = j | X_n = i, X_{n-1}, \ldots, X_0) = P(X_{n+1} = j | X_n = i). \qquad (2.1)$$

A DTMC $\{X_n, n \geq 0\}$ is said to be time homogeneous if, for all $n = 0, 1, \ldots,$

$$P(X_{n+1} = j | X_n = i) = P(X_1 = j | X_0 = i). \qquad (2.2)$$

Note that (2.1) implies that the conditional probability on the left-hand side is the same no matter what values $X_0, X_1, \ldots, X_{n-1}$ take. Sometimes this property is described in words as follows: given the present state of the system (namely X_n), the future state of the DTMC (namely X_{n+1}) is independent of its past (namely $X_0, X_1, \ldots, X_{n-1}$). The quantity $P(X_{n+1}=j | X_n=i)$ is called a *one-step transition probability* of the DTMC at time n. Equation (2.2) implies that, for time-homogeneous DTMCs, the one-step transition probability depends on i and j but is the same at all times n; hence the terminology *time homogeneous*.

In this chapter we shall consider only time-homogeneous DTMCs with *finite* state space $S = \{1, 2, \ldots, N\}$. We shall always mean time-homogeneous DTMC

V.G. Kulkarni, *Introduction to Modeling and Analysis of Stochastic Systems*,
Springer Texts in Statistics, DOI 10.1007/978-1-4419-1772-0_2,
© Springer Science+Business Media, LLC 2011

when we say DTMC. For such DTMCs, we introduce a shorthand notation for the one-step transition probability:

$$p_{i,j} = \mathsf{P}(X_{n+1} = j \mid X_n = i), \quad i, j = 1, 2, \ldots, N. \tag{2.3}$$

Note the absence of n in the notation. This is because the right-hand side is independent of n for time-homogeneous DTMCs. Note that there are N^2 one-step transition probabilities $p_{i,j}$. It is convenient to arrange them in an $N \times N$ matrix form as shown below:

$$P = \begin{bmatrix} p_{1,1} & p_{1,2} & p_{1,3} & \cdots & p_{1,N} \\ p_{2,1} & p_{2,2} & p_{2,3} & \cdots & p_{2,N} \\ p_{3,1} & p_{3,2} & p_{3,3} & \cdots & p_{3,N} \\ \vdots & \vdots & \vdots & \ddots & \vdots \\ p_{N,1} & p_{N,2} & p_{N,3} & \cdots & p_{N,N} \end{bmatrix}. \tag{2.4}$$

The matrix P in the equation above is called the *one-step transition probability matrix*, or transition matrix for short, of the DTMC. Note that the rows correspond to the starting state and the columns correspond to the ending state of a transition. Thus the probability of going from state 2 to state 3 in one step is stored in row number 2 and column number 3.

The information about the transition probabilities can also be represented in a graphical fashion by constructing a *transition diagram* of the DTMC. A transition diagram is a directed graph with N nodes, one node for each state of the DTMC. There is a directed arc going from node i to node j in the graph if $p_{i,j}$ is positive; in this case, the value of $p_{i,j}$ is written next to the arc for easy reference. We can use the transition diagram as a tool to visualize the dynamics of the DTMC as follows. Imagine a particle on a given node, say i, at time n. At time $n + 1$, the particle moves to node 2 with probability $p_{i,2}$, node 3 with probability $p_{i,3}$, etc. X_n can then be thought of as the position (node index) of the particle at time n.

Example 2.1. (Transition Matrix and Transition Diagram). Suppose $\{X_n, n \geq 0\}$ is a DTMC with state space $\{1, 2, 3\}$ and transition matrix

$$P = \begin{bmatrix} .20 & .30 & .50 \\ .10 & .00 & .90 \\ .55 & .00 & .45 \end{bmatrix}. \tag{2.5}$$

If the DTMC is in state 3 at time 17, what is the probability that it will be in state 1 at time 18? The required probability is $p_{3,1}$ and is given by the element in the third row and the first column of the matrix P. Hence the answer is .55.

If the DTMC is in state 2 at time 9, what is the probability that it will be in state 3 at time 10? The required probability can be read from the element in the second row and third column of P. It is $p_{2,3} = .90$.

The transition diagram for this DTMC is shown in Figure 2.1. Note that it has no arc from node 3 to node 2, representing the fact that $p_{3,2} = 0$. ∎

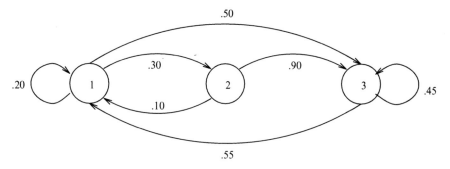

Fig. 2.1 Transition diagram of the DTMC in Example 2.1.

Next we present two main characteristics of a transition probability matrix in the following theorem.

Theorem 2.1. (*Properties of a Transition Probability Matrix*). *Let* $P = [p_{i,j}]$ *be an* $N \times N$ *transition probability matrix of a DTMC* $\{X_n, n \geq 0\}$ *with state space* $S = \{1, 2, \ldots, N\}$. *Then*

1. $p_{i,j} \geq 0,\ 1 \leq i, j \leq N$;
2. $\sum_{j=1}^{N} p_{i,j} = 1,\ 1 \leq i \leq N$.

Proof. The nonnegativity of $p_{i,j}$ follows since it is a (conditional) probability. To prove the second assertion, we have

$$\sum_{j=1}^{N} p_{i,j} = \sum_{j=1}^{N} \mathsf{P}(X_{n+1} = j \mid X_n = i)$$
$$= \mathsf{P}(X_{n+1} \in S \mid X_n = i). \tag{2.6}$$

Since X_{n+1} must take some value in the state space S, regardless of the value of X_n, it follows that the last quantity is 1. Hence the theorem follows. ∎

Any square matrix possessing the two properties of the theorem above is called a *stochastic matrix* and can be thought of as a transition probability matrix of a DTMC.

2.2 Examples of Markov Models

Discrete-time Markov chains appear as appropriate models in many fields: biological systems, inventory systems, queueing systems, computer systems, telecommunication systems, manufacturing systems, manpower systems, economic systems, and so on. The following examples will give some evidence of this diversity. In each of these examples, we derive the transition probability matrix for the appropriate DTMC.

Example 2.2. (Machine Reliability). The Depend-On-Us company manufactures a
machine that is either up or down. If it is up at the beginning of a day, then it is
up at the beginning of the next day with probability .98 (regardless of the history
of the machine), or it fails with probability .02. Once the machine goes down, the
company sends a repair person to repair it. If the machine is down at the beginning
of a day, it is down at the beginning of the next day with probability .03 (regardless
of the history of the machine), or the repair is completed and the machine is up with
probability .97. A repaired machine is as good as new. Model the evolution of the
machine as a DTMC.

Let X_n be the state of the machine at the beginning of day n, defined as follows:

$$X_n = \begin{cases} 0 \text{ if the machine is down at the beginning of day } n, \\ 1 \text{ if the machine is up at the beginning of day } n. \end{cases}$$

The description of the system shows that $\{X_n, n \geq 0\}$ is a DTMC with state space
$\{0, 1\}$ and the transition probability matrix

$$P = \begin{bmatrix} .03 & .97 \\ .02 & .98 \end{bmatrix}. \tag{2.7}$$

Now suppose the company maintains two such machines that are identical and
behave independently of each other, and each has its own repair person. Let Y_n be
the number of machines in the "up" state at the beginning of day n. Is $\{Y_n, n \geq 0\}$ a
DTMC?

First we identify the state space of $\{Y_n, n \geq 0\}$ to be $\{0, 1, 2\}$. Next we see if
the Markov property holds; that is, we check if $P(Y_{n+1} = j | Y_n = i, Y_{n-1}, \ldots, Y_0)$
depends only on i and j for $i, j = 0, 1, 2$. For example, consider the case $Y_n = i = 1$ and $Y_{n+1} = j = 0$. Thus, one machine is up (and one down) at time n. Since
both machines are identical, it does not matter which is up and which is down. In
order to move to state 0 at time $n + 1$, the down machine must stay down and the
up machine must go down at the beginning of the next day. Since the machines are
independent, the probability of this happening is .03*.02, independent of the history
of the two machines. Hence we get

$$P(Y_{n+1} = 0 | Y_n = 1, Y_{n-1}, \ldots, Y_0) = .03 * .02 = .0006 = p_{1,0}.$$

Proceeding in this fashion, we construct the following transition probability matrix:

$$P = \begin{bmatrix} .0009 & .0582 & .9409 \\ .0006 & .0488 & .9506 \\ .0004 & .0392 & .9604 \end{bmatrix}. \blacksquare \tag{2.8}$$

Example 2.3. (Weather Model). The weather in the city of Heavenly is classified as
sunny, cloudy, or rainy. Suppose that tomorrow's weather depends only on today's
weather as follows: if it is sunny today, it is cloudy tomorrow with probability .3 and

rainy with probability .2; if it is cloudy today, it is sunny tomorrow with probability .5 and rainy with probability .3; and finally, if it is rainy today, it is sunny tomorrow with probability .4 and cloudy with probability .5. Model the weather process as a DTMC.

Let X_n be the weather conditions in Heavenly on day n, defined as follows:

$$X_n = \begin{cases} 1 \text{ if it is sunny on day } n, \\ 2 \text{ if it is cloudy on day } n, \\ 3 \text{ if it is rainy on day } n. \end{cases}$$

Then we are told that $\{X_n, n \geq 0\}$ is a DTMC with state space $\{1, 2, 3\}$. We next compute its transition matrix. We are given that $p_{1,2} = .3$ and $p_{1,3} = .2$. We are not explicitly given $p_{1,1}$. We use

$$p_{1,1} + p_{1,2} + p_{1,3} = 1$$

to obtain $p_{1,1} = .5$. Similarly, we can obtain $p_{2,2}$ and $p_{3,3}$. This yields the transition probability matrix

$$P = \begin{bmatrix} .50 & .30 & .20 \\ .50 & .20 & .30 \\ .40 & .50 & .10 \end{bmatrix}. \quad \blacksquare \tag{2.9}$$

Example 2.4. (Inventory System). Computers-R-Us stocks a wide variety of PCs for retail sale. It is open for business Monday through Friday 8:00 a.m. to 5:00 p.m. It uses the following operating policy to control the inventory of PCs. At 5:00 p.m. Friday, the store clerk checks to see how many PCs are still in stock. If the number is less than two, then he orders enough PCs to bring the total in stock up to five at the beginning of the business day Monday. If the number in stock is two or more, no action is taken. The demand for the PCs during the week is a Poisson random variable with mean 3. Any demand that cannot be immediately satisfied is lost. Develop a stochastic model of the inventory of the PCs at Computers-R-Us.

Let X_n be the number of PCs in stock at Computers-R-Us at 8:00 a.m. Monday of the nth week. Let D_n be the number of PCs demanded during the nth week. Then the number of PCs left in the store at the end of the week is $\max(X_n - D_n, 0)$. If $X_n - D_n \geq 2$, then there are two or more PCs left in the store at 5:00 p.m. Friday of the nth week. Hence no more PCs will be ordered that weekend, and we will have $X_{n+1} = X_n - D_n$. On the other hand, if $X_n - D_n \leq 1$, there are 1 or 0 PCs left in the store at the end of the week. Hence enough will be ordered over the weekend so that $X_{n+1} = 5$. Putting these observations together, we get

$$X_{n+1} = \begin{cases} X_n - D_n & \text{if } X_n - D_n \geq 2, \\ 5 & \text{if } X_n - D_n \leq 1. \end{cases}$$

Table 2.1 Demand
distribution for Example 2.4.

$k \rightarrow$	0	1	2	3	4
$P(D_n = k)$	.0498	.1494	.2240	.2240	.1680
$P(D_n \geq k)$	1.000	.9502	.8008	.5768	.3528

It then follows that the state space of $\{X_n, n \geq 0\}$ is $\{2, 3, 4, 5\}$. Now assume that
the demands from week to week are independent of each other and the inventory in
the store. Then we can show that $\{X_n, n \geq 0\}$ is a DTMC. We have, for $j = 2, 3, 4$
and $i = j, j + 1, \ldots, 5$,

$$
\begin{aligned}
P(X_{n+1} &= j \mid X_n = i, X_{n-1}, \ldots, X_0) \\
&= P(X_n - D_n = j \mid X_n = i, X_{n-1}, \ldots, X_0) \\
&= P(i - D_n = j \mid X_n = i, X_{n-1}, \ldots, X_0) \\
&= P(D_n = i - j).
\end{aligned}
$$

Similarly,

$$
P(X_{n+1} = 5 \mid X_n = i) = P(X_n - D_n \leq 1 \mid X_n = i) = P(D_n \geq i - 1)
$$

for $i = 2, 3, 4$. Using the fact that D_n is a P(3) random variable, we get Table 2.1.
Using the data in this table, we can compute the transition probability matrix of the
DTMC $\{X_n, n \geq 0\}$ as follows:

$$
P = \begin{bmatrix}
.0498 & 0 & 0 & .9502 \\
.1494 & .0498 & 0 & .8008 \\
.2240 & .1494 & .0498 & .5768 \\
.2240 & .2240 & .1494 & .4026
\end{bmatrix}.
\tag{2.10}
$$

Note the entry for $p_{5,5}$. The inventory can move from 5 to 5 if the demand is
either 0 or at least 4. Hence we have

$$
p_{5,5} = P(D_n = 0) + P(D_n \geq 4). \blacksquare
$$

Example 2.5. (Manufacturing). The Gadgets-R-Us company has a manufacturing
setup consisting of two distinct machines, each producing one component per hour.
Each component can be tested instantly to be identified as defective or nondefective.
Let a_i be the probability that a component produced by machine i is nondefective,
$i = 1, 2$. (Obviously $1 - a_i$ is the probability that a component produced by machine
i is defective.) The defective components are discarded and the nondefective components produced by each machine are stored in two separate bins. When a component
is present in each bin, the two are instantly assembled together and shipped out.
Each bin can hold at most two components. When a bin is full, the corresponding
machine is turned off. It is turned on again when the bin has space for at least one
component. Model this system by a DTMC.

Let A_n be the number of components in the bin of machine 1 and let B_n be the number of components in the bin of machine 2 at the end of hour n, after accounting for the production and assembly during the hour. The bin-size restrictions imply that $0 \leq A_n, B_n \leq 2$. Note that both bins cannot be nonempty simultaneously since assembly is instantaneous. Thus, $A_n > 0$ implies that $B_n = 0$, and $B_n > 0$ implies that $A_n = 0$. Let $X_n = A_n - B_n$. Then X_n can take values in $S = \{-2, -1, 0, 1, 2\}$. Note that we can recover A_n and B_n from X_n as follows:

$$X_n \leq 0 \Rightarrow A_n = 0; B_n = -X_n,$$
$$X_n \geq 0 \Rightarrow A_n = X_n; B_n = 0.$$

If the successive components are independent of each other, then $\{X_n, n \geq 0\}$ is a DTMC. For example, suppose the history $\{X_0, \ldots, X_{n-1}\}$ is known and $X_n = -1$. That is, the bin for machine 1 is empty, while the bin for machine 2 has one component in it. During the next hour, each machine will produce one new component. If both of these components are nondefective, then they will be assembled instantaneously. If both are defective, they will be discarded. In either case, X_{n+1} will remain -1. If the newly produced component 1 is nondefective and the newly produced component 2 is defective, then the new component 2 will be discarded and the old component from bin 2 will be assembled with the newly produced nondefective component 1 and shipped out. Hence X_{n+1} will become 0. If the new component 1 is defective while new component 2 is nondefective, X_{n+1} will become -2. Bin 2 will now be full, and machine 2 will be turned off. In the next hour, only machine 1 will produce an item. From this analysis, we get

$$P(X_{n+1} = j | X_n = -1, X_{n-1}, \ldots, X_0) = \begin{cases} a_1(1 - a_2) & \text{if } j = 0, \\ a_1 a_2 + (1 - a_1)(1 - a_2) & \text{if } j = -1, \\ (1 - a_1)a_2 & \text{if } j = -2, \\ 0 & \text{otherwise.} \end{cases}$$

Proceeding in this fashion, we can show that $\{X_n, n \geq 0\}$ is a DTMC on $\{-2, -1, 0, 1, 2\}$ with the following transition probability matrix (we use $a = (1-a_1)a_2, b = a_1 a_2 + (1-a_1)(1-a_2)$, and $c = a_1(1-a_2)$ for compactness):

$$P = \begin{bmatrix} 1 - a_1 & a_1 & 0 & 0 & 0 \\ a & b & c & 0 & 0 \\ 0 & a & b & c & 0 \\ 0 & 0 & a & b & c \\ 0 & 0 & 0 & a_2 & 1 - a_2 \end{bmatrix}. \blacksquare \qquad (2.11)$$

Example 2.6. (Manpower Planning). Paper Pushers, Inc., is an insurance company that employs 100 workers organized into four grades, labeled 1, 2, 3, and 4. For the sake of simplicity, we shall assume that workers may get promoted from one grade to another, or leave the company, only at the beginning of a week. A worker in grade 1 at the beginning of a week gets promoted to grade 2 with probability .03, leaves the company with probability .02, or continues in the same grade at the beginning of the next week. A worker in grade 2 at the beginning of a week gets promoted to grade 3 with probability .01, leaves the company with probability .008, or continues in the same grade at the beginning of the next week. A worker in grade 3 at the beginning of a week gets promoted to grade 4 with probability .005, leaves the company with probability .02, or continues in the same grade at the beginning of the next week. A worker in grade 4 at the beginning of a week either leaves the company with probability .01 or continues in the same grade at the beginning of the next week. If a worker leaves the company, he is instantly replaced by a new one in grade 1. Model the worker movement in this company using a DTMC.

We shall assume that all worker promotions are decided in an independent manner. This simplifies our model considerably. Instead of keeping track of all 100 workers, we keep track of a single worker, say worker number k, where $k = 1, 2, \ldots, 100$. We think of k as a worker ID, and if and when this worker leaves the company, it gets assigned to the new replacement. Let X_n^k be the grade that the kth worker is in at the beginning of the nth week. Now, if we assume that the worker promotions are determined independently of the history of the worker so far (meaning the time spent in a given grade does not affect one's chances of promotion), we see that, for $k = 1, 2, \ldots, 100$, $\{X_n^k, n \geq 0\}$ is a DTMC with state space $\{1, 2, 3, 4\}$. We illustrate the computation of the transition probabilities with an example. Suppose $X_n^k = 3$; i.e., employee number k is in grade 3 at the beginning of week n. If he gets promoted (which happens with probability .005), we see that $X_{n+1}^k = 4$. Hence $P(X_{n+1}^k = 4 | X_n^k = 3) = .005$. If he leaves the company (which happens with probability .02), he is replaced by a new employee in grade 1 carrying the ID k, making $X_{n+1}^k = 1$. Hence $P(X_{n+1}^k = 1 | X_n^k = 3) = .02$. With the remaining probability, .975, the employee continues in the same grade in the next week, making $X_{n+1}^k = 3$. Hence $P(X_{n+1}^k = 3 | X_n^k = 3) = .975$. Proceeding in a similar way, we obtain the following transition probability matrix:

$$P = \begin{bmatrix} .9700 & .0300 & 0 & 0 \\ .0080 & .9820 & .0100 & 0 \\ .0200 & 0 & .9750 & .0050 \\ .0100 & 0 & 0 & .9900 \end{bmatrix}. \tag{2.12}$$

Note that the 100 DTMCs $\{X_n^1, n \geq 0\}$ through $\{X_n^{100}, n \geq 0\}$ are independent of each other and have the same transition probability matrix. ∎

Example 2.7. (Stock Market). The common stock of the Gadgets-R-Us company is traded in the stock market. The chief financial officer of Gadgets-R-Us buys and sells the stock in his own company so that the price never drops below \$2 and never goes above \$10. For simplicity, we assume that X_n, the stock price at the end of day n, takes only integer values; i.e., the state space of $\{X_n, n \geq 0\}$ is $\{2, 3, \ldots, 9, 10\}$. Let I_{n+1} be the potential movement in the stock price on day $n + 1$ in the absence of any intervention from the chief financial officer. Thus, we have

$$X_{n+1} = \begin{cases} 2 & \text{if } X_n + I_{n+1} \leq 2, \\ X_n + I_{n+1} & \text{if } 2 < X_n + I_{n+1} < 10, \\ 10 & \text{if } X_n + I_{n+1} \geq 10. \end{cases}$$

Statistical analysis of past data suggests that the potential movements $\{I_n, n \geq 1\}$ form an iid sequence of random variables with common pmf given by .

$$P(I_n = k) = .2, \ k = -2, -1, 0, 1, 2.$$

This implies that $\{X_n, n \geq 0\}$ is a DTMC on $\{2, 3, \ldots, 9, 10\}$. We illustrate the computation of the transition probabilities with three cases:

$$\begin{aligned} P(X_{n+1} = 2 | X_n = 3) &= P(X_n + I_{n+1} \leq 2 | X_n = 3) \\ &= P(I_{n+1} \leq -1) \\ &= .4, \end{aligned}$$

$$\begin{aligned} P(X_{n+1} = 6 | X_n = 5) &= P(X_n + I_{n+1} = 6 | X_n = 5) \\ &= P(I_{n+1} = 1) \\ &= .2, \end{aligned}$$

$$\begin{aligned} P(X_{n+1} = 10 | X_n = 10) &= P(X_n + I_{n+1} \geq 10 | X_n = 10) \\ &= P(I_{n+1} \geq 0) \\ &= .6. \end{aligned}$$

Proceeding in this fashion, we get the following transition probability matrix for the DTMC $\{X_n, n \geq 0\}$:

$$P = \begin{bmatrix} .6 & .2 & .2 & 0 & 0 & 0 & 0 & 0 & 0 \\ .4 & .2 & .2 & .2 & 0 & 0 & 0 & 0 & 0 \\ .2 & .2 & .2 & .2 & .2 & 0 & 0 & 0 & 0 \\ 0 & .2 & .2 & .2 & .2 & .2 & 0 & 0 & 0 \\ 0 & 0 & .2 & .2 & .2 & .2 & .2 & 0 & 0 \\ 0 & 0 & 0 & .2 & .2 & .2 & .2 & .2 & 0 \\ 0 & 0 & 0 & 0 & .2 & .2 & .2 & .2 & .2 \\ 0 & 0 & 0 & 0 & 0 & .2 & .2 & .2 & .4 \\ 0 & 0 & 0 & 0 & 0 & 0 & .2 & .2 & .6 \end{bmatrix}. \ \blacksquare \tag{2.13}$$

Example 2.8. (Telecommunications). The Tel-All Switch Corporation manufactures switching equipment for communication networks. Communication networks move data from switch to switch at lightning-fast speed in the form of packets; i.e., strings of zeros and ones (called bits). The Tel-All switches handle data packets of constant length; i.e., the same number of bits in each packet. At a conceptual level, we can think of the switch as a storage device where packets arrive from network users according to a random process. They are stored in a buffer with the capacity to store K packets and are removed from the buffer one-by-one according to a pre-specified protocol. Under one such protocol, time is slotted into intervals of fixed length, say a microsecond. If there is a packet in the buffer at the beginning of a slot, it is removed instantaneously. If there is no packet at the beginning of a slot, no packet is removed during the slot even if more packets arrive during the slot. If a packet arrives during a slot and there is no space for it, it is discarded. Model this as a DTMC.

Let A_n be the number of packets that arrive at the switch during the nth slot. (Some of these may be discarded.) Let X_n be the number of packets in the buffer at the end of the nth slot. Now, if $X_n = 0$, then there are no packets available for transmission at the beginning of the $(n + 1)$st slot. Hence all the packets that arrive during that slot, namely A_{n+1}, are in the buffer at the end of that slot unless $A_{n+1} > K$, in which case the buffer is full at the end of the $(n + 1)$st slot. Hence $X_{n+1} = \min\{A_{n+1}, K\}$. If $X_n > 0$, one packet is removed at the beginning of the $(n + 1)$st slot and A_{n+1} packets are added during that slot, subject to capacity limitations. Combining these cases, we get

$$X_{n+1} = \begin{cases} \min\{A_{n+1}, K\} & \text{if } X_n = 0, \\ \min\{X_n + A_{n+1} - 1, K\} & \text{if } 0 < X_n \le K. \end{cases}$$

Assume that $\{A_n, n \ge 1\}$ is a sequence of iid random variables with common pmf

$$P(A_n = k) = a_k, \ k \ge 0.$$

Under this assumption, $\{X_n, n \ge 0\}$ is a DTMC on state space $\{0, 1, 2, \ldots, K\}$. The transition probabilities can be computed as follows. For $0 \le j < K$,

$$\begin{aligned} P(X_{n+1} = j | X_n = 0) &= P(\min\{A_{n+1}, K\} = j | X_n = 0) \\ &= P(A_{n+1} = j) \\ &= a_j, \\ P(X_{n+1} = K | X_n = 0) &= P(\min\{A_{n+1}, K\} = K | X_n = 0) \\ &= P(A_{n+1} \ge K) \\ &= \sum_{k=K}^{\infty} a_k. \end{aligned}$$

Similarly, for $1 \le i \le K$ and $i - 1 \le j < K$,

$$\begin{aligned} P(X_{n+1} = j | X_n = i) &= P(\min\{X_n + A_{n+1} - 1, K\} = j | X_n = i) \\ &= P(A_{n+1} = j - i + 1) \\ &= a_{j-i+1}. \end{aligned}$$

Finally, for $1 \leq i \leq K$,

$$
\begin{aligned}
P(X_{n+1} = K | X_n = i) &= P(\min\{X_n + A_{n+1} - 1, K\} = K | X_n = i) \\
&= P(A_{n+1} \geq K - i + 1) \\
&= \sum_{k=K-i+1}^{\infty} a_k.
\end{aligned}
$$

Combining all these cases and using the notation

$$
b_j = \sum_{k=j}^{\infty} a_k,
$$

we get the transition probability matrix

$$
P = \begin{bmatrix}
a_0 & a_1 & \cdots & a_{K-1} & b_K \\
a_0 & a_1 & \cdots & a_{K-1} & b_K \\
0 & a_0 & \cdots & a_{K-2} & b_{K-1} \\
\vdots & \vdots & \ddots & \vdots & \vdots \\
0 & 0 & \cdots & a_0 & b_1
\end{bmatrix} . \blacksquare
\tag{2.14}
$$

Armed with the collection of examples above, we set out to "analyze" them in the next section.

2.3 Transient Distributions

Let $\{X_n, n \geq 0\}$ be a time-homogeneous DTMC on state space $S = \{1, 2, \ldots, N\}$ with transition probability matrix P and initial distribution $a = [a_1, \ldots, a_N]$, where

$$
a_i = P(X_0 = i), \ 1 \leq i \leq N.
$$

In this section, we concentrate on the *transient distribution*; i.e., the distribution of X_n for a fixed $n \geq 0$. In other words, we are interested in $P(X_n = j)$ for all $j \in S$ and $n \geq 0$. We have

$$
\begin{aligned}
P(X_n = j) &= \sum_{i=1}^{N} P(X_n = j | X_0 = i) P(X_0 = i) \\
&= \sum_{i=1}^{N} a_i P(X_n = j | X_0 = i).
\end{aligned}
\tag{2.15}
$$

Thus it suffices to study the conditional probability $P(X_n = j|X_0 = i)$. This quantity is called the *n-step transition probability* of the DTMC. We use the notation

$$a_j^{(n)} = P(X_n = j), \tag{2.16}$$

$$p_{i,j}^{(n)} = P(X_n = j|X_0 = i). \tag{2.17}$$

Analogous to the one-step transition probability matrix $P = [p_{i,j}]$, we build an *n-step transition probability matrix* as follows:

$$P^{(n)} = \begin{bmatrix} p_{1,1}^{(n)} & p_{1,2}^{(n)} & p_{1,3}^{(n)} & \cdots & p_{1,N}^{(n)} \\ p_{2,1}^{(n)} & p_{2,2}^{(n)} & p_{2,3}^{(n)} & \cdots & p_{2,N}^{(n)} \\ p_{3,1}^{(n)} & p_{3,2}^{(n)} & p_{3,3}^{(n)} & \cdots & p_{3,N}^{(n)} \\ \vdots & \vdots & \vdots & \ddots & \vdots \\ p_{N,1}^{(n)} & p_{N,2}^{(n)} & p_{N,3}^{(n)} & \cdots & p_{N,N}^{(n)} \end{bmatrix}. \tag{2.18}$$

We discuss two cases, $P^{(0)}$ and $P^{(1)}$, below. We have

$$p_{i,j}^{(0)} = P(X_0 = j|X_0 = i) = \begin{cases} 1 & \text{if } i = j, \\ 0 & \text{if } i \neq j. \end{cases}$$

This implies that

$$P^{(0)} = I,$$

an $N \times N$ identity matrix. From (2.17) and (2.3), we see that

$$p_{i,j}^{(1)} = P(X_1 = j|X_0 = i) = p_{i,j}. \tag{2.19}$$

Hence, from (2.19), (2.4), and (2.18), we get

$$P^{(1)} = P.$$

Now, construct the transient pmf vector

$$a^{(n)} = \left[a_1^{(n)}, a_2^{(n)}, \ldots, a_N^{(n)} \right], \tag{2.20}$$

so that (2.15) can be written in matrix form as

$$a^{(n)} = a * P^{(n)}. \tag{2.21}$$

Note that

$$a^{(0)} = a * P^{(0)} = a * I = a,$$

the initial distribution of the DTMC.

In this section, we develop methods of computing the n-step transition probability matrix $P^{(n)}$. The main result is given in the following theorem.

Theorem 2.2. (*n-Step Transition Probability Matrix*).

$$P^{(n)} = P^n, \tag{2.22}$$

where P^n is the nth power of the matrix P.

Proof. Since $P^0 = I$ and $P^1 = P$, the theorem is true for $n = 0, 1$. Hence, let $n \geq 2$. We have

$$
\begin{aligned}
p_{i,j}^{(n)} &= P(X_n = j | X_0 = i) \\
&= \sum_{k=1}^{N} P(X_n = j | X_{n-1} = k, X_0 = i) P(X_{n-1} = k | X_0 = i) \\
&= \sum_{k=1}^{N} p_{i,k}^{(n-1)} P(X_n = j | X_{n-1} = k, X_0 = i) \text{ (from (2.17))} \\
&= \sum_{k=1}^{N} p_{i,k}^{(n-1)} P(X_n = j | X_{n-1} = k) \text{ (due to the Markov property)} \\
&= \sum_{k=1}^{N} p_{i,k}^{(n-1)} P(X_1 = j | X_0 = k) \text{ (due to time homogeneity)} \\
&= \sum_{k=1}^{N} p_{i,k}^{(n-1)} p_{k,j}. \tag{2.23}
\end{aligned}
$$

The last sum can be recognized as a matrix multiplication operation, and the equation above, which is valid for all $1 \leq i, j \leq N$, can be written in a more succinct fashion in matrix terminology as

$$P^{(n)} = P^{(n-1)} * P. \tag{2.24}$$

Using the equation above for $n = 2$, we get

$$P^{(2)} = P^{(1)} * P = P * P.$$

Similarly, for $n = 3$, we get

$$P^{(3)} = P^{(2)} * P = P * P * P.$$

In general, using P^n as the nth power of the matrix P, we get (2.22). ∎

From the theorem above, we get the following two corollaries.

Corollary 2.1.
$$a^{(n)} = a * P^n.$$

Proof. This is left to the reader as Conceptual Problem 2.5. ■

Corollary 2.2.

$$P(X_{n+m} = j | X_n = i, X_{n-1}, \ldots, X_0)$$
$$= P(X_{n+m} = j | X_n = i) = p_{ij}^{(m)}.$$

Proof. Use induction on m. ■

A more general version of (2.23) is given in the following theorem.

Theorem 2.3. (Chapman–Kolmogorov Equation). *The n-step transition probabilities satisfy the following equation, called the Chapman–Kolmogorov equation:*

$$p_{i,j}^{(n+m)} = \sum_{k=1}^{N} p_{i,k}^{(n)} p_{k,j}^{(m)}. \tag{2.25}$$

Proof.

$$P(X_{n+m} = j | X_0 = i) = \sum_{k=1}^{N} P(X_{n+m} = j | X_n = k, X_0 = i)P(X_n = k | X_0 = i)$$

$$= \sum_{k=1}^{N} P(X_{n+m} = j | X_n = k)P(X_n = k | X_0 = i)$$
(due to Corollary 2.2)

$$= \sum_{k=1}^{N} P(X_m = j | X_0 = k)P(X_n = k | X_0 = i)$$
(due to time homogeneity)

$$= \sum_{k=1}^{N} P(X_n = k | X_0 = i)P(X_m = j | X_0 = k)$$

$$= \sum_{k=1}^{N} p_{i,k}^{(n)} p_{k,j}^{(m)}. \blacksquare$$

In matrix form, (2.25) can be expressed as

$$P^{(n+m)} = P^{(n)} * P^{(m)}. \tag{2.26}$$

Interchanging the roles of n and m, we get

$$P^{(n+m)} = P^{(m)} * P^{(n)}.$$

The equations above imply that the matrices $P^{(n)}$ and $P^{(m)}$ commute for all n and m. This is an unusual property for matrices. Theorem 2.2 makes it especially easy to compute the transient distributions in DTMCs since it reduces the computations to matrix powers and multiplication. Several matrix-oriented computer packages are available that make these computations easy to perform.

Example 2.9. Consider the three-state DTMC with the transition matrix as given in Example 2.1 and the initial distribution $a = [.1 \ .4 \ .5]$. Compute the pmf of X_3.

Using Corollary 2.1, we see that the pmf of X_3 can be computed as

$$a^{(3)} = a * P^3$$

$$= [.1 \ .4 \ .5] * \begin{bmatrix} .20 & .30 & .50 \\ .10 & .00 & .90 \\ .55 & .00 & .45 \end{bmatrix}^3$$

$$= [0.3580 \ 0.1258 \ 0.5162]. \quad \blacksquare$$

Example 2.10. (Weather Model). Mr. and Mrs. Happy have planned to celebrate their 25th wedding anniversary in Honeymooners' Paradise, a popular resort in the city of Heavenly. Counting today as the first day, they are supposed to be there on the seventh and eighth days. They are thinking about buying vacation insurance that promises to reimburse them for the entire vacation package cost of $2500 if it rains on both days, and nothing is reimbursed otherwise. The insurance costs $200. Suppose the weather in the city of Heavenly changes according to the model in Example 2.3. Assuming that it is sunny today in the city of Heavenly, should Mr. and Mrs. Happy buy the insurance?

Let R be the reimbursement the couple gets from the insurance company. Letting X_n be the weather on the nth day, we see that $X_1 = 1$ and

$$R = \begin{cases} 2500 & \text{if } X_7 = X_8 = 3, \\ 0 & \text{otherwise.} \end{cases}$$

Hence

$$E(R) = 2500P(X_7 = X_8 = 3 | X_1 = 1)$$

$$= 2500P(X_8 = 3 | X_7 = 3, X_1 = 1)P(X_7 = 3 | X_1 = 1)$$

$$= 2500P(X_1 = 3 | X_0 = 3)P(X_6 = 3 | X_0 = 1)$$

$$= (2500)(.10) \left[\begin{bmatrix} .50 & .30 & .20 \\ .50 & .20 & .30 \\ .40 & .50 & .10 \end{bmatrix}^6 \right]_{1,3}$$

$$= 52.52.$$

Since this is less than $200, the insurance is not worth it. $\blacksquare$

Example 2.11. (Manufacturing). Consider the manufacturing operation described in Example 2.5. Suppose that both bins are empty at time 0 at the beginning of an 8-hour shift. What is the probability that both bins are empty at the end of the 8-hour shift? (Assume $a_1 = .99$ and $a_2 = .995$.)

Let $\{X_n, n \geq 0\}$ be the DTMC described in Example 2.5. We see that we are interested in $p_{0,0}^{(8)}$. Using the data given above, we see that the transition probability matrix is given by

$$P = \begin{bmatrix} .0100 & .9900 & 0 & 0 & 0 \\ .00995 & .9851 & .00495 & 0 & 0 \\ 0 & .00995 & .9851 & .00495 & 0 \\ 0 & 0 & .00995 & .9851 & .00495 \\ 0 & 0 & 0 & .9950 & .0050 \end{bmatrix}. \qquad (2.27)$$

Computing P^8, we get

$$p_{0,0}^{(8)} = .88938.$$

What is the probability that machine 2 is shut down at the end of the shift? It is given by

$$p_{0,-2}^{(8)} = .0006533. \quad \blacksquare$$

Example 2.12. (Telecommunications). Consider the model of the Tel-All data switch as described in Example 2.8. Let Y_n be the number of packets *lost* during the nth time slot. Show how to compute $E(Y_n)$ assuming that the buffer is initially empty. Tabulate $E(Y_n)$ as a function of n if the buffer size is seven packets and the number of packets that arrive at the switch during one time slot is a Poisson random variable with mean 1, packet arrivals in successive time slots being iid.

Let X_n be the number of packets in the buffer at the end of the nth time slot. Then $\{X_n, n \geq 0\}$ is a DTMC as described in Example 2.8. Let A_n be the number of packet arrivals during the nth slot. Then

$$Y_{n+1} = \begin{cases} \max\{0, A_{n+1} - K\} & \text{if } X_n = 0, \\ \max\{0, X_n - 1 + A_{n+1} - K\} & \text{if } X_n > 0. \end{cases}$$

Hence

$$E(Y_{n+1}) = E(\max\{0, A_{n+1} - K\})p_{0,0}^{(n)}$$
$$+ \sum_{k=1}^{K} E(\max\{0, k - 1 + A_{n+1} - K\})p_{0,k}^{(n)}.$$

Using $a_r = P(A_n = r)$, we get

$$E(Y_{n+1}) = p_{0,0}^{(n)} \sum_{r=K}^{\infty} (r - K)a_r$$

$$+ \sum_{k=1}^{K} p_{0,k}^{(n)} \sum_{r=K+1-k}^{\infty} (r - K - 1 + k)a_r. \qquad (2.28)$$

We are given that A_n is $P(1)$. The pmf and complementary cdf of A_n are as given in Table 2.2.

Using the data given there, and using the analysis of Example 2.8, we see that $\{X_n, n \geq 0\}$ is a DTMC with state space $\{0, 1, \ldots, 7\}$ and transition probability matrix

$$P = \begin{bmatrix} .3679 & .3679 & .1839 & .0613 & .0153 & .0031 & .0005 & .0001 \\ .3679 & .3679 & .1839 & .0613 & .0153 & .0031 & .0005 & .0001 \\ 0 & .3679 & .3679 & .1839 & .0613 & .0153 & .0031 & .0006 \\ 0 & 0 & .3679 & .3679 & .1839 & .0613 & .0153 & .0037 \\ 0 & 0 & 0 & .3679 & .3679 & .1839 & .0613 & .0190 \\ 0 & 0 & 0 & 0 & .3679 & .3679 & .1839 & .0803 \\ 0 & 0 & 0 & 0 & 0 & .3679 & .3679 & .2642 \\ 0 & 0 & 0 & 0 & 0 & 0 & .3679 & .6321 \end{bmatrix}. \qquad (2.29)$$

The matrix P^n is computed numerically from P, and the n-step transition probabilities $p_{0,k}^{(n)}$ are obtained from it by using Theorem 2.2. These are then used in (2.28) to compute the expected number of packets lost in each slot. The final answer is given in Table 2.3.

Note that even if the switch removes one packet per time slot and one packet arrives at the switch on average per time slot, the losses are not zero. This is due to the randomness in the arrival process. Another interesting feature to note is that, as n becomes large, $E(Y_n)$ seems to tend to a limiting value. We present more on this in Section 2.5. ∎

Table 2.2 Data for Example 2.12.

$r \rightarrow$	0	1	2	3	4	5	6	7
$P(A_n = r)$	.3679	.3679	.1839	.0613	.0153	.0031	.0005	.0001
$P(A_n \geq r)$	1.000	.6321	.2642	.0803	.0190	.0037	.0006	.0001

Table 2.3 Expected packet loss in Example 2.12.

$n \rightarrow$	1	2	5	10	20	30	40	80
$E(Y_n)$	.0000	.0003	.0063	.0249	.0505	.0612	.0654	.0681

2.4 Occupancy Times

Let $\{X_n, n \geq 0\}$ be a time-homogeneous DTMC on state space $S = \{1, 2, \ldots, N\}$ with transition probability matrix P and initial distribution $a = [a_1, \ldots, a_N]$. In this section, we study occupancy times; i.e., the expected amount of time the DTMC spends in a given state during a given interval of time. Since the DTMC undergoes one transition per unit time, the occupancy time is the same as the expected number of times it visits a given state in a finite number of transitions. We define this quantity formally below.

Let $N_j(n)$ be the number of times the DTMC visits state j over the time span $\{0, 1, \ldots, n\}$, and let

$$m_{i,j}(n) = E(N_j(n)|X_0 = i).$$

The quantity $m_{i,j}(n)$ is called the *occupancy time* up to n of state j starting from state i. Let

$$M(n) = \begin{bmatrix} m_{1,1}(n) & m_{1,2}(n) & m_{1,3}(n) & \cdots & m_{1,N}(n) \\ m_{2,1}(n) & m_{2,2}(n) & m_{2,3}(n) & \cdots & m_{2,N}(n) \\ m_{3,1}(n) & m_{3,2}(n) & m_{3,3}(n) & \cdots & m_{3,N}(n) \\ \vdots & \vdots & \vdots & \ddots & \vdots \\ m_{N,1}(n) & m_{N,2}(n) & m_{N,3}(n) & \cdots & m_{N,N}(n) \end{bmatrix} \qquad (2.30)$$

be the *occupancy times matrix*. The next theorem gives a simple method of computing the occupancy times.

Theorem 2.4. (Occupancy Times). *Let $\{X_n, n \geq 0\}$ be a time-homogeneous DTMC on state space $S = \{1, 2, \ldots, N\}$, with transition probability matrix P. The occupancy times matrix is given by*

$$M(n) = \sum_{r=0}^{n} P^r. \qquad (2.31)$$

Proof. Fix i and j. Let $Z_n = 1$ if $X_n = j$ and $Z_n = 0$ if $X_n \neq j$. Then,

$$N_j(n) = Z_0 + Z_1 + \cdots + Z_n.$$

Hence

$$m_{i,j}(n) = E(N_j(n)|X_0 = i)$$
$$= E(Z_0 + Z_1 + \cdots + Z_n|X_0 = i)$$
$$= \sum_{r=0}^{n} E(Z_r|X_0 = i)$$

$$= \sum_{r=0}^{n} P(Z_r = 1 | X_0 = i)$$

$$= \sum_{r=0}^{n} P(X_r = j | X_0 = i)$$

$$= \sum_{r=0}^{n} p_{i,j}^{(r)}. \tag{2.32}$$

We get (2.31) by writing the equation above in matrix form. ∎

Example 2.13. (Three-State DTMC). Consider the DTMC in Example 2.1. Compute the occupancy time matrix $M(10)$.

We have, from Theorem 2.4,

$$M(10) = \sum_{r=0}^{10} \begin{bmatrix} .20 & .30 & .50 \\ .10 & .00 & .90 \\ .55 & .00 & .45 \end{bmatrix}^r$$

$$= \begin{bmatrix} 4.5317 & 1.2484 & 5.2198 \\ 3.5553 & 1.9555 & 5.4892 \\ 3.8583 & 1.0464 & 6.0953 \end{bmatrix}.$$

Thus the expected number of visits to state 1 starting from state 1 over the first ten transitions is 4.5317. Note that this includes the visit at time 0. ∎

Example 2.14. (Weather Model). Consider the three-state weather model described in Example 2.3. Suppose it is sunny in Heavenly today. Compute the expected number of rainy days in the week starting today.

Let X_n be the weather in Heavenly on the nth day. Then, from Example 2.3, $\{X_n, n \geq 0\}$ is a DTMC with transition matrix P given in (2.9). The required quantity is given by $m_{1,3}(6)$ (why 6 and not 7?). The occupancy matrix $M(6)$ is given by

$$M(6) = \sum_{r=0}^{6} \begin{bmatrix} .5 & .3 & .2 \\ .5 & .2 & .3 \\ .4 & .5 & .1 \end{bmatrix}^r$$

$$= \begin{bmatrix} 3.8960 & 1.8538 & 1.2503 \\ 2.8876 & 2.7781 & 1.3343 \\ 2.8036 & 2.0218 & 2.1747 \end{bmatrix}.$$

Hence $m_{1,3}(6) = 1.25$. ∎

2.5 Limiting Behavior

Let $\{X_n, n \geq 0\}$ be a DTMC on state space $S = \{1, 2, \ldots, N\}$ with transition probability matrix P. In this section, we study the limiting behavior of X_n as n tends to infinity. We start with the most obvious question:

Does the pmf of X_n approach a limit as n tends to infinity?

If it does, we call it the *limiting or steady-state distribution* and denote it by

$$\pi = [\pi_1, \pi_2, \ldots, \pi_N], \tag{2.33}$$

where

$$\pi_j = \lim_{n \to \infty} P(X_n = j), \ j \in S. \tag{2.34}$$

The next question is a natural follow-up:

If the limiting distribution exists, is it unique?

This question makes sense since it is conceivable that the limit may depend upon the starting state, or the initial distribution of the DTMC. Finally, the question of practical importance is:

If there is a unique limiting distribution, how do we compute it?

It so happens that the answers to the first two questions are complex but the answer to the last question is easy. Hence we give that first in the following theorem.

Theorem 2.5. (Limiting Distributions). *If a limiting distribution π exists, it satisfies*

$$\pi_j = \sum_{i=1}^{N} \pi_i \, p_{i,j}, \ j \in S, \tag{2.35}$$

and

$$\sum_{j=1}^{N} \pi_j = 1. \tag{2.36}$$

Proof. Conditioning on X_n and using the Law of Total Probability, we get

$$P(X_{n+1} = j) = \sum_{i=1}^{N} P(X_n = i) p_{i,j}, \ j \in S. \tag{2.37}$$

Now, let n tend to infinity on both the right- and left-hand sides. Then, assuming that the limiting distribution exists, we see that

$$\lim_{n \to \infty} P(X_n = j) = \lim_{n \to \infty} P(X_{n+1} = j) = \pi_j.$$

Substituting in (2.37), we get (2.35). Equation (2.36) follows since π is a pmf. ∎

Equations (2.35) can be written in matrix form as

$$\pi = \pi P \tag{2.38}$$

and are called *the balance equations* or *the steady-state equations*. Equation (2.36) is called the *normalizing equation*. We illustrate Theorem 2.5 with an example.

Example 2.15. (A DTMC with a Unique Limiting Distribution). Suppose $\{X_n, n \geq 0\}$ is a DTMC with state space $\{1, 2, 3\}$ and the following transition matrix (see Example 2.1):

$$P = \begin{bmatrix} .20 & .30 & .50 \\ .10 & .00 & .90 \\ .55 & .00 & .45 \end{bmatrix}.$$

We give below the n-step transition probability matrix for various values of n:

$$n = 2: P^2 = \begin{bmatrix} .3450 & .0600 & .5950 \\ .5150 & .0300 & .4550 \\ .3575 & .1650 & .4775 \end{bmatrix},$$

$$n = 4: P^4 = \begin{bmatrix} .3626 & .1207 & .5167 \\ .3558 & .1069 & .5373 \\ .3790 & .1052 & .5158 \end{bmatrix},$$

$$n = 10: P^{10} = \begin{bmatrix} .3704 & .1111 & .5185 \\ .3703 & .1111 & .5186 \\ .3704 & .1111 & .5185 \end{bmatrix},$$

$$n \geq 11: P^n = \begin{bmatrix} .3704 & .1111 & .5185 \\ .3704 & .1111 & .5185 \\ .3704 & .1111 & .5185 \end{bmatrix},$$

From this we see that the pmf of X_n approaches

$$\pi = [.3704, .1111, .5185].$$

It can be checked that π satisfies (2.35) and (2.36). Furthermore, all the rows of P^n are the same in the limit, implying that the limiting distribution of X_n is the same regardless of the initial distribution. ∎

Example 2.16. (A DTMC with No Limiting Distribution). Consider a DTMC $\{X_n, n \geq 0\}$ with state space $\{1, 2, 3\}$ and transition matrix

$$P = \begin{bmatrix} 0 & 1 & 0 \\ .10 & 0 & .90 \\ 0 & 1 & 0 \end{bmatrix}. \tag{2.39}$$

We can check numerically that

$$P^{2n} = \begin{bmatrix} .1000 & 0 & .9000 \\ 0 & 1.0000 & 0 \\ .1000 & 0 & .9000 \end{bmatrix}, \ n \geq 1,$$

$$P^{2n-1} = \begin{bmatrix} 0 & 1.0000 & 0 \\ .1000 & 0 & .9000 \\ 0 & 1.0000 & 0 \end{bmatrix}, \ n \geq 1.$$

Let a be the initial distribution of the DTMC. We see that the pmf of $X_n, n \geq 1$, is $[.1(a_1 + a_3), a_2, .9(a_1 + a_3)]$ if n is even and $[.1a_2, a_1 + a_3, .9a_2]$ if n is odd. Thus the pmf of X_n does not approach a limit. It fluctuates between two pmfs that depend on the initial distribution. The DTMC has no limiting distribution. ■

Although there is no limiting distribution for the DTMC of the example above, we can still solve the balance equations and the normalizing equation. The solution is unique and is given by $[.05 \ .50 \ .45]$. To see what this solution means, suppose the initial distribution of the DTMC is $[.05 \ .50 \ .45]$. Then, using (2.37) for $n = 0$, we can show that the pmf of X_1 is also given by $[.05 \ .50 \ .45]$. Proceeding this way, we see that the pmf of X_n is $[.05 \ .50 \ .45]$ for all $n \geq 0$. Thus the pmf of X_n remains the same for all n if the initial distribution is chosen to be $[.05 \ .50 \ .45]$. We call this initial distribution a *stationary distribution*. Formally, we have the following definition.

Definition 2.2. (Stationary Distribution). A distribution

$$\pi^* = [\pi_1^*, \pi_2^*, \ldots, \pi_N^*] \tag{2.40}$$

is called a stationary distribution if

$$P(X_0 = i) = \pi_i^* \text{ for all } 1 \leq i \leq N \Rightarrow$$
$$P(X_n = i) = \pi_i^* \text{ for all } 1 \leq i \leq N, \text{and } n \geq 0.$$

The questions about the limiting distribution (namely existence, uniqueness, and method of computation) can be asked about the stationary distribution as well. We have a slightly stronger result for the stationary distribution, as given in the following theorem.

Theorem 2.6. (Stationary Distributions). $\pi^* = [\pi_1^*, \pi_2^*, \ldots, \pi_N^*]$ *is a stationary distribution if and only if it satisfies*

$$\pi_j^* = \sum_{i=1}^{N} \pi_i^* p_{i,j}, \ j \in S, \tag{2.41}$$

and

$$\sum_{j=1}^{N} \pi_j^* = 1. \tag{2.42}$$

Proof. First suppose π^* is a stationary distribution. This implies that if

$$P(X_0 = j) = \pi_j^*, \quad j \in S,$$

then

$$P(X_1 = j) = \pi_j^*, \quad j \in S.$$

But, using $n = 0$ in (2.37), we have

$$P(X_1 = j) = \sum_{i=1}^{N} P(X_0 = i) p_{i,j}.$$

Substituting the first two equations into the last, we get (2.41). Equation (2.42) holds because π^* is a pmf.

Now suppose π^* satisfies (2.41) and (2.42). Suppose

$$P(X_0 = j) = \pi_j^*, \quad j \in S.$$

Then, from (2.37), we have

$$P(X_1 = j) = \sum_{i=1}^{N} P(X_0 = i) p_{i,j}$$

$$= \sum_{i=1}^{N} \pi_i^* p_{i,j}$$

$$= \pi_j^* \text{ due to (2.41)}.$$

Thus the pmf of X_1 is π^*. Using (2.37) repeatedly, we can show that the pmf of X_n is π^* for all $n \geq 0$. Hence π^* is a stationary distribution. ∎

Theorem 2.6 implies that, if there is a solution to (2.41) and (2.42), it is a stationary distribution. Also, note that the stationary distribution π^* satisfies the same balance equations and normalizing equation as the limiting distribution π. This yields the following corollary.

Corollary 2.3. *A limiting distribution, when it exists, is also a stationary distribution.*

Proof. Let π be a limiting distribution. Then, from Theorem 2.5, it satisfies (2.35) and (2.36). But these are the same as (2.41) and (2.42). Hence, from Theorem 2.6, $\pi^* = \pi$ is a stationary distribution. ∎

Example 2.17. Using Theorem 2.6 or Corollary 2.3, we see that $\pi^* = [.3704, .1111, .5185]$ is a (unique) limiting, as well as stationary distribution to the DTMC of Example 2.15. Similarly, $\pi^* = [.05, .50, .45]$ is a stationary distribution for the DTMC in Example 2.16, but there is no limiting distribution for this DTMC. ∎

The next example shows that the limiting or stationary distributions need not be unique.

Example 2.18. (A DTMC with Multiple Limiting and Stationary Distributions). Consider a DTMC $\{X_n, n \geq 0\}$ with state space $\{1, 2, 3\}$ and transition matrix

$$P = \begin{bmatrix} .20 & .80 & 0 \\ .10 & .90 & 0 \\ 0 & 0 & 1 \end{bmatrix}. \tag{2.43}$$

Computing the matrix powers P^n for increasing values of n, we get

$$\lim_{n \to \infty} P^n = \begin{bmatrix} .1111 & .8889 & 0 \\ .1111 & .8889 & 0 \\ 0 & 0 & 1.0000 \end{bmatrix}.$$

This implies that the limiting distribution exists. Now let the initial distribution be $a = [a_1, a_2, a_3]$, and define

$$\pi_1 = .1111(a_1 + a_2),$$
$$\pi_2 = .8889(a_1 + a_2),$$
$$\pi_3 = a_3.$$

We see that π is a limiting distribution of $\{X_n, n \geq 0\}$. Thus the limiting distribution exists but is not unique. It depends on the initial distribution. From Corollary 2.3, it follows that any of the limiting distributions is also a stationary distribution of this DTMC. ∎

It should be clear by now that we need to find a normalized solution (i.e., a solution satisfying the normalizing equation) to the balance equations in order to study the limiting behavior of the DTMC. There is another important interpretation of the normalized solution to the balance equations, as discussed below.

Let $N_j(n)$ be the number of times the DTMC visits state j over the time span $\{0, 1, \ldots, n\}$. We studied the expected value of this quantity in Section 2.4. The *occupancy* of state j is defined as

$$\hat{\pi}_j = \lim_{n \to \infty} \frac{E(N_j(n)|X_0 = i)}{n + 1}. \tag{2.44}$$

Thus occupancy of state j is the same as the long-run fraction of the time the DTMC spends in state j. The next theorem shows that the *occupancy distribution*

$$\hat{\pi} = [\hat{\pi}_1, \hat{\pi}_2, \ldots, \hat{\pi}_N],$$

if it exists, satisfies the same balance and normalizing equations.

Theorem 2.7. (Occupancy Distribution). *If the occupancy distribution $\hat{\pi}$ exists, it satisfies*

$$\hat{\pi}_j = \sum_{i=1}^{N} \hat{\pi}_i \, p_{i,j}, \quad j \in S, \tag{2.45}$$

and

$$\sum_{j=1}^{N} \hat{\pi}_j = 1. \tag{2.46}$$

Proof. From Theorem 2.4, we have

$$\mathsf{E}(N_j(n)|X_0 = i) = m_{i,j}(n) = \sum_{r=0}^{n} p_{i,j}^{(r)}.$$

Hence,

$$\frac{\mathsf{E}(N_j(n)|X_0 = i)}{n+1} = \frac{1}{n+1} \sum_{r=0}^{n} p_{i,j}^{(r)}$$

$$= \frac{1}{n+1} \left(p_{i,j}^{(0)} + \sum_{r=1}^{n} p_{i,j}^{(r)} \right)$$

$$= \frac{1}{n+1} \left(p_{i,j}^{(0)} + \sum_{r=1}^{n} \sum_{k=1}^{N} p_{i,k}^{(r-1)} p_{k,j} \right)$$

(using Chapman–Kolmogorov equations)

$$= \frac{1}{n+1} \left(p_{i,j}^{(0)} + \sum_{k=1}^{N} \sum_{r=1}^{n} p_{i,k}^{(r-1)} p_{k,j} \right)$$

$$= \frac{1}{n+1} (p_{i,j}^{(0)}) + \frac{n}{n+1} \sum_{k=1}^{N} \frac{1}{n} \left(\sum_{r=0}^{n-1} p_{i,k}^{(r)} p_{k,j} \right).$$

Now, let n tend to ∞,

$$\lim_{n \to \infty} \frac{\mathsf{E}(N_j(n)|X_0 = i)}{n+1} = \lim_{n \to \infty} \frac{1}{n+1} (p_{i,j}^{(0)})$$

$$+ \lim_{n \to \infty} \left(\frac{n}{n+1} \right) \sum_{k=1}^{N} \lim_{n \to \infty} \frac{1}{n} \left(\sum_{r=0}^{n-1} p_{i,k}^{(r)} \right) p_{k,j}.$$

Assuming the limits exist and using (2.44) and (2.32), we get

$$\hat{\pi}_j = \sum_{k=1}^{N} \hat{\pi}_k \, p_{k,j},$$

which is (2.45). The normalization equation (2.46) follows because

$$\sum_{j=1}^{N} N_j(n) = n + 1. \quad \blacksquare$$

Thus the normalized solution of the balance equations can have as many as three interpretations: limiting distribution, stationary distribution, or occupancy distribution. The questions are: Will there always be a solution? Will it be unique? When can this solution be interpreted as a limiting distribution, stationary distribution, or occupancy distribution? Although these questions can be answered strictly in terms of solutions of linear systems of equations, it is more useful to develop the answers in terms of the DTMC framework. That is what we do below.

Definition 2.3. (Irreducible DTMC). A DTMC $\{X_n, n \geq 0\}$ on state space $S = \{1, 2, \ldots, N\}$ is said to be irreducible if, for every i and j in S, there is a $k > 0$ such that

$$P(X_k = j \,|\, X_0 = i) > 0. \tag{2.47}$$

A DTMC that is not irreducible is called reducible.

Note that the condition in (2.47) holds if and only if it is possible to go from any state i to any state j in the DTMC in one or more steps or, alternatively, there is a directed path from any node i to any node j in the transition diagram of the DTMC. It is in general easy to check if the DTMC is irreducible.

Example 2.19. (Irreducible DTMCs).

(a) The DTMC of Example 2.15 is irreducible since the DTMC can visit any state from any other state in two or fewer steps.

(b) The DTMC of Example 2.3 is irreducible since it can go from any state to any other state in one step.

(c) The five-state DTMC of Example 2.5 is irreducible since it can go from any state to any other state in four steps or less.

(d) The nine-state DTMC of Example 2.7 is irreducible since it can go from any state to any other state in seven steps or less.

(e) The $(K + 1)$-state DTMC of Example 2.8 is irreducible since it can go from any state to any other state in K steps or less. $\blacksquare$

Example 2.20. (Reducible DTMCs). The DTMC of Example 2.18 is reducible since this DTMC cannot visit state 3 from state 1 or 2. $\blacksquare$

The usefulness of the concept of irreducibility arises from the following two theorems, whose proofs are beyond the scope of this book.

Theorem 2.8. (Unique Stationary Distribution). *A finite-state irreducible DTMC has a unique stationary distribution; i.e., there is a unique normalized solution to the balance equation.*

Theorem 2.9. (Unique Occupancy Distribution). *A finite-state irreducible DTMC has a unique occupancy distribution and is equal to the stationary distribution.*

Next we introduce the concept of periodicity. This will help us decide when the limiting distribution exists.

Definition 2.4. (Periodicity). Let $\{X_n, n \geq 0\}$ be an irreducible DTMC on state space $S = \{1, 2, \ldots, N\}$, and let d be the largest integer such that

$$P(X_n = i | X_0 = i) > 0 \Rightarrow n \text{ is an integer multiple of } d \qquad (2.48)$$

for all $i \in S$. The DTMC is said to be periodic with period d if $d > 1$ and aperiodic if $d = 1$.

A DTMC with period d can return to its starting state only at times d, $2d, 3d, \ldots$. It is an interesting fact of irreducible DTMCs that it is sufficient to find the largest d satisfying (2.48) for any one state $i \in S$. All other states are guaranteed to produce the same d. This makes it easy to establish the periodicity of an irreducible DTMC. In particular, if $p_{i,i} > 0$ for any $i \in S$ for an irreducible DTMC, then d must be 1 and the DTMC must be aperiodic!

Periodicity is also easy to spot from the transition diagrams. First, define a directed cycle in the transition diagram as a directed path from any node to itself. If all the directed cycles in the transition diagram of the DTMC are multiples of some integer d and this is the largest such integer, then this is the d of the definition above.

Example 2.21. (Aperiodic DTMCs). All the irreducible DTMCs mentioned in Example 2.19 are aperiodic since each of them has at least one state i with $p_{i,i} > 0$. ∎

Example 2.22. (Periodic DTMCs).

(a) Consider a DTMC on state space $\{1, 2\}$ with the transition matrix

$$P = \begin{bmatrix} 0 & 1 \\ 1 & 0 \end{bmatrix}. \qquad (2.49)$$

This DTMC is periodic with period 2.

(b) Consider a DTMC on state space $\{1, 2, 3\}$ with the transition matrix

$$P = \begin{bmatrix} 0 & 1 & 0 \\ 0 & 0 & 1 \\ 1 & 0 & 0 \end{bmatrix}. \qquad (2.50)$$

This DTMC is periodic with period 3.

(c) Consider a DTMC on state space $\{1, 2, 3\}$ with the transition matrix

$$P = \begin{bmatrix} 0 & 1 & 0 \\ .5 & 0 & .5 \\ 0 & 1 & 0 \end{bmatrix}. \tag{2.51}$$

This DTMC is periodic with period 2. ∎

The usefulness of the concept of irreducibility arises from the following main theorem, whose proof is beyond the scope of this book.

Theorem 2.10. (Unique Limiting Distribution). *A finite-state irreducible aperiodic DTMC has a unique limiting distribution.*

Theorem 2.10, along with Theorem 2.5, shows that the (unique) limiting distribution of an irreducible aperiodic DTMC is given by the solution to (2.35) and (2.36). From Corollary 2.3, this is also the stationary distribution of the DTMC, and from Theorem 2.9 this is also the occupancy distribution of the DTMC.

We shall restrict ourselves to irreducible and aperiodic DTMCs in our study of limiting behavior. The limiting behavior of periodic and/or reducible DTMCs is more involved. For example, the pmf of X_n eventually cycles with period d if $\{X_n, n \geq 0\}$ is an irreducible periodic DTMC with period d. The stationary/limiting distribution of a reducible DTMC is not unique and depends on the initial state of the DTMC. We refer the reader to an advanced text for a more complete discussion of these cases.

We end this section with several examples.

Example 2.23. (Three-State DTMC). Consider the DTMC of Example 2.15. This DTMC is irreducible and aperiodic. Hence the limiting distribution, the stationary distribution, and the occupancy distribution all exist and are given by the unique solution to

$$[\pi_1 \ \pi_2 \ \pi_3] = [\pi_1 \ \pi_2 \ \pi_3] * \begin{bmatrix} .20 & .30 & .50 \\ .10 & .00 & .90 \\ .55 & .00 & .45 \end{bmatrix}$$

and the normalizing equation

$$\pi_1 + \pi_2 + \pi_3 = 1.$$

Note that although we have four equations in three unknowns, one of the balance equations is redundant. Solving the equations above simultaneously yields

$$\pi_1 = .3704, \pi_2 = .1111, \pi_3 = .5185.$$

This matches our answer in Example 2.15. Thus we have

$$\pi = \pi^* = \hat{\pi} = [.3704, .1111, .5185].$$

Now consider the three-state DTMC from Example 2.16. This DTMC is irreducible but periodic. Hence there is no limiting distribution. However, the stationary distribution exists and is given by the solution to

$$[\pi_1^* \; \pi_2^* \; \pi_3^*] = [\pi_1^* \; \pi_2^* \; \pi_3^*] * \begin{bmatrix} 0 & 1 & 0 \\ .10 & 0 & .90 \\ 0 & 1 & 0 \end{bmatrix}$$

and

$$\pi_1^* + \pi_2^* + \pi_3^* = 1.$$

The solution is

$$[\pi_1^* \; \pi_2^* \; \pi_3^*] = [.0500, \; .5000, \; .4500].$$

This matches with the numerical analysis presented in Example 2.16. Since the DTMC is irreducible, the occupancy distribution is also given by $\hat{\pi} = \pi^*$. Thus the DTMC spends 45% of the time in state 3 in the long run. ∎

Example 2.24. (Telecommunications). Consider the DTMC model of the Tel-All data switch described in Example 2.12.

(a) Compute the long-run fraction of the time that the buffer is full.
   Let X_n be the number of packets in the buffer at the beginning of the nth time slot. Then $\{X_n, n \geq 0\}$ is a DTMC on state space $\{0, 1, \ldots, 7\}$ with transition probability matrix P given in (2.29). We want to compute the long-run fraction of the time the buffer is full; i.e., the occupancy of state 7. Since this is an irreducible aperiodic DTMC, the occupancy distribution exists and is given by the solution to

$$\hat{\pi} = \hat{\pi} * P$$

and

$$\sum_{i=0}^{7} \hat{\pi}_i = 1.$$

The solution is given by

$$\hat{\pi} = [.0681, .1171, .1331, .1361, .1364, .1364, .1364, .1364].$$

The occupancy of state 7 is .1364. Hence the buffer is full 13.64% of the time.
(b) Compute the expected number of packets waiting in the buffer in steady state.
   Note that the DTMC has a limiting distribution and is given by $\pi = \hat{\pi}$. Hence the expected number of packets in the buffer in steady state is given by

$$\lim_{n \to \infty} E(X_n) = \sum_{i=0}^{7} i\pi_i = 3.7924.$$

Thus the buffer is a little more than half full on average in steady state. ∎

Example 2.25. (Manufacturing). Consider the manufacturing operation of the Gadgets-R-Us company as described in Example 2.11. Compute the long-run fraction of the time that both the machines are operating.

Let $\{X_n, n \geq 0\}$ be the DTMC described in Example 2.5 with state space $\{-2, -1, 0, 1, 2\}$ and the transition probability matrix given in (2.27). We are interested in computing the long-run fraction of the time that the DTMC spends in states $-1, 0, 1$. This is an irreducible and aperiodic DTMC. Hence the occupancy distribution exists and is given by the solution to

$$\hat{\pi} = \hat{\pi} * \begin{bmatrix} .0100 & .9900 & 0 & 0 & 0 \\ .00995 & .9851 & .00495 & 0 & 0 \\ 0 & .00995 & .9851 & .00495 & 0 \\ 0 & 0 & .00995 & .9851 & .00495 \\ 0 & 0 & 0 & .9950 & .0050 \end{bmatrix}$$

and

$$\sum_{i=-2}^{2} \hat{\pi}_i = 1.$$

Solving, we get

$$\hat{\pi} = [.0057, .5694, .2833, .1409, .0007].$$

Hence the long-run fraction of the time that both machines are working is given by

$$\hat{\pi}_{-1} + \hat{\pi}_0 + \hat{\pi}_1 = 0.9936. \quad \blacksquare$$

2.6 Cost Models

Recall the inventory model of Example 1.3(c), where we were interested in computing the total cost of carrying the inventory over 10 weeks. In this section, we develop methods of computing such costs. We start with a simple cost model first.

Let X_n be the state of a system at time n. Assume that $\{X_n, n \geq 0\}$ is a DTMC on state space $\{1, 2, \ldots, N\}$ with transition probability matrix P. Suppose the system incurs a random cost of $C(i)$ dollars every time it visits state i. Let $c(i) = \mathsf{E}(C(i))$ be the expected cost incurred at every visit to state i. Although we think of $c(i)$ as a cost per visit, it need not be so. It may be any other quantity, like reward per visit, loss per visit, profit per visit, etc. We shall consider two cost-performance measures in the two subsections below.

2.6.1 Expected Total Cost over a Finite Horizon

In this subsection, we shall develop methods of computing *expected total cost* (ETC) up to a given finite time n, called the horizon. The actual cost incurred at time r is $C(X_r)$. Hence the actual total cost up to time n is given by

$$\sum_{r=0}^{n} C(X_r),$$

and the ETC is given by

$$E\left(\sum_{r=0}^{n} C(X_r)\right).$$

For $1 \leq i \leq N$, define

$$g(i, n) = E\left(\sum_{r=0}^{n} C(X_r) \middle| X_0 = i\right) \tag{2.52}$$

as the ETC up to time n starting from state i. Next, let

$$c = \begin{bmatrix} c(1) \\ c(2) \\ \vdots \\ c(N) \end{bmatrix}$$

and

$$g(n) = \begin{bmatrix} g(1, n) \\ g(2, n) \\ \vdots \\ g(N, n) \end{bmatrix}.$$

Let $M(n)$ be the occupancy time matrix of the DTMC as defined in (2.30). The next theorem gives a method of computing $g(n)$ in terms of $M(n)$.

Theorem 2.11. (ETC: Finite Horizon).

$$g(n) = M(n) * c. \tag{2.53}$$

Proof. We have

$$g(i, n) = E\left(\sum_{r=0}^{n} C(X_r) \middle| X_0 = i\right)$$

$$= \sum_{r=0}^{n} \sum_{j=1}^{N} E(C(X_r)|X_r = j)P(X_r = j|X_0 = i)$$

$$= \sum_{r=0}^{n} \sum_{j=1}^{N} c(j) p_{i,j}^{(r)}$$

$$= \sum_{j=1}^{N} \left[\sum_{r=0}^{n} p_{i,j}^{(r)} \right] c(j)$$

$$= \sum_{j=1}^{N} m_{i,j}(n) c(j), \tag{2.54}$$

where the last equation follows from (2.32). This yields (2.53) in matrix form. ∎

We illustrate the theorem with several examples.

Example 2.26. (Manufacturing). Consider the manufacturing model of Example 2.11. Assume that both bins are empty at the beginning of a shift. Compute the expected total number of assembled units produced during an 8-hour shift.

Let $\{X_n, n \geq 0\}$ be the DTMC described in Example 2.5. The transition probability matrix is given by (see (2.27))

$$P = \begin{bmatrix} .0100 & .9900 & 0 & 0 & 0 \\ .00995 & .9851 & .00495 & 0 & 0 \\ 0 & .00995 & .9851 & .00495 & 0 \\ 0 & 0 & .00995 & .9851 & .00495 \\ 0 & 0 & 0 & .9950 & .0050 \end{bmatrix}. \tag{2.55}$$

Recall that a_i is the probability that a component produced by machine i is nondefective, $i = 1, 2$. Let $c(i)$ be the expected number of assembled units produced in 1 hour if the DTMC is in state i at the beginning of the hour. (Note that $c(i)$ as defined here is not a cost but can be treated as such!) Thus, if $i = 0$, both the bins are empty and a unit is assembled in the next hour if both machines produce nondefective components. Hence the expected number of assembled units produced per visit to state 0 is $a_1 a_2 = .99 * .995 = .98505$. A similar analysis for other states yields

$$c(-2) = .99, c(-1) = .99, c(1) = .995, c(2) = .995.$$

We want to compute $g(0, 7)$. (Note that the production during the eighth hour is counted as production at time 7.) Using Theorem 2.4 and (2.53), we get

$$g(7) = \begin{bmatrix} 7.9195 \\ 7.9194 \\ 7.8830 \\ 7.9573 \\ 7.9580 \end{bmatrix}.$$

Hence the expected production during an 8-hour shift starting with both bins empty is 7.8830 units. If there were no defectives, the production would be 8 units. Thus the loss due to defective production is .1170 units on this shift! ∎

Example 2.27. (Inventory Systems). Consider the DTMC model of the inventory system as described in Example 2.4. Suppose the store buys the PCs for $1500 and sells them for $1750. The weekly storage cost is $50 per PC that is in the store at the beginning of the week. Compute the net revenue the store expects to get over the next 10 weeks, assuming that it begins with five PCs in stock at the beginning of the week.

Following Example 2.4, let X_n be the number of PCs in the store at the beginning of the nth week. $\{X_n, n \geq 0\}$ is a DTMC on state space $\{2, 3, 4, 5\}$ with the transition probability matrix given in (2.10). We are given $X_0 = 5$. If there are i PCs at the beginning of the nth week, the expected storage cost during that week is $50i$. Let D_n be the demand during the nth week. Then the expected number of PCs sold during the nth week is $E(\min(i, D_n))$. Hence the expected net revenue is given as

$$c(i) = -50i + (1750 - 1500)E(\min(i, D_n)), \quad 2 \leq i \leq 5.$$

Computing the expectations above, we get

$$c = \begin{bmatrix} 337.7662 \\ 431.9686 \\ 470.1607 \\ 466.3449 \end{bmatrix}.$$

Note that the expected total net revenue over the next n weeks, starting in state i, is given by $g(i, n-1)$. Hence we need to compute $g(5, 9)$. Using Theorem 2.4 and (2.53), we get

$$g(9) = \begin{bmatrix} 4298.65 \\ 4381.17 \\ 4409.41 \\ 4404.37 \end{bmatrix}.$$

Hence the expected total net revenue over the next 10 weeks, starting with five PCs, is $4404.37. Note that the figure is higher if the initial inventory is 4! This is the result of storage costs. ∎

2.6.2 Long-Run Expected Cost Per Unit Time

The ETC $g(i, n)$ computed in the previous subsection tends to ∞ as n tends to ∞ in many examples. In such cases, it makes more sense to compute the expected long-run cost rate, defined as

$$g(i) = \lim_{n \to \infty} \frac{g(i, n)}{n + 1}.$$

The following theorem shows that this long-run cost rate is independent of i when the DTMC is irreducible and gives an easy method of computing it.

Theorem 2.12. (Long-Run Cost Rate). *Suppose $\{X_n, n \geq 0\}$ is an irreducible DTMC with occupancy distribution $\hat{\pi}$. Then*

$$g = g(i) = \sum_{j=1}^{N} \hat{\pi}_j c(j). \tag{2.56}$$

Proof. From Theorem 2.9, we have

$$\lim_{n \to \infty} \frac{m_{i,j}(n)}{n+1} = \hat{\pi}_j.$$

Using this and Theorem 2.11, we get

$$g(i) = \lim_{n \to \infty} \frac{g(i,n)}{n+1}$$

$$= \lim_{n \to \infty} \frac{1}{n+1} \sum_{j=1}^{N} m_{i,j}(n) c(j)$$

$$= \sum_{j=1}^{N} \left[\lim_{n \to \infty} \frac{m_{i,j}(n)}{n+1} \right] c(j)$$

$$= \sum_{j=1}^{N} \hat{\pi}_j c(j).$$

This yields the desired result. ■

The theorem is intuitive: in the long run, among all the visits to all the states, $\hat{\pi}_j$ is the fraction of the visits made by the DTMC to state j. The DTMC incurs a cost of $c(j)$ dollars for every visit to state j. Hence the expected cost per visit in the long run must be $\sum c(j)\hat{\pi}_j$. We can use Theorem 2.7 to compute the occupancy distribution $\hat{\pi}$. We illustrate this with two examples below.

Example 2.28. (Manpower Planning). Consider the manpower planning model of Paper Pushers Insurance Company, Inc., as described in Example 2.6. Suppose the company has 70 employees and this level does not change with time. Suppose the per person weekly payroll expenses are $400 for grade 1, $600 for grade 2, $800 for grade 3, and $1000 for grade 4. Compute the long-run weekly payroll expenses for the company.

We shall compute the long-run weekly payroll expenses for each employee slot and multiply that figure by 70 to get the final answer since all employees behave identically. The grade of an employee evolves according to a DTMC with state space $\{1, 2, 3, 4\}$ and transition probability matrix as given in (2.12). Since this is an irreducible DTMC, the unique occupancy distribution is obtained using (2.45) and (2.46) as

$$\hat{\pi} = [.2715, .4546, .1826, .0913].$$

The cost vector is

$$c = [400\ 600\ 800\ 1000]'.$$

Hence the long-run weekly payroll expense for a single employee is

$$\sum_{j=1}^{4} \hat{\pi}_j c(j) = 618.7185.$$

For the 70 employees, we get as the total weekly payroll expense $\$70 * 618.7185 = \$43,310.29$. ∎

Example 2.29. (Telecommunications). Consider the model of the data switch as described in Examples 2.8 and 2.12. Compute the long-run packet-loss rate if the parameters of the problem are as in Example 2.12.

Let $c(i)$ be the expected number of packets lost during the $(n + 1)$st slot if there were i packets in the buffer at the end of the nth slot. Following the analysis of Example 2.12, we get

$$c(i) = \sum_{r=K}^{\infty} (r - K)a_r, \text{ if } i = 0$$

$$= \sum_{r=K+1-i}^{\infty} (r - K - 1 + i)a_r, \text{ if } 0 < i \leq K,$$

where a_r is the probability that a Poisson random variable with parameter 1 takes a value r. Evaluating these sums, we get

$$c = [.0000, .0000, .0001, .0007, .0043, .0233, .1036, .3679].$$

The occupancy distribution of this DTMC has already been computed in Example 2.24. It is given by

$$\hat{\pi} = [.0681, .1171, .1331, .1361, .1364, .1364, .1364, .1364].$$

Hence the long-run rate of packet loss per slot is

$$\sum_{j=0}^{7} \hat{\pi}_j c(j) = .0682.$$

Since the arrival rate of packets is one packet per slot, this implies that the loss fraction is 6.82%. This is too high in practical applications. This loss can be reduced by either increasing the buffer size or reducing the input packet rate. Note that the expected number of packets lost during the nth slot, as computed in Example 2.12, was .0681 for $n = 80$. This agrees quite well with the long-run loss rate computed in this example. ∎

2.7 First-Passage Times

We saw in Example 1.3(a) that one of the questions of interest in weather prediction was "How long will the current heat wave last?" If the heat wave is declared to be over when the temperature falls below 90°F, the problem can be formulated as "When will the stochastic process representing the temperature visit a state below 90°F?" Questions of this sort lead us to study the *first-passage time*; i.e., the random time at which a stochastic process "first passes into" a given subset of the state space. In this section, we study the first-passage times in DTMCs.

Let $\{X_n, n \geq 0\}$ be a DTMC on state space $S = \{1, 2, \ldots, N\}$ with transition probability matrix P. We shall first study a simple case, first-passage time into state N, defined as

$$T = \min\{n \geq 0 : X_n = N\}. \tag{2.57}$$

Note that T is *not* the minimum number of steps in which the DTMC can reach state N. It is the (random) number of steps that it takes to actually visit state N. Typically T can take values in $\{0, 1, 2, 3, \ldots\}$. We shall study the expected value of this random variable in detail below.

Let

$$m_i = \mathsf{E}(T | X_0 = i). \tag{2.58}$$

Clearly, $m_N = 0$. The next theorem gives a method of computing $m_i, 1 \leq i \leq N - 1$.

Theorem 2.13. (Expected First-Passage Times). $\{m_i, 1 \leq i \leq N - 1\}$ *satisfy*

$$m_i = 1 + \sum_{j=1}^{N-1} p_{i,j} m_j, \quad 1 \leq i \leq N - 1. \tag{2.59}$$

Proof. We condition on X_1. Suppose $X_0 = i$ and $X_1 = j$. If $j = N$, then $T = 1$, and if $j \neq N$, then the DTMC has already spent one time unit to go to state j and the expected time from then on to reach state N is now given by m_j. Hence we get

$$\mathsf{E}(T | X_0 = i, X_1 = j) = \begin{cases} 1 & \text{if } j = N, \\ 1 + m_j & \text{if } j \neq N. \end{cases}$$

Unconditioning with respect to X_1 yields

$$m_i = E(T|X_0 = i)$$

$$= \sum_{j=1}^{N} E(T|X_0 = i, X_1 = j)P(X_1 = j|X_0 = i)$$

$$= \sum_{j=1}^{N-1} (1 + m_j)p_{i,j} + (1)(p_{i,N})$$

$$= \sum_{j=1}^{N} (1)p_{i,j} + \sum_{j=1}^{N-1} p_{i,j}m_j$$

$$= 1 + \sum_{j=1}^{N-1} p_{i,j}m_j$$

as desired. ∎

The following examples illustrate the theorem above.

Example 2.30. (Machine Reliability). Consider the machine shop with two independent machines as described by the three-state DTMC $\{Y_n, n \geq 0\}$ in Example 2.2. Suppose both machines are up at time 0. Compute the expected time until both machines are down for the first time.

Let Y_n be the number of machines in the "up" state at the beginning of day n. From Example 2.2, we see that $\{Y_n, n \geq 0\}$ is a DTMC with state space $\{0, 1, 2\}$ and transition probability matrix given by

$$P = \begin{bmatrix} .0009 & .0582 & .9409 \\ .0006 & .0488 & .9506 \\ .0004 & .0392 & .9604 \end{bmatrix}. \tag{2.60}$$

Let T be the first-passage time into state 0 (both machines down). We are interested in $m_2 = E(T|Y_0 = 2)$. Equations (2.59) become

$$m_2 = 1 + .9604m_2 + .0392m_1,$$
$$m_1 = 1 + .9506m_2 + .0488m_1.$$

Solving simultaneously, we get

$$m_1 = 2451 \text{ days}, m_2 = 2451.5 \text{ days}.$$

Thus the expected time until both machines are down is $2451.5/365 = 6.71$ years! ∎

Example 2.31. (Manpower Planning). Consider the manpower model of Example 2.6. Compute the expected amount of time a new recruit spends with the company.

Note that the new recruit starts in grade 1. Let X_n be the grade of the new recruit at the beginning of the nth week. If the new recruit has left the company by the beginning of the nth week, we set $X_n = 0$. Then, using the data in Example 2.6, we see that $\{X_n, n \geq 0\}$ is a DTMC on state space $\{0, 1, 2, 3, 4\}$ with the following transition probability matrix:

$$P = \begin{bmatrix} 1 & 0 & 0 & 0 & 0 \\ .020 & .950 & .030 & 0 & 0 \\ .008 & 0 & .982 & .010 & 0 \\ .020 & 0 & 0 & .975 & .005 \\ .010 & 0 & 0 & 0 & .990 \end{bmatrix}. \tag{2.61}$$

Note that state 0 is absorbing since once the new recruit leaves the company the problem is finished. Let T be the first-passage time into state 0. We are interested in $m_1 = E(T|X_0 = 1)$. Equations (2.59) can be written as follows:

$$m_1 = 1 + .950m_1 + .030m_2,$$
$$m_2 = 1 + .982m_2 + .010m_3,$$
$$m_3 = 1 + .975m_3 + .005m_4,$$
$$m_4 = 1 + .990m_4.$$

Solving simultaneously, we get

$$m_1 = 73.33, m_2 = 88.89, m_3 = 60, m_4 = 100.$$

Thus the new recruit stays with the company for 73.33 weeks, or about 1.4 years.
∎

So far we have dealt with a first-passage time into a single state. What if we are interested in a first-passage time into a set of states? We consider such a case next.

Let A be a subset of states in the state space, and define

$$T = \min\{n \geq 0 : X_n \in A\}. \tag{2.62}$$

Theorem 2.13 can be easily extended to the case of the first-passage time defined above. Let $m_i(A)$ be the expected time to reach the set A starting from state i. Clearly, $m_i(A) = 0$ if $i \in A$. Following the same argument as in the proof of Theorem 2.13, we can show that

$$m_i(A) = 1 + \sum_{j \notin A} p_{i,j} m_j(A), \ i \notin A. \tag{2.63}$$

In matrix form, the equations above can be written as

$$m(A) = e + P(A)m(A), \tag{2.64}$$

where $m(A)$ is a column vector $[m_i(A)]_{i \notin A}$, e is a column vector of ones, and $P(A) = [p_{i,j}]_{i,j \notin A}$ is a submatrix of P. A matrix language package can be used to solve this equation easily. We illustrate this with an example.

Example 2.32. (Stock Market). Consider the model of stock movement as described in Example 2.7. Suppose Mr. Jones buys the stock when it is trading for \$5 and decides to sell it as soon as it trades at or above \$8. What is the expected amount of time that Mr. Jones will end up holding the stock?

Let X_n be the value of the stock in dollars at the end of the nth day. From Example 2.7, $\{X_n, n \geq 0\}$ is a DTMC on state space $\{2, 3, \ldots, 9, 10\}$. We are given that $X_0 = 5$. Mr. Jones will sell the stock as soon as X_n is 8 or 9 or 10. Thus we are interested in the first-passage time T into the set $A = \{8, 9, 10\}$, in particular in $m_5(A)$. Equations (2.64) are

$$
\begin{bmatrix} m_2(A) \\ m_3(A) \\ m_4(A) \\ m_5(A) \\ m_6(A) \\ m_7(A) \end{bmatrix} = \begin{bmatrix} 1 \\ 1 \\ 1 \\ 1 \\ 1 \\ 1 \end{bmatrix} + \begin{bmatrix} .6 & .2 & .2 & 0 & 0 & 0 \\ .4 & .2 & .2 & .2 & 0 & 0 \\ .2 & .2 & .2 & .2 & .2 & 0 \\ 0 & .2 & .2 & .2 & .2 & .2 \\ 0 & 0 & .2 & .2 & .2 & .2 \\ 0 & 0 & 0 & .2 & .2 & .2 \end{bmatrix} \begin{bmatrix} m_2(A) \\ m_3(A) \\ m_4(A) \\ m_5(A) \\ m_6(A) \\ m_7(A) \end{bmatrix}.
$$

Solving the equation above, we get

$$
\begin{bmatrix} m_2(A) \\ m_3(A) \\ m_4(A) \\ m_5(A) \\ m_6(A) \\ m_7(A) \end{bmatrix} = \begin{bmatrix} 24.7070 \\ 23.3516 \\ 21.0623 \\ 17.9304 \\ 13.2601 \\ 9.0476 \end{bmatrix}.
$$

Thus the expected time for the stock to reach \$8 or more, starting from \$5, is about 18 days.

∎

Example 2.33. (Gambler's Ruin). Two gamblers, A and B, bet on successive independent tosses of a coin that lands heads up with probability p. If the coin turns up heads, gambler A wins a dollar from gambler B, and if the coin turns up tails, gambler B wins a dollar from gambler A. Thus the total number of dollars among the two gamblers stays fixed, say N. The game stops as soon as either gambler is ruined; i.e., is left with no money! Compute the expected duration of the game, assuming that the game stops as soon as one of the two gamblers is ruined. Assume the initial fortune of gambler A is i.

Let X_n be the amount of money gambler A has after the nth toss. If $X_n = 0$, then gambler A is ruined and the game stops. If $X_n = N$, then gambler B is ruined and the game stops. Otherwise the game continues. We have

$$X_{n+1} = \begin{cases} X_n & \text{if } X_n \text{ is } 0 \text{ or } N, \\ X_n + 1 & \text{if } 0 < X_n < N \text{ and the coin turns up heads,} \\ X_n - 1 & \text{if } 0 < X_n < N \text{ and the coin turns up tails.} \end{cases}$$

Since the successive coin tosses are independent, we see that $\{X_n, n \geq 0\}$ is a DTMC on state space $\{0, 1, \ldots, N-1, N\}$ with the following transition probability matrix (with $q = 1 - p$):

$$P = \begin{bmatrix} 1 & 0 & 0 & \cdots & 0 & 0 & 0 \\ q & 0 & p & \cdots & 0 & 0 & 0 \\ 0 & q & 0 & \cdots & 0 & 0 & 0 \\ \vdots & \vdots & \vdots & \ddots & \vdots & \vdots & \vdots \\ 0 & 0 & 0 & \cdots & 0 & p & 0 \\ 0 & 0 & 0 & \cdots & q & 0 & p \\ 0 & 0 & 0 & \cdots & 0 & 0 & 1 \end{bmatrix}. \tag{2.65}$$

The game ends when the DTMC visits state 0 or N. Thus we are interested in $m_i(A)$, where $A = \{0, N\}$. Equations (2.63) are

$$m_0(A) = 0,$$
$$m_i(A) = 1 + q m_{i-1}(A) + p m_{i+1}(A), \quad 1 \leq i \leq N - 1,$$
$$m_N(A) = 0.$$

We leave it to the reader to verify that the solution, whose derivation is rather tedious, is given by

$$m_i(A) = \begin{cases} \dfrac{i}{q-p} - \dfrac{N}{q-p} \cdot \dfrac{1 - (q/p)^i}{1 - (q/p)^N} & \text{if } q \neq p, \\ (N-i)(i) & \text{if } q = p. \end{cases} \tag{2.66}$$

∎

2.8 Case Study: Passport Credit Card Company

This case study, is inspired by a paper by P. E. Pfeifer and R. L. Carraway (2000).

Passport is a consumer credit card company that has a large number of customers (or accounts). These customers charge some of their purchases on their Passport cards. The charges made in one month are due by the end of the next month. If a customer fails to make the minimum payment in a given month, the company flags the account as delinquent. The company keeps track of the payment history of each customer so that it can identify customers who are likely to default on their obligations and not pay their debt to the company.

Here we describe the simplest method by which passport tracks its accounts. A customer is said to be in state (or delinquency stage) k if he or she has missed making the minimum payment for the last k consecutive months. A customer in state k has four possible futures: make a minimum payment (or more) and move to stage 0, make no payment (or less than the minimum payment) and move to stage $k + 1$, default by declaring bankruptcy, thus moving to stage D, or the company can cancel the customer's card and terminate the account, in which case the customer moves to stage C. Currently the company has a simple policy: it terminates an account as soon as it misses seven minimum payments in a row and writes off the remaining outstanding balance on that account as a loss.

To make the discussion above more precise, let p_k be the probability that a customer in state k fails to make the minimum payment in the current period and thus moves to state $k + 1$. Let q_k be the probability that a customer in state k declares bankruptcy in the current period and thus moves to state D. Also, let b_k be the average outstanding balance of a customer in state k.

From its experience with its customers over the years, the company has estimated the parameters above for $0 \le k \le 6$ as given in Table 2.4. Note that the company has no data for $k > 6$ since it terminates an account as soon as it misses the seventh payment in a row.

We will build stochastic models using DTMCs to help the management of Passport analyze the performance of this policy in a rational way. First we assume that the state of an account changes in a Markov fashion. Also, when a customer account is terminated or the customer declares bankruptcy, we shall simply replace that account with an active one, so that the number of accounts does not change. This is the same modeling trick we used in the manpower planning model of Example 2.6.

Let X_n be the state of a particular customer account at time n (i.e., during the nth month). When the customer goes bankrupt or the account is closed, we start a new account in state 0. Thus $\{X_n, n \ge 0\}$ is a stochastic process on state space $S = \{0, 1, 2, 3, 4, 5, 6\}$. We assume that it is a DTMC. In this case, the dynamics of $\{X_n, n \ge 0\}$ are given by

$$
X_{n+1} = \begin{cases} X_n + 1 & \text{if the customer misses the minimum payment in the } n\text{th} \\ & \text{month} \\ 0 & \text{if the customer makes the minimum payment in the } n\text{th} \\ & \text{month, declares bankruptcy, or the account is terminated.} \end{cases}
$$

Table 2.4 Data for Passport account holders.

k	0	1	2	3	4	5	6
p_k	.033	.048	.090	.165	.212	.287	.329
q_k	.030	.021	.037	.052	.075	.135	.182
b_k	1243.78	2090.33	2615.16	3073.13	3502.99	3905.77	4280.26

With the interpretation above, we see that $\{X_n, n \geq 0\}$ is a DTMC on state space $\{0, 1, 2, 3, 4, 5, 6\}$ with the following transition probabilities:

$$p_{k,k+1} = P(X_{n+1} = k + 1 | X_n = k) = p_k, \quad 0 \leq k \leq 5,$$
$$p_{k,0} = 1 - p_k, \quad 0 \leq k \leq 6.$$

Using the data from Table 2.4, we get the following transition probability matrix:

$$P = \begin{bmatrix} .967 & .033 & 0 & 0 & 0 & 0 & 0 \\ .952 & 0 & .048 & 0 & 0 & 0 & 0 \\ .910 & 0 & 0 & .090 & 0 & 0 & 0 \\ .835 & 0 & 0 & 0 & .165 & 0 & 0 \\ .788 & 0 & 0 & 0 & 0 & .212 & 0 \\ .713 & 0 & 0 & 0 & 0 & 0 & .287 \\ 1 & 0 & 0 & 0 & 0 & 0 & 0 \end{bmatrix}. \tag{2.67}$$

We are now ready to analyze the current policy (P_c), which terminates an account as soon as it misses the seventh minimum payment in a row. To analyze the performance of the policy, we need a performance measure. Although one can devise many performance measures, here we concentrate on the expected annual loss due to bankruptcies and account closures. Let l_k be the expected loss from an account in state k in one month. Now, for $0 \leq k \leq 6$, a customer in state k declares bankruptcy with probability q_k, and that leads to a loss of b_k, the outstanding balance in that account. Additionally, in state 6, a customer fails to make the minimum payment with probability p_6, in which case the account is terminated with probability 1 and creates a loss of b_6. Thus

$$l_k = \begin{cases} q_k b_k & \text{if } 0 \leq k \leq 5, \\ (p_6 + q_6) b_6 & \text{if } k = 6. \end{cases}$$

Let $l = [l_0, l_1, \cdots, l_6]$. Using the data in Table 2.4, we get

$$l = [37.31 \quad 43.90 \quad 96.76 \quad 159.80 \quad 262.72 \quad 527.28 \quad 779.01].$$

Now note that the transition probability matrix P of (2.67) is irreducible and aperiodic. From Theorem 2.10 we see that the (unique) limiting distribution of such a DTMC exists and is given as the solution to (2.35) and (2.36). Solving these, we get

$$\pi = [0.9664 \quad 0.0319 \quad 0.0015 \quad 0.0001 \quad 0.0000 \quad 0.0000 \quad 0.0000].$$

Hence the long-run net loss per account is given by

$$L_a = \sum_{k=0}^{6} \pi_k l_k = 37.6417$$

dollars per month. Of course, the company must generate far more than this in revenue on average from each account to stay in business.

Now Passport has been approached by a debt collection agency, We Mean Business, or WMB for short. If a customer declares bankruptcy, Passport loses the entire outstanding balance as before. However, if a customer does not declare bankruptcy, the company can decide to terminate the account and turn it over to the WMB company. If Passport decides to do this, WMB pays Passport 75% of the current outstanding balance on that account. When an account is turned over to WMB, it collects the outstanding balance on the account from the account holder by (barely) legal means. Passport also has to pay WMB an annual retainer fee of $50,000 for this service. Passport management wants to decide if they should hire WMB and, if they do, when they should turn over an account to them.

Passport management has several possible policy options if it decides to retain WMB's services. We study six such policy options, denoted by P_m ($2 \leq m \leq 7$). Under P_m, Passport turns over an account to WMB as soon as it misses m minimum payments in a row. Clearly, under P_m, $\{X_n, n \geq 0\}$ is a DTMC with state space $\{0, 1, \cdots, m-1\}$. The expected cost l_k in state k is the same as under the policy P_c for $0 \leq k \leq m-2$. In state $m-1$, we have

$$l_{m-1} = .25p_{m-1}b_{m-1} + q_{m-1}b_{m-1}.$$

We give the main steps in the performance evaluation of P_2. In this case, $\{X_n, n \geq 0\}$ is a DTMC on state space $\{0, 1\}$ with transition probability matrix

$$P = \begin{bmatrix} .967 & .033 \\ 1 & 0 \end{bmatrix} \tag{2.68}$$

and limiting distribution

$$\pi = [0.96805 \quad 0.03195].$$

The expected loss vector is

$$l = [37.3134 \quad 94.06485].$$

Hence the long-run net loss per account is given by

$$L_a = \sum_{k=0}^{1} \pi_k l_k = 38.3250$$

dollars per month.

Table 2.5 Annual losses per account for different policies.

Policy	Annual Loss \$/Year
P_2	459.9005
P_3	452.4540
P_4	451.7845
P_5	451.6870
P_6	451.6809
P_7	451.6828
P_c	451.7003

Similar analyses can be done for the other policies. Table 2.5 gives the annual loss rate for all of them. For comparison, we have also included the annual loss rate of the current policy P_c.

It is clear that among the policies above it is best to follow the policy P_6; that is, turn over the account to WMB as soon as the account holder fails to make six minimum payments in a row. This policy saves Passport

$$451.7003 - 451.6809 = .0194$$

dollars per year per account over the current Passport policy P_c. Since Passport also has to pay the annual fee of \$50,000/year, the services of WMB are worth it if the number of Passport accounts is at least

$$50,000/.0194 = 2,577,319.$$

We are told that Passport has 14 million accounts, which is much larger than the number above. So our analysis suggests that the management of Passport should hire WMB and follow policy P_6. This will save Passport

$$.0194 * 14,000,000 - 50,000 = 221,600$$

dollars per year.

At this point, we should see what assumptions we have made that may not be accurate. First of all, what we have presented here is an enormously simplified version of the actual problem faced by a credit card company. We have assumed that all accounts are stochastically similar and independent. Both these assumptions are patently untrue. In practice, the company will classify the accounts into different classes so that accounts within a class are similar. The independence assumption might be invalidated if a large fraction of the account holders work in a particular sector of the economy (such as real estate), and if that sector suffers, then the bankruptcy rates can be affected by the health of that sector. The Markov nature of account evolution is another assumption that may or may not hold. This can be validated by further statistical analysis. Another important assumption is that the data in Table 2.5 remain unaffected by the termination policy that Passport follows. This is probably not true, and there is no easy way to verify it short of implementing a new policy and studying how customer behavior changes in response. One needs to be aware of all such pitfalls before trusting the results of such an exercise.

We end this section with the Matlab function that we used to do the computations to produce Table 2.5.

```
*************************
function rpa = consumercreditcase(p,q,b,m,r)
%consumer credit case study
%p(k) = probability that the customer in state k makes a minimum payment
%q(k) = probability that the customer in state k declares bankruptcy
%b(k) = average outstanding balance owed by the customer in state k
%m = the account is terminated as soon as it misses m payments in a row
%r = WMB buys an account in state k for r*b(k) 0 ≤ r ≤ 1
%Set r = 0 if WMB is not being used
%Output: rpa = annual expected loss from a single account
l = zeros(1,m); %l(k) = loss in state k for k = 1:m
l(k) = q(k)*b(k);
end;
l(m) = l(m) + p(m)*(1−r)*b(m);
P = zeros(m,m);
for k = 1:m−1
P(k,k+1) = p(k);
P(k,1) = 1−p(k);
end
P(m,1) = 1;
P100 = P^100;
piv = P100(1,:);
rpa = 12*piv*l(1:m)'  %annual loss per account
```

2.9 Problems

CONCEPTUAL PROBLEMS

2.1. Let $\{X_n, n \geq 0\}$ be a time-homogeneous DTMC on state space $S = \{1, 2, \ldots, N\}$ with transition probability matrix P. Then, for $i_0, i_1, \ldots, i_{k-1}$, $i_k \in S$, show that

$$P(X_1 = i_1, \ldots, X_{k-1} = i_{k-1}, X_k = i_k | X_0 = i_0) = p_{i_0,i_1} \cdots p_{i_{k-1},i_k}.$$

2.2. Let $\{X_n, n \geq 0\}$ be a time-homogeneous DTMC on state space $S = \{1, 2, \ldots, N\}$ with transition probability matrix P. Prove or disprove by counterexample

$$P(X_1 = i, X_2 = j, X_3 = k) = P(X_2 = i, X_3 = j, X_4 = k).$$

2.3. Consider the machine reliability model of Example 2.2. Now suppose that there are three independent and identically behaving machines in the shop. If a machine is up at the beginning of a day, it stays up at the beginning of the next day with probability p, and if it is down at the beginning of a day, it stays down at the beginning of the next day with probability q, where $0 < p, q < 1$ are fixed numbers. Let X_n be the number of working machines at the beginning of the nth day. Show that $\{X_n, n \geq 0\}$ is a DTMC, and display its transition probability matrix.

2.4. Let P be an $N \times N$ stochastic matrix. Using the probabilistic interpretation, show that P^n is also a stochastic matrix.

2.5. Prove Corollaries 2.1 and 2.2.

2.6. Let $\{X_n, n \geq 0\}$ be a DTMC on state space $S = \{1, 2, \ldots, N\}$ with transition probability matrix P. Let $Y_n = X_{2n}, n \geq 0$. Show that $\{Y_n, n \geq 0\}$ is a DTMC on S with transition matrix P^2.

2.7. Let $\{X_n, n \geq 0\}$ be a DTMC on state space $S = \{1, 2, \ldots, N\}$ with transition probability matrix P. Suppose $X_0 = i$ with probability 1. The sojourn time T_i of the DTMC in state i is said to be k if $\{X_0 = X_1 = \cdots = X_{k-1} = i, X_k \neq i\}$. Show that T_i is a $G(1 - p_{i,i})$ random variable.

2.8. Consider a machine that works as follows. If it is up at the beginning of a day, it stays up at the beginning of the next day with probability p and fails with probability $1 - p$. It takes exactly 2 days for the repairs, at the end of which the machine is as good as new. Let X_n be the state of the machine at the beginning of day n, where the state is 0 if the machine has just failed, 1 if 1 day's worth of repair work is done on it, and 2 if it is up. Show that $\{X_n, n \geq 0\}$ is a DTMC, and display its transition probability matrix.

2.9. Items arrive at a machine shop in a deterministic fashion at a rate of one per minute. Each item is tested before it is loaded onto the machine. An item is found to be nondefective with probability p and defective with probability $1 - p$. If an item is found defective, it is discarded. Otherwise, it is loaded onto the machine. The machine takes exactly 1 minute to process the item, after which it is ready to process the next one. Let X_n be 0 if the machine is idle at the beginning of the nth minute and 1 if it is starting the processing of an item. Show that $\{X_n, n \geq 0\}$ is a DTMC, and display its transition probability matrix.

2.10. Consider the system of Conceptual Problem 2.9. Now suppose the machine can process two items simultaneously. However, it takes 2 minutes to complete the processing. There is a bin in front of the machine where there is room to store two nondefective items. As soon as there are two items in the bin, they are loaded onto the machine and the machine starts processing them. Model this system as a DTMC.

2.11. Consider the system of Conceptual Problem 2.10. However, now suppose that the machine starts working on whatever items are waiting in the bin when it becomes idle. It takes 2 minutes to complete the processing whether the machine is processing one or two items. Processing on a new item cannot start unless the machine is idle. Model this as a DTMC.

2.12. The weather at a resort city is either sunny or rainy. The weather tomorrow depends on the weather today and yesterday as follows. If it was sunny yesterday and today, it will be sunny tomorrow with probability .9. If it was rainy yesterday but sunny today, it will be sunny tomorrow with probability .8. If it was sunny yesterday but rainy today, it will be sunny tomorrow with probability .7. If it was rainy yesterday and today, it will be sunny tomorrow with probability .6. Define today's state of the system as the pair (weather yesterday, weather today). Model this system as a DTMC, making appropriate independence assumptions.

2.13. Consider the following weather model. The weather normally behaves as in Example 2.3. However, when the cloudy spell lasts for two or more days, it continues to be cloudy for another day with probability .8 or turns rainy with probability .2. Develop a four-state DTMC model to describe this behavior.

2.14. N points, labeled $1, 2, \ldots, N$, are arranged clockwise on a circle and a particle moves on it as follows. If the particle is on point i at time n, it moves one step in clockwise fashion with probability p or one step in counterclockwise fashion with probability $1 - p$ to move to a new point at time $n + 1$. Let X_n be the position of the particle at time n. Show that $\{X_n, n \geq 0\}$ is a DTMC, and display its transition probability matrix.

2.15. Let $\{X_n, n \geq 0\}$ be an irreducible DTMC on state space $\{1, 2, \ldots, N\}$. Let u_i be the probability that the DTMC visits state 1 before it visits state N, starting from state i. Using the first-step analysis, show that

$$u_1 = 1,$$
$$u_i = \sum_{j=1}^{N} p_{i,j} u_j, \quad 2 \leq i \leq N - 1,$$
$$u_N = 0.$$

2.16. A total of N balls are put in two urns, so that initially urn A has i balls and urn B has $N - i$ balls. At each step, one ball is chosen at random from the N balls. If it is from urn A, it is moved to urn B, and vice versa. Let X_n be the number of balls in urn A after n steps. Show that $\{X_n, n \geq 0\}$ is a DTMC, assuming that the successive random drawings of the balls are independent. Display the transition probability matrix of the DTMC.

2.17. The following selection procedure is used to select one of two brands, say A and B, of infrared light bulbs. Suppose the brand A light bulb life-times are iid

$\exp(\lambda)$ random variables and brand B light bulb lifetimes are iid $\exp(\mu)$ random variables. One light bulb from each brand is turned on simultaneously, and the experiment ends when one of the two light bulbs fails. Brand A wins a point if the brand A light bulb outlasts brand B, and vice versa. (The probability that the bulbs fail simultaneously is zero.) The experiment is repeated with new light bulbs until one of the brands accumulates five points more than the other, and that brand is selected as the better brand. Let X_n be the number of points for brand A minus the number of points accumulated by brand B after n experiments. Show that $\{X_n, n \geq 0\}$ is a DTMC, and display its transition probability matrix. (*Hint*: Once X_n takes a value of 5 or -5, it stays there forever.)

2.18. Let $\{X_n, n \geq 0\}$ be a DTMC on state space $\{1, 2, \ldots, N\}$. Suppose it incurs a cost of $c(i)$ dollars every time it visits state i. Let $g(i)$ be the total expected cost incurred by the DTMC until it visits state N starting from state i. Derive the following equations:

$$g(N) = 0,$$
$$g(i) = c(i) + \sum_{j=1}^{N} p_{i,j} g(j), 1 \leq j \leq N - 1.$$

2.19. Another useful cost model is when the system incurs a random cost of $C(i, j)$ dollars whenever it undergoes a transition from state i to j in one step. This model is called the *cost per transition model*. Define

$$c(i) = \sum_{j=1}^{N} E(C(i, j)) p_{i,j}, 1 \leq i \leq N.$$

Show that $g(i, T)$, the total cost over a finite horizon T, under this cost model satisfies Theorem 2.11 with $c(i)$ as defined above. Also show that $g(i)$, the long-run cost rate, satisfies Theorem 2.12.

COMPUTATIONAL PROBLEMS

2.1. Consider the telecommunications model of Example 2.12. Suppose the buffer is full at the end of the third time slot. Compute the expected number of packets in the buffer at the end of the fifth time slot.

2.2. Consider the DTMC in Conceptual Problem 2.14 with $N = 5$. Suppose the particle starts on point 1. Compute the probability distribution of its position at time 3.

2.3. Consider the stock market model of Example 2.7. Suppose Mr. BigShot has bought 100 shares of stock at $5 per share. Compute the expected net change in the value of his investment in 5 days.

2.4. Mr. Smith is a coffee addict. He keeps switching between three brands of coffee, say A, B, and C, from week to week according to a DTMC with the following transition probability matrix:

$$P = \begin{bmatrix} .10 & .30 & .60 \\ .10 & .50 & .40 \\ .30 & .20 & .50 \end{bmatrix}. \tag{2.69}$$

If he is using brand A this week (i.e., week 1), what is the probability distribution of the brand he will be using in week 10?

2.5. Consider the telecommunications model of Example 2.8. Suppose the buffer is full at the beginning. Compute the expected number of packets in the buffer at time n for $n = 1, 2, 5,$ and 10, assuming that the buffer size is 10 and that the number of packets arriving during one time slot is a binomial random variable with parameters $(5, .2)$.

2.6. Consider the machine described in Conceptual Problem 2.8. Suppose the machine is initially up. Compute the probability that the machine is up at times $n = 5, 10, 15,$ and 20. (Assume $p = .95$.)

2.7. Consider Paper Pushers Insurance Company, Inc., of Example 2.6. Suppose it has 100 employees at the beginning of week 1, distributed as follows: 50 in grade 1, 25 in grade 2, 15 in grade 3, and 10 in grade 4. If employees behave independently of each other, compute the expected number of employees in each grade at the beginning of week 4.

2.8. Consider the machine reliability model in Example 2.2 with one machine. Suppose the machine is up at the beginning of day 0. Compute the probability that the state of the machine at the beginning of the next three days is up, down, down, in that order.

2.9. Consider the machine reliability model in Example 2.2 with two machines. Suppose both machines are up at the beginning of day 0. Compute the probability that the number of working machines at the beginning of the next three days is two, one, and two, in that order.

2.10. Consider the weather model of Example 2.3. Compute the probability that once the weather becomes sunny, the sunny spell lasts for at least 3 days.

2.11. Compute the expected length of a rainy spell in the weather model of Example 2.3.

2.12. Consider the inventory system of Example 2.4 with a starting inventory of five PCs on a Monday. Compute the probability that the inventory trajectory over the next four Mondays is as follows: 4, 2, 5, and 3.

2.13. Consider the inventory system of Example 2.4 with a starting inventory of five PCs on a Monday. Compute the probability that an order is placed at the end of the first week for more PCs.

2.14. Consider the manufacturing model of Example 2.5. Suppose both bins are empty at time 0. Compute the probability that both bins stay empty at times $n = 1, 2$, and then machine 1 is shut down at time $n = 4$.

2.15. Compute the occupancy matrix $M(10)$ for the DTMCs with transition matrices as given below:

(a)

$$P = \begin{bmatrix} .10 & .30 & .20 & .40 \\ .10 & .30 & .40 & .20 \\ .30 & .30 & .10 & .30 \\ .15 & .25 & .35 & .25 \end{bmatrix},$$

(b)

$$P = \begin{bmatrix} 0 & 1 & 0 & 0 \\ 0 & 0 & 1 & 0 \\ 0 & 0 & 0 & 1 \\ 1 & 0 & 0 & 0 \end{bmatrix},$$

(c)

$$P = \begin{bmatrix} .10 & 0 & .90 & 0 \\ 0 & .30 & 0 & .70 \\ .30 & 0 & .70 & 0 \\ 0 & .25 & 0 & .75 \end{bmatrix},$$

(d)

$$P = \begin{bmatrix} .10 & .30 & 0 & .60 \\ .10 & .30 & 0 & .60 \\ .30 & .10 & .10 & .50 \\ .5 & .25 & 0 & .25 \end{bmatrix}.$$

2.16. Consider the inventory system of Example 2.4. Compute the occupancy matrix $M(52)$. Using this, compute the expected number of weeks that Computers-R-Us starts with a full inventory (i.e., five PCs) during a year given that it started the first week of the year with an inventory of five PCs.

2.17. Consider the manufacturing model of Example 2.11. Suppose that at time 0 there is one item in bin 1 and bin 2 is empty. Compute the expected amount of time that machine 1 is turned off during an 8-hour shift.

2.18. Consider the telecommunications model of Example 2.12. Suppose the buffer is empty at time 0. Compute the expected number of slots that have an empty buffer at the end during the next 50 slots.

2.19. Classify the DTMCs with the transition matrices given in Computational Problem 2.15 as irreducible or reducible.

2.20. Classify the irreducible DTMCs with the transition matrices given below as periodic or aperiodic:

(a)

$$P = \begin{bmatrix} .10 & .30 & .20 & .40 \\ .10 & .30 & .40 & .20 \\ .30 & .10 & .10 & .50 \\ .15 & .25 & .35 & .25 \end{bmatrix},$$

(b)

$$P = \begin{bmatrix} 0 & 1 & 0 & 0 \\ 0 & 0 & 1 & 0 \\ 0 & 0 & 0 & 1 \\ 1 & 0 & 0 & 0 \end{bmatrix},$$

(c)

$$P = \begin{bmatrix} 0 & .20 & .30 & .50 \\ 1 & 0 & 0 & 0 \\ 1 & 0 & 0 & 0 \\ 1 & 0 & 0 & 0 \end{bmatrix},$$

(d)

$$P = \begin{bmatrix} 0 & 0 & .40 & .60 \\ 1 & 0 & 0 & 0 \\ 0 & 1 & 0 & 0 \\ 0 & 1 & 0 & 0 \end{bmatrix}.$$

2.21. Compute a normalized solution to the balance equations for the DTMC in Computational Problem 2.20(a). When possible, compute:

1. the limiting distribution;
2. the stationary distribution;
3. the occupancy distribution.

2.22. Do Computational Problem 2.21 for Computational Problem 2.20(b).

2.23. Do Computational Problem 2.21 for Computational Problem 2.20(c).

2.24. Do Computational Problem 2.21 for Computational Problem 2.20(d).

2.25. Consider the DTMC of Computational Problem 2.5. Compute:

1. the long-run fraction of the time that the buffer is full;
2. the expected number of packets in the buffer in the long run.

2.26. Consider Computational Problem 2.7. Compute the expected number of employees in each grade in steady state.

2.27. Consider the weather model of Conceptual Problem 2.12. Compute the long-run fraction of days that are rainy.

2.28. Consider the weather model of Conceptual Problem 2.13. Compute the long-run fraction of days that are sunny.

2.29. What fraction of the time does the coffee addict of Computational Problem 2.4 consume brand A coffee?

2.30. Consider the machine described in Conceptual Problem 2.8. What is the long-run fraction of the time that this machine is up? (Assume $p = .90$.)

2.31. Consider the manufacturing model of Example 2.11. Compute the expected number of components in bins A and B in steady state.

2.32. Consider the stock market model of Example 2.7. What fraction of the time does the chief financial officer have to interfere in the stock market to control the price of the stock?

2.33. Consider the single-machine production system of Conceptual Problem 2.10. Compute the expected number of items processed by the machine in 10 minutes, assuming that the bin is empty and the machine is idle to begin with. (Assume $p = .95$.)

2.34. Do Computational Problem 2.33 for the production system of Conceptual Problem 2.11. (Assume $p = .95$.)

2.35. Which one of the two production systems described in Conceptual Problems 2.10 and 2.11 has a higher per minute rate of production in steady state?

2.36. Consider the three-machine workshop described in Conceptual Problem 2.3. Suppose each working machine produces revenue of $500 per day, while repairs cost $300 per day per machine. What is the net rate of revenue per day in steady state? (*Hint*: Can we consider the problem with one machine to obtain the answer for three machines?)

2.37. Consider the inventory system of Example 2.27. Compute the long-run expected cost per day of operating this system.

2.38. Consider the manufacturing system of Example 2.11. Compute the expected number of assemblies produced per hour in steady state.

2.39. (Computational Problem 2.38 continued). What will be the increase in the production rate (in number of assemblies per hour) if we provide bins of capacity 3 to the two machines in Example 2.11?

2.40. Compute the long-run expected number of packets transmitted per unit time by the data switch of Example 2.12. How is this connected to the packet-loss rate computed in Example 2.29?

2.41. Consider the brand-switching model of Computational Problem 2.4. Suppose the per pound cost of coffee is $6, $8, and $15 for brands A, B, and C, respectively. Assuming Mr. Smith consumes one pound of coffee per week, what is his long-run expected coffee expense per week?

2.42. Compute the expected time to go from state 1 to 4 in the DTMCs of Computational Problems 2.20(a) and (c).

2.43. Compute the expected time to go from state 1 to 4 in the DTMCs of Computational Problems 2.20(b) and (d).

2.44. Consider the selection procedure of Conceptual Problem 2.17. Suppose the mean lifetime of Brand A light bulbs is 1, while that of Brand B light bulbs is 1.25. Compute the expected number of experiments done before the selection procedure ends. (*Hint*: Use the Gambler's ruin model of Example 2.33.)

2.45. Consider the DTMC model of the data switch described in Example 2.12. Suppose the buffer is full to begin with. Compute the expected amount of time (counted in number of time slots) before the buffer becomes empty.

2.46. Do Computational Problem 2.45 for the data buffer described in Computational Problem 2.5.

2.47. Consider the manufacturing model of Example 2.11. Compute the expected time (in hours) before one of the two machines is shut down, assuming that both bins are empty at time 0.

Case Study Problems. You may use the Matlab program of Section 2.8 to solve the following problems.

2.48. Suppose Passport has decided not to employ the services of WMB. However, this has generated discussion within the company about whether it should terminate accounts earlier. Let T_m ($1 \leq m \leq 7$) be the policy of terminating the account as soon as it misses m payments in a row. Which policy should Passport follow?

2.49. Consider the current policy P_c. One of the managers wants to see if it would help to alert the customers of their impending account termination in a more dire form by a phone call when the customer has missed six minimum payments in a row. This will cost a dollar per call. The manager estimates that this will decrease the missed payment probability from the current $p_6 = .329$ to $.250$. Is this policy cost-effective?

2.50. The company has observed over the past year that the downturn in the economy has increased the bankruptcy rate by 50%. In this changed environment, should Passport engage the services of WMB? When should it turn over the accounts to WMB?

2.51. Passport has been approached by another collection agency, which is willing to work with no annual service contract fee. However, it pays only 60% of the outstanding balance of any account turned over to them. Is this option better than hiring WMB?

Chapter 3
Poisson Processes

In the previous chapter, we studied a discrete-time stochastic process $\{X_n, n \geq 0\}$ on finite state space with Markov property at times $n = 0, 1, 2 \cdots$. Now we would like to study a continuous-time stochastic process $\{X(t), t \geq 0\}$ on a finite state space with Markov property at each time $t \geq 0$. We shall call such a process continuous-time Markov Chain (CTMC). We shall see in the next chapter that a finite-state CTMC spends an exponentially distributed amount of time in a given state before jumping out of it. Thus exponential distributions play an important role in CTMCs. In addition, the Poisson distribution and Poisson processes also form the foundation of many CTMC models. Hence we study these topics in this chapter.

3.1 Exponential Random Variables

Consider a nonnegative random variable with parameter $\lambda > 0$ with the following pdf:

$$f(x) = \lambda e^{-\lambda x}, \quad x \geq 0. \tag{3.1}$$

The corresponding cdf can be computed using Equation (B.3) as

$$F(x) = 1 - e^{-\lambda x}, \quad x \geq 0. \tag{3.2}$$

Note that if the random variable X has units of time, the parameter λ has units of time^{-1}. From Equation (3.2), we see that

$$P(X > x) = e^{-\lambda x}, \quad x \geq 0.$$

Definition 3.1. (Exponential Distribution). The pdf given by Equation (3.1) is called the exponential density. The cdf given by Equation (3.2) is called the

V.G. Kulkarni, *Introduction to Modeling and Analysis of Stochastic Systems*, Springer Texts in Statistics, DOI 10.1007/978-1-4419-1772-0_3, © Springer Science+Business Media, LLC 2011

exponential distribution. A random variable with the cdf given in Equation (3.2) is called an exponential random variable. All three are denoted by $Exp(\lambda)$.

We leave it to the reader to verify that

$$E(X) = \frac{1}{\lambda} \tag{3.3}$$

and

$$Var(X) = \frac{1}{\lambda^2}. \tag{3.4}$$

Example 3.1. (Time to Failure). Suppose a new machine is put into operation at time zero. Its lifetime is known to be an $Exp(\lambda)$ random variable with $\lambda = .1/\text{hour}$. What are the mean and variance of the lifetime of the machine?

Using Equation (3.3), we see that the mean lifetime is $1/\lambda = 10$ hours. Similarly, Equation (3.4) yields the variance as $1/\lambda^2 = 100$ hours^2.

What is the probability that the machine will give trouble-free service continuously for 1 day?

To use consistent units, we use 24 hours instead of 1 day. We compute the required probability as

$$P(X > 24) = e^{-(.1)(24)} = e^{-2.4} = .0907.$$

Suppose the machine has not failed by the end of the first day. What is the probability that it will give trouble-free service for the whole of the next day?

The required probability is given by

$$P(X > 48 | X > 24) = \frac{P(X > 48; X > 24)}{P(X > 24)}$$
$$= \frac{P(X > 48)}{P(X > 24)}$$
$$= \frac{e^{-(.1)(48)}}{e^{-(.1)(24)}}$$
$$= e^{-(.1)(24)}$$
$$= .0907.$$

But this is the same as the earlier answer. Thus, given that the machine has not failed by day 1, it is as good as new! ∎

The property discovered in the example above is called the *memoryless property* and is one of the most important properties of the exponential random variable. We define it formally below.

Definition 3.2. (Memoryless Property). A random variable X on $[0, \infty)$ is said to have the memoryless property if

$$P(X > t + s | X > s) = P(X > t), \quad s, t \geq 0. \tag{3.5}$$

The unique feature of an exponential distribution is that it is the only continuous nonnegative random variable with the memoryless property, as stated in the next theorem.

Theorem 3.1. (Memoryless Property of $Exp(\lambda)$). *Let X be a continuous random variable taking values in $[0, \infty)$. It has the memoryless property if and only if it is an $Exp(\lambda)$ random variable for some $\lambda > 0$.*

Proof. Suppose $X \sim Exp(\lambda)$. Then, for all $s, t \geq 0$,

$$P(X > s + t | X > s) = \frac{P(X > s + t; X > s)}{P(X > s)}$$

$$= \frac{e^{-\lambda(s+t)}}{e^{-\lambda s}}$$

$$= e^{-\lambda t}$$

$$= P(X > t).$$

Thus X has the memoryless property. Next, suppose X is a continuous random variable with pdf $f(\cdot)$. Define

$$G(x) = \int_x^\infty f(x) dx = P(X > x).$$

Assume that

$$G(0) = 1. \tag{3.6}$$

Suppose X has the memoryless property. Then

$$G(t) = P(X > t)$$

$$= P(X > t + s | X > s)$$

$$= \frac{P(X > t + s, X > s)}{P(X > s)}$$

$$= \frac{P(X > t + s)}{P(X > s)}$$

$$= \frac{G(t + s)}{G(s)}.$$

Hence we must have

$$G(t + s) = G(t)G(s), \quad s, t \geq 0.$$

Thus

$$\frac{G(t+s) - G(t)}{s} = -\frac{G(t)(1 - G(s))}{s}. \tag{3.7}$$

We have

$$\lim_{s \to 0} \frac{1 - G(s)}{s} = f(0) = \lambda \ (\text{say}).$$

Taking limits on both sides in Equation (3.7), we get

$$G'(t) = -\lambda G(t).$$

Hence,

$$G(t) = e^{-\lambda t}, \quad t \geq 0.$$

This implies that $X \sim \text{Exp}(\lambda)$. ∎

Next consider a random variable with parameters $k = 1, 2, 3, \ldots$ and $\lambda > 0$ taking values in $[0, \infty)$ with the following pdf:

$$f(x) = \lambda e^{-\lambda x} \frac{(\lambda x)^{k-1}}{(k-1)!}, \quad x \geq 0. \tag{3.8}$$

Computing F from f of the equation above by using Equation (B.3) is a tedious exercise in integration by parts, and we omit the details. The final expression is

$$F(x) = 1 - \sum_{r=0}^{k-1} e^{-\lambda x} \frac{(\lambda x)^r}{r!}, \quad x \geq 0. \tag{3.9}$$

If the random variable X has units of time, the parameter λ has units of time^{-1}. From Equation (3.9), we see that

$$P(X > x) = e^{-\lambda x} \sum_{r=0}^{k-1} \frac{(\lambda x)^r}{r!}, \quad x \geq 0.$$

Definition 3.3. (Erlang Distribution). The pdf given by Equation (3.8) is called the Erlang density. The cdf given by Equation (3.9) is called the Erlang distribution. A random variable with cdf given in Equation (3.9) is called an Erlang random variable. All three are denoted by $\text{Erl}(k, \lambda)$.

We leave it to the reader to verify that

$$E(X) = \frac{k}{\lambda} \tag{3.10}$$

and

$$\text{Var}(X) = \frac{k}{\lambda^2}. \tag{3.11}$$

Example 3.2. (Time to Failure). We shall redo Example 3.1 under the assumption that the lifetime of the machine is an $\mathrm{Erl}(k, \lambda)$ random variable with parameters $k = 2$ and $\lambda = .2/\mathrm{hr}$.

Using Equation (3.10), we see that the mean lifetime is $2/\lambda = 10$ hours. Similarly, Equation (3.11) yields the variance as $2/\lambda^2 = 50$ hours^2.

From Equation (3.9), we have

$$P(\text{no failure in the first 24 hours}) = P(X > 24)$$
$$= e^{-(.2)(24)}(1 + (.2)(24))$$
$$= .0477$$

and

$$P(\text{no failure in the second 24 hours} \mid \text{no failure in the first 24 hours})$$
$$= P(X > 48 | X > 24)$$
$$= \frac{P(X > 48)}{P(X > 24)}$$
$$= \frac{e^{-(.2)(48)}(1 + (.2)(48))}{e^{-(.2)(24)}(1 + (.2)(24))}$$
$$= .0151.$$

The second probability is lower than the first, indicating that the machine deteriorates with age. This is what we would expect. ■

The next theorem shows that Erlang random variables appear naturally as the sum of iid exponential random variables.

Theorem 3.2. (Sums of Exponentials). *Suppose $\{X_i, i = 1, 2, \cdots, n\}$ are iid* $\mathrm{Exp}(\lambda)$ *random variables, and let*

$$Z_n = X_1 + X_2 + \cdots + X_n.$$

Then Z_n is an $\mathrm{Erl}(n, \lambda)$ random variable.

Proof. Note that $\mathrm{Erl}(1, \lambda)$ is the same as $\mathrm{Exp}(\lambda)$. Hence the result is true for $n = 1$. We shall show that it holds for a general n by induction. So suppose the result is true for $n = k$. We shall show that it holds for $n = k + 1$. Using

$$Z_{k+1} = Z_k + X_{k+1}$$

in Equation (C.12), we get

$$f_{Z_{k+1}}(z) = \int_{-\infty}^{\infty} f_{X_{k+1}}(z - x) f_{Z_k}(x) dx$$

$$= \int_0^z \lambda e^{-\lambda(z-x)} \lambda e^{-\lambda x} \frac{(\lambda x)^{k-1}}{(k-1)!} dx$$

$$= \lambda^2 e^{-\lambda z} \int_0^z \frac{(\lambda x)^{k-1}}{(k-1)!} dx$$

$$= \lambda e^{-\lambda z} \frac{(\lambda z)^k}{k!}.$$

The last expression can be recognized as the pdf of an $Erl(k+1, \lambda)$ random variable. Hence the result follows by induction. ∎

Next we study the minimum of independent exponential random variables. Let X_i be an $Exp(\lambda_i)$ random variable $(1 \le i \le k)$, and suppose $X_1, X_2, \ldots, X_k$ are independent. Let

$$X = \min\{X_1, X_2, \ldots, X_k\}. \tag{3.12}$$

We can think of X_i as the time when an event of type i occurs. Then X is the time when the first of these k events occurs. The main result is given in the next theorem.

Theorem 3.3. (Minimum of Exponentials). X of Equation (3.12) is an $Exp(\lambda)$ random variable, where

$$\lambda = \sum_{i=1}^k \lambda_i. \tag{3.13}$$

Proof. If $X_i > x$ for all $1 \le i \le k$, then $X > x$ and vice versa. Using this and the independence of $X_i, 1 \le i \le k$, we get

$$P(X > x) = P(\min\{X_1, X_2, \ldots, X_k\} > x)$$
$$= P(X_1 > x, X_2 > x, \ldots, X_k > x)$$
$$= P(X_1 > x)P(X_2 > x) \cdots P(X_k > x)$$
$$= e^{-\lambda_1 x} e^{-\lambda_2 x} \cdots e^{-\lambda_k x}$$
$$= e^{-\lambda x}. \tag{3.14}$$

This proves that X is an $Exp(\lambda)$ random variable. ∎

Next, let X be as in Equation (3.12), and define $Z = i$ if $X_i = X$; i.e., if event i is the first of the k events to occur. Note that X_i's are continuous random variables, and hence the probability that two or more of them will be equal to each other is zero. Thus there is no ambiguity in defining Z. The next theorem gives the joint distribution of Z and X.

Theorem 3.4. (Distribution of (Z, X)). Z and X are independent random variables with

$$P(Z = i; X > x) = P(Z = i)P(X > x) = \frac{\lambda_i}{\lambda} e^{-\lambda x}, \quad 1 \le i \le k, \ x \ge 0. \tag{3.15}$$

Proof. We have

$$
\begin{aligned}
P(Z = i; X > x) &= \int_x^\infty P(Z = i; X > x | X_i = y)\lambda_i e^{-\lambda_i y} dy \\
&= \int_x^\infty P(X_j > y, \ j \neq i | X_i = y)\lambda_i e^{-\lambda_i y} dy \\
&= \int_x^\infty \prod_{j \neq i} e^{-\lambda_j y} \lambda_i e^{-\lambda_i y} dy \\
&= \lambda_i \int_x^\infty e^{-\lambda y} dy \\
&= \frac{\lambda_i}{\lambda} e^{-\lambda x}.
\end{aligned}
$$

Setting $x = 0$ in the above and using the fact that X is a continuous random variable, we get

$$
P(Z = i) = P(Z = i; X > 0) = P(Z = i; X \geq 0) = \frac{\lambda_i}{\lambda}. \tag{3.16}
$$

Hence we see that

$$
P(Z = i; X > x) = P(Z = i)P(X > x),
$$

proving the independence of Z and X. ∎

It is counterintuitive that Z and X are independent random variables because it implies that the time until the occurrence of the first of the k events is independent of which of the k events occurs first! In yet other terms, the conditional distribution of X, given that $Z = i$, is Exp(λ) and not Exp(λ_i) as we might have (incorrectly) guessed.

Example 3.3. (Boy or a Girl?). A maternity ward at a hospital currently has seven pregnant women waiting to give birth. Three of them are expected to give birth to boys, and the remaining four are expected to give birth to girls. From prior experience, the hospital staff knows that a mother spends on average 6 hours in the hospital before delivering a boy and 5 hours before delivering a girl. Assume that these times are independent and exponentially distributed. What is the probability that the first baby born is a boy and is born in the next hour?

Let X_i be the time to delivery of the ith expectant mother. Assume mothers 1, 2, and 3 are expecting boys and mothers 4, 5, 6, and 7 are expecting girls. Then our assumption implies that $\{X_i, 1 \leq i \leq 7\}$ are independent exponential random variables with parameters $1/6$, $1/6$, $1/6$, $1/5$, $1/5$, $1/5$, and $1/5$, respectively. Let B be the time when the first boy is born, and G be the time when the first girl is born. Then Theorem 3.3 implies

$$B = \min\{X_1, X_2, X_3\} \sim \text{Exp}(3/6),$$

$$G = \min\{X_4, X_5, X_6, X_7\} \sim \text{Exp}(4/5).$$

Also, B and G are independent. Then the probability that the first baby is a boy is given by

$$P(B < G) = \frac{3/6}{3/6 + 4/5} = \frac{5}{13} = 0.3846.$$

The time of the first birth is

$$\min\{B, G\} \sim \text{Exp}(3/6 + 4/5) = \text{Exp}(1.3).$$

Thus the probability that the first baby is born within the hour is given by

$$P(\min\{B, G\} \leq 1) = 1 - \exp(-1.3) = 0.7275.$$

Using Equation (3.15), we get the probability that the first baby born is a boy and is born in the next hour as $0.3846 * 0.7275 = 0.2798.$ ∎

In the next section, we study another random variable that is quite useful in building stochastic models.

3.2 Poisson Random Variables

Suppose we conduct n independent trials of an experiment. Each trial is successful with probability p or unsuccessful with probability $1 - p$. Let X be the number of successes among the n trials. The state space of X is $\{0, 1, 2, \ldots, n\}$. The random variable X is called a binomial random variable with parameters n and p and is denoted as $\text{Bin}(n, p)$. The pmf of X is given by

$$p_k = P(X = k) = \binom{n}{k} p^k (1 - p)^{n-k}, \quad 0 \leq k \leq n. \tag{3.17}$$

The pmf above is called the binomial distribution. Computing this distribution for large values of n is difficult since the term $\binom{n}{k}$ becomes very large, while the term $p^k(1-p)^{n-k}$ becomes very small. To avoid such difficulties, it is instructive to look at the limit of the binomial distribution as $n \to \infty$ and $p \to 0$ in such a way that np approaches a fixed number, say $\lambda \in (0, \infty)$. We shall denote this limiting region as

D to avoid clutter in the following derivation. We have

$$\lim_{D} \binom{n}{k} p^k (1-p)^{n-k} \tag{3.18}$$

$$= \lim_{D} \frac{n!}{k!(n-k)!} p^k (1-p)^{n-k}$$

$$= \lim_{D} \frac{(np)^k}{k!} \left(\prod_{r=0}^{k} \left(\frac{n-r}{n} \right) \right) \left(1 - \frac{np}{n} \right)^{n-k}$$

$$= \lim_{D} \frac{(np)^k}{k!} \left(\lim_{D} \prod_{r=0}^{k} \left(\frac{n-r}{n} \right) \right) \lim_{D} \left(1 - \frac{np}{n} \right)^{n} \lim_{D} \left(1 - \frac{np}{n} \right)^{-k}$$

$$= \left(\frac{\lambda^k}{k!} \right) (1) \left(e^{-\lambda} \right) (1)$$

$$= e^{-\lambda} \frac{\lambda^k}{k!}.$$

Here we have used

$$\lim_{D} \left(1 - \frac{np}{n} \right)^{n} = \lim_{n \to \infty} \left(1 - \frac{\lambda}{n} \right)^{n} = e^{-\lambda}.$$

Thus, in the limit, the Bin(n, p) random variable approaches a random variable with state space $S = \{0, 1, 2, ...\}$ and pmf

$$p_k = e^{-\lambda} \frac{\lambda^k}{k!}, \quad k \in S. \tag{3.19}$$

The pmf above plays an important part in probability models and hence has been given the special name Poisson distribution.

Definition 3.4. (Poisson Distribution). A random variable with pmf given by Equation (3.19) is called a Poisson random variable with parameter λ and is denoted by P(λ). The pmf given in Equation (3.19) is called a Poisson distribution with parameter λ and is also denoted by P(λ).

We see that Equation (3.19) defines a proper pmf since

$$\sum_{k=0}^{\infty} e^{-\lambda} \frac{\lambda^k}{k!} = e^{-\lambda} \sum_{k=0}^{\infty} \frac{\lambda^k}{k!}$$

$$= e^{-\lambda} e^{\lambda} = 1.$$

We leave it to the reader to verify that, if X is a $P(\lambda)$,

$$E(X) = \lambda \tag{3.20}$$

and

$$\mathrm{Var}(X) = \lambda. \tag{3.21}$$

Example 3.4. (Counting Accidents). Suppose accidents occur one at a time at a dangerous traffic intersection during a 24-hour day. We divide the day into 1440 minute-long intervals and assume that there can be only 0 or 1 accidents during each 1-minute interval. Let E_k be the event that there is an accident during the kth minute-long interval, $1 \le k \le 1440$. Suppose that the event E_k's are mutually independent and that $P(E_k) = .001$. Compute the probability that there are exactly k accidents during one day.

Let X be the number of accidents during a 24-hour period. Using the assumptions above, we see that X is a Bin(1440, .001) random variable. Hence

$$P(X = k) = \binom{1440}{k}(.001)^k(.999)^{1440-k}, \quad 0 \le k \le 1440.$$

This is numerically difficult to compute even for moderate values of k. Hence we approximate X as a $P(1440 * .0001) = P(1.440)$ random variable. This yields

$$P(X = k) = e^{-1.440}\frac{1.440^k}{k!}, \quad k \ge 0.$$

For example, for $k = 0$, the binomial formula produces a value .2368, while the Poisson formula produces .2369. ∎

Example 3.5. (Service Facility). Customers arrive at a service facility one at a time. Suppose that the total number of arrivals during a 1-hour period is a Poisson random variable with parameter 8. Compute the probability that at least three customers arrive during 1 hour.

Let X be the number of arrivals during 1 hour. Then we are given that X is a $P(8)$ random variable. The required probability is given by

$$P(X \ge 3) = 1 - P(X = 0) - P(X = 1) - P(X = 2)$$
$$= 1 - e^{-8}(1 + 8 + 64/2)$$
$$= 1 - 41e^{-8} = .9862.$$

What are the mean and variance of the number of arrivals in a 1-hour period? Using Equations (3.20) and (3.21), we see that the mean and variance are $\lambda = 8$. ∎

Next we study the sums of independent Poisson random variables. The main result is given in the next theorem.

Theorem 3.5. (Sums of Poissons). *Suppose $\{X_i, i = 1, 2, \ldots, n\}$ are independent random variables with $X_i \sim P(\lambda_i)$, $1 \le i \le n$. Let*

$$Z_n = X_1 + X_2 + \cdots + X_n.$$

Then Z_n is a $P(\lambda)$ random variable, where

$$\lambda = \sum_{i=1}^{n} \lambda_i.$$

Proof. We shall show the result for $n = 2$. The general result then follows by induction. Using Equation (C.11), we get

$$p_Z(k) = \sum_{k_2=0}^{k} e^{-\lambda_1} \frac{\lambda_1^{k-k_2}}{(k-k_2)!} e^{-\lambda_2} \frac{\lambda_2^{k_2}}{k_2!}$$

$$= e^{-\lambda_1-\lambda_2} \frac{1}{k!} \sum_{k_2=0}^{k} \frac{k!}{(k-k_2)!k_2!} \lambda_1^{k-k_2} \lambda_2^{k_2}$$

$$= e^{-\lambda_1-\lambda_2} \frac{(\lambda_1+\lambda_2)^k}{k!}. \tag{3.22}$$

The last equality follows from the binomial identity

$$\sum_{r=0}^{k} \frac{k!}{(k-r)!r!} \lambda_1^{k-r} \lambda_2^{r} = (\lambda_1+\lambda_2)^k.$$

This completes the proof. ∎

Since adding two Poisson random variables generates another Poisson random variable, it is natural to ask: can we construct a reverse process? That is, can we produce two Poisson random variables starting from one Poisson random variable? The answer is affirmative. The reverse process is called thinning and can be visualized as follows. Let X be a $P(\lambda)$ random variable representing the random number of events. An event may or may not be registered. Let $p \in [0, 1]$ be a fixed number representing the probability that any given event is registered, independent of everything else. Let X_1 be the number of registered events and X_2 be the number of unregistered events. Now, from the definition of the random variables, it is clear that $X_1 \sim \text{Bin}(X, p)$ and $X_2 = X - X_1$. Hence

$$P(X_1 = i, X_2 = j)$$
$$= P(X_1 = i, X_2 = j | X = i + j)P(X = i + j)$$
$$= \binom{i+j}{i} p^i (1-p)^j e^{-\lambda} \frac{\lambda^{i+j}}{(i+j)!}$$
$$= e^{-\lambda p} \frac{(\lambda p)^i}{i!} e^{-\lambda(1-p)} \frac{(\lambda(1-p))^j}{j!}, \quad i \geq 0, j \geq 0.$$

From the joint distribution above, we can compute the following marginal distributions:

$$P(X_1 = i) = e^{-\lambda p} \frac{(\lambda p)^i}{i!}, \quad i \geq 0,$$

$$P(X_2 = j) = e^{-\lambda(1-p)} \frac{(\lambda(1-p))^j}{j!}, \quad j \geq 0.$$

Hence we have

$$P(X_1 = i, X_2 = j) = P(X_1 = i)P(X_2 = j). \tag{3.23}$$

Thus X_1 is $P(\lambda p)$ and X_2 is $P(\lambda(1-p))$, and they are mutually independent. The independence is quite surprising!

Using the properties of the exponential and Poisson distributions studied above, we shall build our first continuous-time stochastic process in the next section.

3.3 Poisson Processes

Frequently we encounter systems whose state transitions are triggered by streams of events that occur one at a time, for example, arrivals to a queueing system, shocks to an engineering system, earthquakes in a geological system, biological stimuli in a neural system, accidents in a given city, claims on an insurance company, demands on an inventory system, failures in a manufacturing system, etc. The modeling of these systems becomes more tractable if we assume that the successive inter-event times are iid exponential random variables. For reasons that will become clear shortly, such a stream of events is given a special name: *Poisson process*. We define it formally below.

Let S_n be the occurrence time of the nth event. Assume $S_0 = 0$, and define

$$T_n = S_n - S_{n-1}, \quad n \geq 1.$$

Thus T_n is the time between the occurrence of the nth and the $(n-1)$st event. Let $N(t)$ be the total number of events that occur during the interval $(0, t]$. Thus an event at 0 is not counted, but an event at t, if any, is counted in $N(t)$. One can formally define it as

$$N(t) = \max\{n \geq 0 : S_n \leq t\}, \quad t \geq 0. \tag{3.24}$$

Definition 3.5. (Poisson Process). The stochastic process $\{N(t), t \geq 0\}$, where $N(t)$ is as defined by Equation (3.24), is called a Poisson process with rate λ (denoted by PP(λ)) if $\{T_n, n \geq 1\}$ is a sequence of iid Exp(λ) random variables.

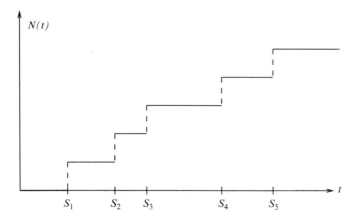

Fig. 3.1 A typical sample path of a Poisson process.

A typical sample path of a Poisson process is shown in Figure 3.1. The next theorem gives the pmf of $N(t)$ for a given t. It also justifies the name Poisson process for this stochastic process.

Theorem 3.6. (Poisson Process). *Let* $\{N(t), t \geq 0\}$ *be a* $\text{PP}(\lambda)$. *For a given* t, $N(t)$ *is a Poisson random variable with parameter* λt; *i.e.,*

$$P(N(t) = k) = e^{-\lambda t} \frac{(\lambda t)^k}{k!}, \quad k \geq 0. \tag{3.25}$$

Proof. We shall first compute $P(N(t) \geq k)$ and then obtain $P(N(t) = k)$ by using

$$P(N(t) = k) = P(N(t) \geq k) - P(N(t) \geq k + 1). \tag{3.26}$$

Suppose $N(t) \geq k$. Then the kth event in the Poisson process must have occurred at or before t; i.e., $S_k \leq t$. On the other hand, suppose $S_k \leq t$. Then the kth event has occurred at or before t. Hence there must be at least k events up to t; i.e., $N(t) \geq k$. Thus the events $\{N(t) \geq k\}$ and $\{S_k \leq t\}$ are identical and must have the same probability. However, S_k is a sum of k iid $\text{Exp}(\lambda)$ random variables. Hence, from Theorem 3.2, it is an $\text{Erl}(k, \lambda)$ random variable. Using Equation (3.9), we get

$$P(N(t) \geq k) = P(S_k \leq t) = 1 - \sum_{r=0}^{k-1} e^{-\lambda t} \frac{(\lambda t)^r}{r!}.$$

Substituting in (3.26), we get

$$P(N(t) = k) = P(N(t) \geq k) - P(N(t) \geq k + 1)$$
$$= 1 - \sum_{r=0}^{k-1} e^{-\lambda t} \frac{(\lambda t)^r}{r!} - \left(1 - \sum_{r=0}^{k} e^{-\lambda t} \frac{(\lambda t)^r}{r!}\right)$$
$$= e^{-\lambda t} \frac{(\lambda t)^k}{k!}.$$

This completes the proof. ∎

As an immediate consequence of the theorem above, we get

$$E(N(t)) = \lambda t, \quad \mathrm{Var}(N(t)) = \lambda t.$$

Thus the expected number of events up to t increases linearly in t with rate λ. This justifies calling λ the rate of the Poisson process. The next theorem gives an important property of the Poisson process.

Theorem 3.7. (Markov Property). *Let $\{N(t), t \geq 0\}$ be a PP(λ). It has the Markov property at each time t; that is,*

$$P(N(t + s) = k | N(s) = j, N(u), 0 \leq u \leq s) = P(N(t + s) = k | N(s) = j).$$
$$(3.27)$$

Idea of Proof. This can be seen as follows. Fix a $t \geq 0$, and suppose $N(t) = k$ is given. Then S_{k+1} is the time of occurrence of the first event after t. Using the (strong) memoryless property (see Conceptual Problem 3.11) of the exponential random variables, we can show that the time until occurrence of this next event, namely $S_{k+1} - t$, is exponentially distributed with parameter λ and is independent of the history of the Poisson process up to t. Thus the time until the occurrence of the next event in a Poisson process is always Exp(λ), regardless of the occurrence of events so far. Equation (3.27) follows as a consequence. ∎

The result above can be stated in another way: the distribution of the increment $N(t + s) - N(s)$ over any interval $(s, t + s]$ is independent of the events occurring outside this interval. In particular, the increments over non-overlapping intervals are independent. In fact, the argument in the proof above shows that the number of events on any time interval $(s, s + t]$ is given by

$$N(t + s) - N(s) \sim P(\lambda t).$$

Thus we have

$$P(N(t + s) = k | N(s) = j) = e^{-\lambda t} \frac{(\lambda t)^{k-j}}{(k - j)!}, \quad 0 \leq j \leq k.$$

Thus the distribution of the increment $N(t + s) - N(s)$ over any interval $(s, t + s]$ depends only on t, the length of the interval. We say that the increments are stationary. We say that the Poisson process has *stationary and independent increments*. We shall see this property again when we study Brownian motion in Chapter 7.

Example 3.6. Let $N(t)$ be the number of births in a hospital over the interval $(0, t]$. Suppose $\{N(t), t \geq 0\}$ is a Poisson process with rate 10 per day.

1. What are the mean and variance of the number of births in an 8-hour shift?

   Using days as the time unit, we see that we want to compute the mean and variance of $N(1/3)$. Now $N(1/3) \sim P(10/3)$. Hence we get

   $$E(N(1/3)) = 10/3, \quad \text{Var}(N(1/3)) = 10/3.$$

2. What is the probability that there are no births from noon to 1 p.m.?

   Using 1 hour $= 1/24$ day, we see that the number of births in one hour window is given by $N(1/24) \sim P(10/24)$. Hence the required probability is given by

   $$P(N(1/24) = 0) = \exp(-10/24) = 0.6592.$$

3. What is the probability that there are three births from 8 a.m. to 12 noon and four from 12 noon to 5 p.m.?

   Assuming that 8 a.m. is $t = 0$, the desired probability is given by

   $$P(N(4/24) - N(0) = 3; N(8/24) - N(4/24) = 4)$$
   $$= P(N(4/24) - N(0) = 3)P(N(8/24) - N(4/24) = 4)$$
   $$\text{(independence of increments)}$$
   $$= P(N(4/24) = 3)P(N(4/24) = 4)$$
   $$\text{(stationarity of increments)}$$
   $$= e^{-10/6}\frac{(10/6)^3}{3!}e^{-10/6}\frac{(10/6)^4}{4!}$$
   $$= .0088.$$

This calculation illustrates the usefulness of the stationary and independent increments property of a Poisson process. ∎

Poisson processes and their variations and generalizations have been extensively studied, and we study a few of them in the rest of this chapter. However, the reader can safely skip these sections since the material on Poisson processes presented so far is sufficient for the purposes of modeling encountered in the rest of this book.

3.4 Superposition of Poisson Processes

Superposition of stochastic processes is the equivalent of addition of random variables. It is a natural operation in the real world. For example, suppose $N_1(t)$ is the number of male arrivals at a store up to time t and $N_2(t)$ is the number of female arrivals at that store up to time t. Then $N(t) = N_1(t) + N_2(t)$ is the total number of arrivals at the store up to time t. The process $\{N(t), t \geq 0\}$ is called the superposition of $\{N_1(t), t \geq 0\}$ and $\{N_2(t), t \geq 0\}$. The next theorem gives the probabilistic structure of the superposed process $\{N(t), t \geq 0\}$.

Theorem 3.8. (Superposition). *Let $\{N_1(t), t \geq 0\}$ be a PP(λ_1) and $\{N_2(t), t \geq 0\}$ be a PP(λ_2). Suppose the two processes are independent of each other. Let $\{N(t), t \geq 0\}$ be the superposition of $\{N_1(t), t \geq 0\}$ and $\{N_2(t), t \geq 0\}$. Then $\{N(t), t \geq 0\}$ is a PP $(\lambda_1 + \lambda_2)$.*

Proof. Let T_i be the time of the first event in the $\{N_i(t), t \geq 0\}$ process ($i = 1, 2$). From the definition of the Poisson process, we see that T_i is an Exp(λ_i) random variable for $i = 1, 2$. The independence of the two Poisson processes implies that T_1 and T_2 are independent. Now let T be the time of the first event in the superposed process. Since the superposed process counts events in both the processes, it follows that $T = \min\{T_1, T_2\}$. From Theorem 3.3 it follows that T is an Exp($\lambda_1 + \lambda_2$) random variable.

Now consider the $\{N_i(t), t \geq 0\}$ processes from time T onward. From the property of the exponential random variables, it follows that $\{N_1(t + T), t \geq 0\}$ and $\{N_2(t + T), t \geq 0\}$ are independent Poisson processes with parameters λ_1 and λ_2, respectively. Thus we can repeat the argument above to see that the time until the first event in the process $\{N(t + T), t \geq 0\}$ is again an Exp($\lambda_1 + \lambda_2$) random variable, and it is independent of the past up to time T. Thus it follows that the inter-event times in the superposed process are iid Exp($\lambda_1 + \lambda_2$) random variables. Hence $\{N(t), t \geq 0\}$ is a PP($\lambda_1 + \lambda_2$). ∎

From Theorem 3.6 we see that, for a fixed t, $N_i(t) \sim$ P($\lambda_i t$) ($i = 1, 2$) are independent. Hence it follows from Theorem 3.8 that $N(t) \sim$ P($\lambda_1 t + \lambda_2 t$). We could have derived this directly from Theorem 3.5.

Example 3.7. (Superposition). Let $B(t)$ be the number of boys born in a hospital over $(0, t]$. Similarly, let $G(t)$ be the number of girls born at this hospital over $(0, t]$. Assume that $\{B(t), t \geq 0\}$ is a Poisson process with rate 10 per day, while $\{G(t), t \geq 0\}$ is an independent Poisson process with rate 9 per day. What are the mean and variance of the number of births in a week?

Let $N(t)$ be the number of births during $(0, t]$. Then $N(t) = B(t) + G(t)$. From Theorem 3.8, we see that $\{N(t), t \geq 0\}$ is a Poisson process with rate 19 per day. The desired answer is given by

$$E(N(7)) = \text{Var}(N(7)) = 19 * 7 = 133.$$
∎

3.5 Thinning of a Poisson Process

Let $\{N(t), t \geq 0\}$ be a PP(λ) and let $0 \leq p \leq 1$ be a given real number. Suppose every event in the Poisson process is registered with probability p, independent of everything else. Let $R(t)$ be the number of registered events over $(0, t]$. The process $\{R(t), t \geq 0\}$ is called a thinning of $\{N(t), t \geq 0\}$. The next theorem describes the process $\{R(t), t \geq 0\}$.

Theorem 3.9. (Thinning). *The thinned process $\{R(t), t \geq 0\}$ as constructed above is a PP(λp).*

Proof. Let $\{T_n, n \geq 1\}$ be inter-event times in the $\{N(t), t \geq 0\}$ process. From the definition of a Poisson process, it follows that $\{T_n, n \geq 1\}$ is a sequence of iid Exp(λ) random variables. Let K be the first event in the N process that is registered. Then K is a geometric random variable with probability mass function

$$p_k = P(K = k) = (1 - p)^{k-1}p, \quad k = 1, 2, 3, \cdots.$$

Let T be the time of the first event in the R process. We have

$$T = \sum_{i=1}^{K} T_i.$$

We see that, for a fixed k, if $K = k$, $T = T_1 + T_2 + \cdots + T_k$. From Theorem 3.2, we see that, given $K = k$, T is an Erl(k, λ) random variable with density

$$f_{T|K=k}(t) = \lambda e^{-\lambda t} \frac{(\lambda t)^{k-1}}{(k-1)!}, \quad t \geq 0.$$

Now we can use conditioning to compute the probability density $f_T(\cdot)$ of T as follows:

$$
\begin{aligned}
f_T(t) &= \sum_{k=1}^{\infty} f_{T|K=k}(t) P(K = k) \\
&= \sum_{k=1}^{\infty} \lambda e^{-\lambda t} \frac{(\lambda t)^{k-1}}{(k-1)!} (1 - p)^{k-1} p \\
&= \lambda p e^{-\lambda t} \sum_{k=1}^{\infty} \frac{(\lambda t)^{k-1}}{(k-1)!} (1 - p)^{k-1} \\
&= \lambda p e^{-\lambda t} \sum_{k=0}^{\infty} \frac{(\lambda(1 - p)t)^k}{k!} \\
&= \lambda p e^{-\lambda t} e^{\lambda(1-p)t} \\
&= \lambda p e^{-\lambda p t}.
\end{aligned}
$$

Thus T is an $\text{Exp}(\lambda p)$ random variable. Since each event is registered independently, it follows that the successive inter-event times in the R process are iid $\text{Exp}(\lambda p)$ random variables. This proves that $\{R(t), t \geq 0\}$ is a $\text{PP}(\lambda p)$ as desired. ∎

We state one more important property of the thinning process. We have defined $R(t)$ as the number of registered events up to time t. That means $N(t) - R(t)$ is the number of unregistered events up to time t. We say the $N(t)$ is split into $R(t)$ and $N(t) - R(t)$.

Theorem 3.10. (Splitting). $\{N(t) - R(t), t \geq 0\}$ is a $\text{PP}(\lambda(1-p))$ and is independent of $\{R(t), t \geq 0\}$.

Proof. Each event of the $\{N(t), t \geq 0\}$ process is counted in the $\{N(t) - R(t), t \geq 0\}$ process with probability $1 - p$. Hence it follows from Theorem 3.9 that $\{N(t) - R(t), t \geq 0\}$ is a $\text{PP}(\lambda(1-p))$. Proving independence is more involved. We shall satisfy ourselves with the independence of the marginal distributions. This follows directly from Equation (3.23), which implies that $R(t)$ and $N(t) - R(t)$ are independent for a fixed t. ∎

Example 3.8. (Thinning). Let $N(t)$ be the number of births in a hospital over $(0, t]$. Suppose $\{N(t), t \geq 0\}$ is a Poisson process with rate 16 per day. The hospital experience shows that 48% of the births are boys and 52% are girls. What is the probability that we see six boys and eight girls born on a given day?

We assume that each baby is a boy with probability .48 and a girl with probability .52, independent of everything else. Let $B(t)$ be the number of boys born in a hospital over $(0, t]$. Similarly, let $G(t)$ be the number of girls born at this hospital over $(0, t]$. Then, from Theorem 3.10, we see that $\{B(t), t \geq 0\}$ is a $\text{PP}(16*.48) = \text{PP}(7.68)$ and is independent of $\{G(t), t \geq 0\}$, which is a $\text{PP}(16*.52) = \text{PP}(8.32)$. Thus $B(1) \sim P(7.68)$ is independent of $G(1) \sim P(8.32)$. Hence the desired probability is given by

$$P(B(1) = 6, G(1) = 8) = e^{-7.68}\frac{7.68^6}{6!}e^{-8.32}\frac{8.32^8}{8!} = 0.0183.$$ ∎

3.6 Compound Poisson Processes

From Definition 3.5 and Figure 3.1, we see that events occur one at a time in a Poisson process. Now we define a process that is useful in modeling a situation when multiple events can occur at the same time. For example, customers arrive in batches at a restaurant. Thus, if we define $X(t)$ to be the total number of arrivals at this restaurant over $(0, t]$, it cannot be adequately modeled as a Poisson process. We introduce the compound Poisson processes as a model of such a situation.

Definition 3.6. (Compound Poisson Process). Let $\{N(t), t \geq 0\}$ be a Poisson process with rate λ. Let $\{C_n, n \geq 1\}$ be a sequence of iid random variables with mean τ and second moment s^2. Suppose it is also independent of the Poisson process. Let

$$C(t) = \sum_{n=1}^{N(t)} C_n, \quad t \geq 0. \tag{3.28}$$

The stochastic process $\{C(t), t \geq 0\}$ is called a *compound Poisson process (CPP)*.

The next theorem gives the expressions for the mean and variance of $C(t)$. The computation of the distribution of $C(t)$ is hard.

Theorem 3.11. (Mean and Variance of a CPP). *Let $\{C(t), t \geq 0\}$ be a CPP. Then*

$$\mathsf{E}(C(t)) = \tau\lambda t, \tag{3.29}$$

$$\mathrm{Var}(C(t)) = s^2\lambda t. \tag{3.30}$$

Proof. From Section 3.3 we see that $N(t)$ is a Poisson random variable with parameter λt, and hence,

$$\mathsf{E}(N(t)) = \mathrm{Var}(N(t)) = \lambda t.$$

Note that $C(t)$ is a random sum of random variables. Using Equation (D.9), we get

$$\mathsf{E}(C(t)) = \mathsf{E}\left(\sum_{n=1}^{N(t)} C_n\right)$$
$$= \mathsf{E}(C_1)\mathsf{E}(N(t))$$
$$= \tau\lambda t.$$

From Equation (D.10), we get

$$\mathrm{Var}(C(t)) = \mathrm{Var}\left(\sum_{n=1}^{N(t)} C_n\right)$$
$$= \mathsf{E}(N(t))\mathrm{Var}(C_1) + (\mathsf{E}(C_1))^2\mathrm{Var}(N(t))$$
$$= \lambda t(s^2 - \tau^2 + \tau^2)$$
$$= \lambda t s^2.$$

This proves the theorem. ∎

Equation (3.29) makes intuitive sense since λt is the expected number of batches that arrive until t and each batch brings in τ customers on average. Equation (3.30), on the other hand, is a bit surprising since one would expect the variance of C_1 to appear where s^2 appears. This is the consequence of the variance of $N(t)$. We end with a numerical example.

Example 3.9. (Restaurant Arrival Process). Consider the process of customers arriving at a restaurant. Suppose the customers arrive in parties (or batches) of variable sizes. The successive party sizes are iid random variables and are binomially distributed with parameters $n = 10$ and $p = .4$. Thus the mean size of a party is $\tau = np = 4$ and its variance is $np(1 - p) = 2.4$. Hence the second moment is $s^2 = 2.4 + 4 * 4 = 18.4$. Note that a binomial random variable can take a value 0 with probability $(1 - p)^n$. What does it mean to get a batch arrival of size 0? The parties themselves arrive according to a Poisson process with a rate of 10 per hour. Let $C(t)$ be the total number of arrivals up to time t, assuming $C(0) = 0$. Then $\{C(t), t \geq 0\}$ is a compound Poisson process. Equation (3.29) yields

$$\mathsf{E}(C(t)) = \tau\lambda t = (4)(10)t = 40t,$$

and Equation (3.30) yields

$$\mathrm{Var}(C(t)) = s^2\lambda t = (18.4)(10)t = 184t.$$

Suppose the process above is a valid model of the arrival process over the time interval from 7 p.m. to 10 p.m. Then the expected total number of arrivals over those 3 hours will be given by $40 * 3 = 120$, and its variance is $184*3 = 552$. ∎

3.7 Problems

CONCEPTUAL PROBLEMS

3.1. Find the mode of a Bin(n, p) distribution.

3.2. Find the mode of a P(λ) distribution.

3.3. Find the mode of an Erl(k, λ) density.

3.4. Find the median of an Exp(λ) distribution.

3.5. Compute the hazard rate of an Exp(λ) random variable.

3.6. Let h be the hazard rate of a random variable X with cdf F. Show that

$$1 - F(x) = \exp\left(-\int_0^x h(u)du\right).$$

This shows that the hazard rate uniquely determines the cdf.

3.7. Compute $\mathsf{E}(X(X - 1))$, where X is a P(λ) random variable.

3.8. Let X be an Exp(λ) random variable. Show that

$$\mathsf{E}(X^n) = n!/\lambda^n.$$

3.9. Let X be a $P(\lambda)$ random variable. Show that

$$E(X(X-1)(X-2)\cdots(X-k+1)) = \lambda^k, \quad k \geq 1.$$

3.10. Let $X_1 \sim P(\lambda_1)$ and $X_2 \sim P(\lambda_2)$ be two independent random variables. Show that, given $X_1 + X_2 = k$, $X_i \sim \text{Bin}(k, \lambda_i/(\lambda_1 + \lambda_2))$, $i = 1, 2$.

3.11. Suppose $T_i, i = 1, 2$, the lifetimes of machine i, are independent $\text{Exp}(\lambda_i)$ random variables. Suppose machine 1 fails before machine 2. Show that the remaining lifetime of machine 2 from the time of failure of machine 1 is an $\text{Exp}(\lambda_2)$ random variable; i.e., show that

$$P(T_2 > T_1 + t | T_2 > T_1) = e^{-\lambda_2 t}.$$

This is sometimes called the *strong memoryless property.*

3.12. Let T_1 and T_2 be two iid $\text{Exp}(\lambda)$ random variables. Using Conceptual Problem 3.11 or otherwise, show that

$$E(\max(T_1, T_2)) = \frac{1}{2\lambda} + \frac{1}{\lambda}.$$

3.13. Generalize Conceptual Problem 3.12 as follows. Let $T_i, 1 \leq i \leq k$, be k iid $\text{Exp}(\lambda)$ random variables. Define

$$T = \max(T_1, T_2, \cdots, T_k).$$

Show that

$$E(T) = \frac{1}{\lambda} \sum_{i=1}^{k} \frac{1}{i}.$$

3.14. Let T be as in Conceptual Problem 3.13. Compute $P(T \leq t)$.

3.15. A system consists of n independent components. Let X_i be the lifetime of the ith component, and assume that $\{X_i, 1 \leq i \leq n\}$ are iid $\text{Exp}(\lambda)$. The system as a whole functions if at least k components are functioning at time T $(1 \leq k \leq n)$, where T is the mission time of the system. What is the probability that the system functions? (Such a system is called a k out of n system.)

3.16. Let $\{X_n, n = 1, 2, \cdots\}$ be a set of iid $\text{Exp}(\lambda)$ random variables and N be an independent $G(p)$ random variable. Show that

$$Z = X_1 + X_2 + \cdots + X_N$$

is an $\text{Exp}(\lambda p)$ random variable.

3.17. Let $\{N(t), t \geq 0\}$ be a $PP(\lambda)$. Let S_k be the time of occurrence of the kth event in this Poisson process. Show that, given $N(t) = 1$, S_1 is uniformly distributed over $[0, t]$.

3.18. Suppose customers arrive at a service station with two servers according to a PP(λ). An arriving customer is routed to server 1 or server 2 with equal probability. What are the mean and variance of the time between two consecutive arrivals to the first server? Next suppose we route the arriving customers to the two servers alternately. Now what are the mean and variance of the time between two consecutive arrivals to the first server? Is the arrival process of the customers to server 1 a Poisson process?

3.19. Let $\{N(t), t \geq 0\}$ be a PP(λ). Let s, $t \geq 0$. Compute the joint distribution

$$P(N(s) = i, N(s + t) = j), \quad 0 \leq i \leq j < \infty.$$

3.20. Let $\{N(t), t \geq 0\}$ be a PP(λ). Show that

$$\text{Cov}(N(s), N(s + t)) = \lambda s, \quad s, t \geq 0.$$

3.21. A bus with infinite capacity runs on an infinite route with stops indexed $n = 1, 2, \cdots$. The bus arrives empty at stop 1. When the bus arrives at stop n, each passenger on it gets off with probability p, independent of each other. Then all the waiting passengers get on. Let W_n be the number of passengers waiting at stop n, and suppose $\{W_n, n \geq 0\}$ are iid P(λ) random variables. Let R_n be the number of passengers on the bus when it leaves stop n. What is the distribution of R_n?

3.22. In Computational Problem 3.21, compute the expected number of passengers that get off at stop n.

3.23. Let $\{C(t), t \geq 0\}$ be a CPP with batch sizes of mean 1 and second moment 1. Show that $\{C(t), t \geq 0\}$ is in fact a PP(λ).

3.24. Let $\{C(t), t \geq 0\}$ be a CPP of Definition 3.6. Compute $E(C(t + s) - C(s))$ and $\text{Var}(C(t + s) - C(s))$ for $s, t \geq 0$.

3.25. Let $\{C(t), t \geq 0\}$ be a CPP of Definition 3.6. Compute $\text{Cov}(C(t + s), C(s))$ for $s, t \geq 0$.

COMPUTATIONAL PROBLEMS

3.1. The lifetime of a car (in miles) is an Exp(λ) random variable with $1/\lambda = 130{,}000$ miles. If the car is driven 13,000 miles a year, what is the distribution of the lifetime of the car in years?

3.2. The lifetimes of two car batteries (brands A and B) are independent exponential random variables with means 12 hours and 10 hours, respectively. What is the probability that the Brand B battery outlasts the Brand A battery?

3.3. Suppose a machine has three independent components with Exp(.1) lifetimes. Compute the expected lifetime of the machine if it needs all three components to function properly.

3.4. A communications satellite is controlled by two computers. It needs at least one functioning computer to operate properly. Suppose the lifetimes of the computers are iid exponential random variables with mean 5 years. Once a computer fails, it cannot be repaired. What is the cdf of the lifetime of the satellite? What is the expected lifetime of the satellite?

3.5. The run times of two computer programs, A and B, are independent random variables. The run time of program A is an exponential random variable with mean 3 minutes, while that of program B is an Erl$(2, \lambda)$ random variable with mean 3 minutes. If both programs are being executed on two identical computers, what is the probability that program B will finish first?

3.6. A spacecraft is sent to observe Jupiter. It is supposed to take pictures of Jupiter's moons and send them back to Earth. There are three critical systems involved: the camera, the batteries, and the transmission antenna. These three systems fail independently of each other. The mean lifetime of the battery system is 6 years, that of the camera is 12 years, and that of the antenna is 10 years. Assume all lifetimes are exponential random variables. The spacecraft is to reach Jupiter after 3 years to carry out this mission. What is the probability that the mission is successful?

3.7. In Computational Problem 3.6, suppose the monitoring station on Earth conducts a test run after 2 years and finds that no data are received, indicating that one of the three systems has failed. What is the probability that the camera has failed?

3.8. A car comes equipped with one spare tire. The lifetimes of the four tires at the beginning of a long-distance trip are iid exponential random variables with a mean of 5000 miles. The spare tire has an exponential lifetime with a mean of 1000 miles. Compute the expected number of miles that can be covered without having to go to a tire shop. (*Hint:* Compute the expected time until the first tire failure and then that until the second tire failure.)

3.9. A customer enters a bank and finds that all three tellers are occupied and he is the next in line. Suppose the mean transaction times are 5 minutes and are independent exponential random variables. What is the expected amount of time that he has to wait before he is served?

3.10. In Computational Problem 3.9, suppose the bank also has a teller machine outside with one person currently using it and no one waiting behind him. The processing time at the machine is exponentially distributed with mean 2 minutes. Should the customer in Computational Problem 3.9 wait at the machine or at the inside tellers if the aim is to minimize the expected waiting time until the start of service? Assume that once he decides to wait in one place, he does not change his mind.

3.11. The weight of a ketchup bottle is Erlang distributed with $k = 10$ and $\lambda = .25$ Oz^{-1}. The bottle is declared to be underweight if the weight is under 39 Oz. What is the probability that the bottle is declared underweight?

3.12. Two vendors offer functionally identical products. The expected lifetime of both the products is 10 months. However, the distribution of the first is $Exp(\lambda)$, while that of the other is $Erl(2, \mu)$. If the aim is to maximize the probability that the lifetime of the product is greater than 8 months, which of the two products should be chosen?

3.13. The lifetime of an item is an exponential random variable with mean 5 days. We replace this item upon failure or at the end of the fourth day, whichever occurs first. Compute the expected time of replacement.

3.14. Consider the 2 out of 3 system of Conceptual Problem 3.15 with the mean lifetime of each component being 3 years. Suppose that initially all the components are functioning. We visit this system after 3 years and replace all the failed components at a cost of $75.00 each. If the system has failed, it costs us an additional $1000.00. Compute the expected total cost incurred at time 3.

3.15. In Computational Problem 3.9, suppose customers arrive according to a PP with rate 12 per hour. At time zero, all three tellers are occupied and there is no one waiting. What is the probability that the next arriving customer will have to wait for service?

3.16. The number of industrial accidents in a factory is adequately modeled as a PP with a rate of 1.3 per year. What is the probability that there are no accidents in one year? What is the expected time (in days) between two consecutive accidents?

3.17. A computer system is subject to failures and self-repairs. The failures occur according to a PP with rate 2 per day. The self-repairs are instantaneous. The only effect of a failure is that the work done on any program running on the machine at the time of failure is lost and the program has to be restarted from scratch. Now consider a program whose execution time is 3 hours if uninterrupted by failures. What is the probability that the program completes successfully without a restart?

3.18. Redo Computational Problem 3.17 if the program execution time (in hours) is a random variable that is uniformly distributed over (2,4).

3.19. A machine is subject to shocks arriving from two independent sources. The shocks from source 1 arrive according to a PP with rate 3 per day and those from source 2 at rate 4 per day. What are the mean and variance of the total number of shocks from both the sources over an 8-hour shift.

3.20. In Computational Problem 3.19, what is the probability that the machine is subject to a total of exactly two shocks in 1 hour?

3.21. In Computational Problem 3.19, suppose a shock from source i can cause the machine to fail with probability p_i, independent of everything else. Suppose $p_1 = .011$ and $p_2 = .005$. A failed machine is replaced instantaneously. Let $N(t)$ be the number of machine replacements over the interval $(0, t]$. Is $\{N(t), t \geq 0\}$ a Poisson process? If it is, what is its parameter?

3.22. Consider a maternity ward in a hospital. A delivery may result in one, two or three births with probabilities .9, .08, and .02, respectively. The number of deliveries forms a Poisson process with rate 10 per day. Formulate the number of births as a compound Poisson process, and compute the mean number of births during one day.

3.23. In Computational Problem 3.22, compute the probability of at least one twin being born on a given day.

3.24. In Computational Problem 3.22, suppose a twin is born during a 24-hour interval. What is the expected number of deliveries that occurred during that interval?

3.25. Suppose 45% of the cars on the road are domestic brands and 55% are foreign brands. Let $N(t)$ be the number of cars crossing at a given toll booth during $(0, t]$. Suppose $\{N(t), t \geq 0\}$ is a PP with rate 40 per minute. What is the probability that four foreign cars and five domestic cars cross the intersection during a 10-second interval?

3.26. In Computational Problem 3.25, suppose 30 foreign cars have crossed the toll booth during a 1-minute interval. What is the expected number of domestic cars that cross the intersection during this interval given this information?

3.27. Consider the toll booth of Computational Problem 3.25. Suppose the toll paid by a car at the toll booth is $1.00 with probability .3 and $2.00 with probability .7. What are the mean and variance of the toll collected at the toll booth during 1 hour?

3.28. Suppose the distribution of the toll depends on whether the car is domestic or foreign as follows. The domestic cars pay $1.00 with probability .4 and $2.00 with probability .6, while the foreign cars pay $1.00 with probability .25 and $2.00 with probability .75. Now compute the mean and the variance of the total toll collected at the booth over a 1-hour period.

3.29. The number of cars visiting a national park forms a PP with rate 15 per hour. Each car has k occupants with probability p_k as given below:

$$p_1 = .2, \; p_2 = .3, \; p_3 = .3, \; p_4 = .1, \; p_5 = .05, \; p_6 = .05.$$

Compute the mean and variance of the number of visitors to the park during a 10-hour window.

3.30. Now suppose the national park of Computational Problem 3.29 charges $4.00 per car plus $1.00 per occupant as the entry fee. Compute the mean and variance of the total fee collected during a 10-hour window.

Chapter 4
Continuous-Time Markov Models

4.1 Continuous-Time Markov Chains

In this chapter, we consider a system that is observed continuously, with $X(t)$ being the state at time $t, t \geq 0$. Following the definition of DTMCs, we define *continuous-time Markov chains* (CTMCs) below.

Definition 4.1. (Continuous-Time Markov Chain (CTMC)). A stochastic process $\{X(t), t \geq 0\}$ on state space S is called a CTMC if, for all i and j in S and $t, s \geq 0$,

$$P(X(s+t) = j | X(s) = i, X(u), 0 \leq u \leq s) = P(X(s+t) = j | X(s) = i).$$
(4.1)

The CTMC $\{X(t), t \geq 0\}$ is said to be time homogeneous if, for $t, s \geq 0$,

$$P(X(s+t) = j | X(s) = i) = P(X(t) = j | X(0) = i).$$
(4.2)

Equation (4.1) implies that the evolution of a CTMC from a fixed time s onward depends on $X(u), 0 \leq u \leq s$ (the history of the CTMC up to time s) only via $X(s)$ (the state of the CTMC at time s). Furthermore, for time-homogeneous CTMCs, it is also independent of s! In this chapter, we shall assume, unless otherwise mentioned, that all CTMCs are time homogeneous and have a finite state space $\{1, 2, \ldots, N\}$. For such CTMCs, we define

$$p_{i,j}(t) = P(X(t) = j | X(0) = i), \quad 1 \leq i, j \leq N.$$
(4.3)

The N^2 entities $p_{i,j}(t)$ are arranged in matrix form as follows:

$$P(t) = \begin{bmatrix} p_{1,1}(t) & p_{1,2}(t) & p_{1,3}(t) & \cdots & p_{1,N}(t) \\ p_{2,1}(t) & p_{2,2}(t) & p_{2,3}(t) & \cdots & p_{2,N}(t) \\ p_{3,1}(t) & p_{3,2}(t) & p_{3,3}(t) & \cdots & p_{3,N}(t) \\ \vdots & \vdots & \vdots & \ddots & \vdots \\ p_{N,1}(t) & p_{N,2}(t) & p_{N,3}(t) & \cdots & p_{N,N}(t) \end{bmatrix}.$$
(4.4)

V.G. Kulkarni, *Introduction to Modeling and Analysis of Stochastic Systems*,
Springer Texts in Statistics, DOI 10.1007/978-1-4419-1772-0_4,
© Springer Science+Business Media, LLC 2011

The matrix above is called the *transition probability matrix* of the CTMC $\{X(t), t \geq 0\}$. Note that $P(t)$ is a matrix of functions of t; i.e., it has to be specified for each t. The following theorem gives the important properties of the transition probability matrix $P(t)$.

Theorem 4.1. *(Properties of $P(t)$). A transition probability matrix $P(t) = [p_{i,j}(t)]$ of a time-homogeneous CTMC on state space $S = \{1, 2, \ldots, N\}$ satisfies the following:*

1.
$$p_{i,j}(t) \geq 0, \quad 1 \leq i, j \leq N; t \geq 0, \tag{4.5}$$

2.
$$\sum_{j=1}^{N} p_{i,j}(t) = 1, \quad 1 \leq i \leq N; t \geq 0, \tag{4.6}$$

3. $p_{i,j}(s+t) = \displaystyle\sum_{k=1}^{N} p_{i,k}(s) p_{k,j}(t) = \sum_{k=1}^{N} p_{i,k}(t) p_{k,j}(s), \quad 1 \leq i \leq N; t, s \geq 0.$
$$\tag{4.7}$$

Proof. **(1)** The nonnegativity of $p_{i,j}(t)$ follows because it is a (conditional) probability.

(2) To prove the second assertion, we have

$$\sum_{j=1}^{N} p_{i,j}(t) = \sum_{j=1}^{N} \mathsf{P}(X(t) = j | X(0) = i)$$

$$= \mathsf{P}(X(t) \in S | X(0) = i).$$

Since $X(t)$ must take some value in the state space S, regardless of the value of $X(0)$, it follows that the last quantity is 1.

(3) To prove (4.7), we have

$$p_{i,j}(s+t) = \mathsf{P}(X(s+t) = j | X(0) = i)$$

$$= \sum_{k=1}^{N} \mathsf{P}(X(s+t) = j | X(s) = k, X(0) = i) \mathsf{P}(X(s) = k | X(0) = i)$$

$$= \sum_{k=1}^{N} \mathsf{P}(X(s+t) = j | X(s) = k) p_{i,k}(s)$$

(due to the Markov property at s)

$$= \sum_{k=1}^{N} P(X(t) = j | X(0) = k) p_{i,k}(s)$$

(due to time homogeneity)

$$= \sum_{k=1}^{N} p_{i,k}(s) p_{k,j}(t).$$

The other part of (4.7) follows by interchanging s and t. This proves the theorem. ■

Equation (4.7) states the *Chapman–Kolmogorov equations* for CTMCs and can be written in matrix form as

$$P(t + s) = P(s)P(t) = P(t)P(s). \tag{4.8}$$

Thus the transition probability matrices $P(s)$ and $P(t)$ commute! Since AB is generally not the same as BA for square matrices A and B, the property above implies that the transition probability matrices are very special indeed!

Example 4.1. (Two-State CTMC). Suppose the lifetime of a high-altitude satellite is an $\text{Exp}(\mu)$ random variable. Once it fails, it stays failed forever since no repair is possible. Let $X(t)$ be 1 if the satellite is operational at time t and 0 otherwise.

We first show that $\{X(t), t \geq 0\}$ is a CTMC. The state space is $\{0, 1\}$. We directly check (4.1) and (4.2). Consider the case $i = j = 1$. Let the lifetime of the satellite be T. Then $X(s) = 1$ if and only if $T > s$, and if $X(s) = 1$ then $X(u)$ must be 1 for $0 \leq u \leq s$. Hence, using the memoryless property (see Equation (3.5)) of the exponential variable T, we have

$$\begin{aligned}
P(X(s + t) = 1 | X(s) &= 1, X(u) : 0 \leq u \leq s) \\
&= P(T > s + t | T > s) = e^{-\mu t} \\
&= P(X(t) = 1 | X(0) = 1).
\end{aligned}$$

Performing similar calculations for all pairs i, j, we establish that $\{X(t), t \geq 0\}$ is a CTMC. Next we compute its transition probability matrix. We have

$$p_{0,0}(t) = P(\text{satellite is down at } t | \text{satellite is down at } 0) = 1,$$
$$p_{1,1}(t) = P(\text{satellite is up at } t | \text{satellite is up at } 0) = P(T > t) = e^{-\mu t}.$$

Hence the transition probability matrix is given by

$$P(t) = \begin{bmatrix} 1 & 0 \\ 1 - e^{-\mu t} & e^{-\mu t} \end{bmatrix}. \tag{4.9}$$

We can check by direct calculations that (4.8) are satisfied. ■

In Chapter 2, we described a DTMC by giving its one-step transition probability matrix. How do we describe a CTMC? Giving $P(t)$ for each t is too complicated for most CTMCs. We need a simpler method of describing a CTMC. We develop such a method below based on the insight provided by the example above.

The discussion above implies that the stochastic process $\{X(t), t \geq 0\}$ of Example 4.1 can be a CTMC if and only if the lifetime of the satellite has the memoryless property. We have seen in Theorem 3.1 that the only continuous random variable with the memoryless property is the exponential random variable. This suggests that we should build CTMCs using exponential random variables. This is what we do next.

Let $X(t)$ be the state of a system at time t. Suppose the state space of the stochastic process $\{X(t), t \geq 0\}$ is $\{1, 2, \ldots, N\}$. The random evolution of the system occurs as follows. Suppose the system starts in state i. It stays there for an $\text{Exp}(r_i)$ amount of time, called the *sojourn time in state i*. At the end of the sojourn time in state i, the system makes a sudden transition to state $j \neq i$ with probability $p_{i,j}$, independent of how long the system has been in state i. Once in state j, it stays there for an $\text{Exp}(r_j)$ amount of time and then moves to a new state $k \neq j$ with probability $p_{j,k}$, independently of the history of the system so far. And it continues this way forever. A typical sample path of the system is shown in Figure 4.1.

Three remarks are in order at this time. First, the jump probabilities $p_{i,j}$ are not to be confused with the transition probabilities $p_{i,j}(t)$ of (4.3). Think of $p_{i,j}$ as the probability that the system moves to state j when it moves out of state i. Second, $p_{i,i} = 0$. This is because, by definition, the sojourn time in state i is the time the system spends in state i until it moves out of it. Hence, a transition from i to i is not allowed. (Contrast this with the DTMC.) Third, in case state i is absorbing (i.e., the system stays in state i forever once it gets there), we set $r_i = 0$, interpreting an $\text{Exp}(0)$ random variable as taking a value equal to ∞ with probability 1. In such a case, $p_{i,j}$'s have no meaning and can be left undefined.

Theorem 4.2. *The stochastic process $\{X(t), t \geq 0\}$ with parameters $r_i, 1 \leq i \leq N$, and $p_{i,j}, 1 \leq i, j \leq N$, as described above is a CTMC.*

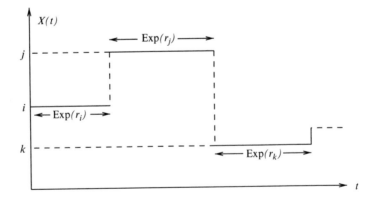

Fig. 4.1 A typical sample path of a CTMC.

Idea of Proof. The rigorous proof of this theorem is tedious, but the idea is simple. Let $s, t \geq 0$ be fixed. Suppose we are given the history of the system up to s, i.e., we are given $X(u), 0 \leq u \leq s$. Also, suppose $X(s) = i$. Then, using the memoryless property of the exponential random variable, we see that the remaining sojourn time of the system in state i from time s onward is again an $\text{Exp}(r_i)$ random variable, independent of the past. At the end of the sojourn time, the system will move to state j with probability $p_{i,j}$. The subsequent sojourn times and the transitions are independent of the history by construction. Hence the evolution of the system from time s onward depends only on i, the state of the system at time s. Hence $\{X(t), t \geq 0\}$ is a CTMC. ∎

What is even more important is that all CTMCs on finite state spaces that have nonzero sojourn times in each state can be described this way. We shall not prove this fact here. In all our applications, this condition will be met and hence we can describe a CTMC by giving the parameters $\{r_i, 1 \leq i \leq N\}$ and $\{p_{i,j}, 1 \leq i, j \leq N\}$. This is a much simpler description than giving the transition probability matrix $P(t)$ for all $t \geq 0$.

Analogously to DTMCs, a CTMC can also be represented graphically by means of a directed graph as follows. The directed graph has one node (or vertex) for each state. There is a directed arc from node i to node j if $p_{i,j} > 0$. The quantity $r_{i,j} = r_i p_{i,j}$, called the *transition rate from i to j*, is written next to this arc. Note that there are no self-loops (arcs from i to i) in this representation. This graphical representation is called the *rate diagram* of the CTMC.

Rate diagrams are useful visual representations. We can understand the dynamics of the CTMC by visualizing a particle that moves from node to node in the rate diagram as follows: it stays on node i for an $\text{Exp}(r_i)$ amount of time and then chooses one of the outgoing arcs from node i with probabilities proportional to the rates on the arcs and moves to the node at the other end of the arc. This motion continues forever. The node occupied by the particle at time t is the state of the CTMC at time t.

Note that we can recover the original parameters $\{r_i, 1 \leq i \leq N\}$ and $\{p_{i,j}, 1 \leq i, j \leq N\}$ from the rates $\{r_{i,j}, 1 \leq i, j \leq N\}$ shown in the rate diagram as follows:

$$r_i = \sum_{j=1}^{N} r_{i,j},$$
$$p_{i,j} = \frac{r_{i,j}}{r_i} \text{ if } r_i \neq 0.$$

It is convenient to put all the rates $r_{i,j}$ in matrix form as follows:

$$R = \begin{bmatrix} r_{1,1} & r_{1,2} & r_{1,3} & \cdots & r_{1,N} \\ r_{2,1} & r_{2,2} & r_{2,3} & \cdots & r_{2,N} \\ r_{3,1} & r_{3,2} & r_{3,3} & \cdots & r_{3,N} \\ \vdots & \vdots & \vdots & \ddots & \vdots \\ r_{N,1} & r_{N,2} & r_{N,3} & \cdots & r_{N,N} \end{bmatrix}. \tag{4.10}$$

Note that $r_{i,i} = 0$ for all $1 \leq i \leq N$, and hence the diagonal entries in R are always zero. R is called the *rate matrix* of the CTMC. It is closely related to $Q = [q_{i,j}]$, called the *generator matrix* of the CTMC, which is defined as

$$q_{i,j} = \begin{cases} -r_i & \text{if } i = j, \\ r_{i,j} & \text{if } i \neq j. \end{cases} \tag{4.11}$$

Thus the generator matrix Q is the same as the rate matrix R with the diagonal elements replaced by $-r_i$'s. It is common in the literature to describe a CTMC by the Q matrix. However, we shall describe it by the R matrix. Clearly, one can be obtained from the other.

Example 4.2. (Two-State CTMC). Consider the CTMC of Example 4.1. The system stays in state 1 (satellite up) for an $\text{Exp}(\mu)$ amount of time and then moves to state 0. Hence, $r_1 = \mu$ and $p_{1,0} = 1$. Once the satellite fails (state 0), it stays failed forever. Thus state 0 is absorbing. Hence $r_0 = 0$, and $p_{0,1}$ is left undefined. The rate matrix is

$$R = \begin{bmatrix} 0 & 0 \\ \mu & 0 \end{bmatrix}. \tag{4.12}$$

The generator matrix is given by

$$Q = \begin{bmatrix} 0 & 0 \\ \mu & -\mu \end{bmatrix}. \tag{4.13}$$

The rate diagram is shown in Figure 4.2. ∎

Example 4.3. (Two-State Machine). Consider a machine that operates for an $\text{Exp}(\mu)$ amount of time and then fails. Once it fails, it gets repaired. The repair time is an $\text{Exp}(\lambda)$ random variable and is independent of the past. The machine is as good as new after the repair is complete. Let $X(t)$ be the state of the machine at time t, 1 if it is up and 0 if it is down. Model this as a CTMC.

The state space of $\{X(t), t \geq 0\}$ is $\{0, 1\}$. The sojourn time in state 0 is the repair time, which is given as an $\text{Exp}(\lambda)$ random variable. Hence $r_0 = \lambda$. After the repair is complete, the machine is up with probability 1; hence $p_{0,1} = 1$. Similarly, $r_1 = \mu$ and $p_{1,0} = 1$. The independence assumptions are satisfied, thus making $\{X(t), t \geq 0\}$ a CTMC. The rate matrix is

$$R = \begin{bmatrix} 0 & \lambda \\ \mu & 0 \end{bmatrix}. \tag{4.14}$$

Fig. 4.2 Rate diagram of the satellite system.

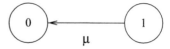

Fig. 4.3 Rate diagram of the
two-state machine.

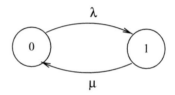

The generator matrix is given by

$$Q = \begin{bmatrix} -\lambda & \lambda \\ \mu & -\mu \end{bmatrix}. \tag{4.15}$$

The rate diagram is shown in Figure 4.3. Note that describing this CTMC by giving its $P(t)$ matrix would be very difficult since the machine can undergo many failures and repairs up to time t. ∎

In the next section, we use the properties of exponential random variables studied in Chapter 3 and develop CTMC models of real-life systems. We give the rate matrix R for the CTMCs and the rate diagrams but omit the generator matrix Q.

4.2 Examples of CTMCs: I

In this section, we consider CTMC models that are built using the exponential random variables and their properties as studied in Chapter 3.

Example 4.4. (Two-Machine Workshop). Consider a workshop with two independent machines, each with its own repair person and each machine behaving as described in Example 4.3. Let $X(t)$ be the number of machines operating at time t. Model $\{X(t), t \geq 0\}$ as a CTMC.

The state space of $\{X(t), t \geq 0\}$ is $\{0, 1, 2\}$. Consider state 0. In this state, both machines are down and hence under repair. Due to the memoryless property, the remaining repair times of the two machines are iid $\text{Exp}(\lambda)$ random variables. The system moves from state 0 to state 1 as soon as one of the two machines finishes service. Hence the sojourn time in state 0 is a minimum of two iid $\text{Exp}(\lambda)$ random variables. Furthermore, by Theorem 3.3, the sojourn time is an $\text{Exp}(2\lambda)$ random variable. Hence $r_0 = 2\lambda$ and $p_{0,1} = 1$. The transition rate from state 0 to state 1 is thus $r_{0,1} = 2\lambda$.

Now consider state 1, where one machine is up and one is down. Let T_1 be the remaining repair time of the down machine, and let T_2 be the remaining life time of the up machine. Again from the memoryless property of the exponential random variables, T_1 is $\text{Exp}(\lambda)$ and T_2 is $\text{Exp}(\mu)$, and they are independent. The sojourn time in state 1 is $\min(T_1, T_2)$, which is an $\text{Exp}(\lambda + \mu)$ random variable. Hence $r_1 = \lambda + \mu$. The system moves to state 2 if $T_1 < T_2$, which has probability $\lambda/(\lambda + \mu)$ and it moves to state 0 if $T_2 < T_1$, which has probability $\mu/(\lambda + \mu)$.

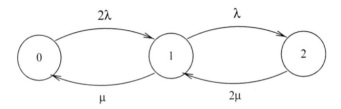

Fig. 4.4 Rate diagram of the two-machine workshop.

Note that the next state is independent of the sojourn time from Theorem 3.4. Hence $p_{1,2} = \lambda/(\lambda + \mu)$ and $p_{1,0} = \mu/(\lambda + \mu)$. The transition rates are $r_{1,2} = \lambda$ and $r_{1,0} = \mu$.

A similar analysis of state 2 yields $r_2 = 2\mu$, $p_{2,1} = 1$, and $r_{2,1} = 2\mu$. Thus $\{X(t), t \geq 0\}$ is a CTMC with rate matrix

$$R = \begin{bmatrix} 0 & 2\lambda & 0 \\ \mu & 0 & \lambda \\ 0 & 2\mu & 0 \end{bmatrix}. \tag{4.16}$$

The rate diagram is as shown in Figure 4.4. ∎

The example above produces the following extremely useful insight. Suppose the state of the system at time t is $X(t) = i$. Suppose there are m different events that can trigger a transition in the system state from state i to state j. Suppose the occurrence time of the kth event ($1 \leq k \leq m$) is an $\text{Exp}(\lambda_k)$ random variable. Then, assuming the m events are independent of each other and of the history of the system up to time t, the transition rate from state i to state j is given by

$$r_{i,j} = \lambda_1 + \lambda_2 + \cdots + \lambda_m.$$

If the transition rates can be computed this way for all i and j in the state space, then $\{X(t), t \geq 0\}$ is a CTMC with transition rates $r_{i,j}$. We use this observation in the remaining examples.

Example 4.5. (General Machine Shop). A machine shop consists of N machines and M repair persons. ($M \leq N$.) The machines are identical, and the lifetimes of the machines are independent $\text{Exp}(\mu)$ random variables. When the machines fail, they are serviced in the order of failure by the M repair persons. Each failed machine needs one and only one repair person, and the repair times are independent $\text{Exp}(\lambda)$ random variables. A repaired machine behaves like a new machine. Let $X(t)$ be the number of machines that are functioning at time t. Model $\{X(t), t \geq 0\}$ as a CTMC.

We shall explain with a special case: four machines and two repair persons. The general case is similar. For the special case, the state space of $\{X(t), t \geq 0\}$ is $S = \{0, 1, 2, 3, 4\}$. We obtain the transition rates below. In state 0, all machines are down, and two are under repair. (The other two are waiting for a repair person.) The repair times are iid $\text{Exp}(\lambda)$, and either of the two repair completions will move the system

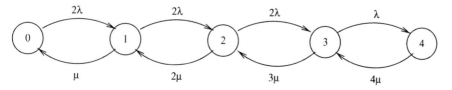

Fig. 4.5 Rate diagram of the general machine shop with four machines and two repair persons.

to state 1. Hence $r_{0,1} = \lambda + \lambda = 2\lambda$. In state 1, one machine is up and three are down, two of which are under repair. Either repair completion will send the system to state 2. Hence $r_{1,2} = 2\lambda$. The failure of the functioning machine, which will happen after an $\text{Exp}(\mu)$ amount of time, will trigger a transition to state 0. Hence $r_{1,0} = \mu$. Similarly, for state 2, we get $r_{2,3} = 2\lambda$ and $r_{2,1} = 2\mu$. In state 3, three machines are functioning and one is down and under repair. (One repair person is idle.) Thus there are three independent failure events that will trigger a transition to state 2, and a repair completion will trigger a transition to state 4. Hence $r_{3,2} = 3\mu$ and $r_{3,4} = \lambda$. Similarly, for state 4, we get $r_{4,3} = 4\mu$.

Thus $\{X(t), t \geq 0\}$ is a CTMC with a rate matrix as follows:

$$R = \begin{bmatrix} 0 & 2\lambda & 0 & 0 & 0 \\ \mu & 0 & 2\lambda & 0 & 0 \\ 0 & 2\mu & 0 & 2\lambda & 0 \\ 0 & 0 & 3\mu & 0 & \lambda \\ 0 & 0 & 0 & 4\mu & 0 \end{bmatrix}. \tag{4.17}$$

The rate diagram is as shown in Figure 4.5. ∎

Example 4.6. (Airplane Reliability). A commercial jet airplane has four engines, two on each wing. Each engine lasts for a random amount of time that is an exponential random variable with parameter λ and then fails. If the failure takes place in flight, there can be no repair. The airplane needs at least one engine on each wing to function properly in order to fly safely. Model this system so that we can predict the probability of a trouble-free flight.

Let $X_L(t)$ be the number of functioning engines on the left wing at time t, and let $X_R(t)$ be the number of functioning engines on the right wing at time t. The state of the system at time t is given by $X(t) = (X_L(t), X_R(t))$. Assuming the engine failures are independent of each other, we see that $\{X(t), t \geq 0\}$ is a CTMC on state space

$$S = \{(0,0), (0,1), (0,2), (1,0), (1,1), (1,2), (2,0), (2,1), (2,2)\}.$$

Note that the flight continues in a trouble-free fashion as long as $X_L(t) \geq 1$ and $X_R(t) \geq 1$; i.e., as long as the CTMC is in states $(1,1), (1,2), (2,1), (2,2)$. The flight crashes as soon as the CTMC visits any state in $\{(0,0), (0,1), (0,2), (1,0), (2,0)\}$. We assume that the CTMC continues to evolve even after the airplane crashes!

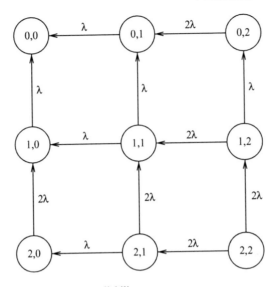

Fig. 4.6 Rate diagram of the airplane reliability system.

Label the states in S 1 through 9 in the order in which they are listed above. Performing the standard triggering event analysis (assuming that no repairs are possible), we get the following rate matrix:

$$R = \begin{bmatrix} 0 & 0 & 0 & 0 & 0 & 0 & 0 & 0 & 0 \\ \lambda & 0 & 0 & 0 & 0 & 0 & 0 & 0 & 0 \\ 0 & 2\lambda & 0 & 0 & 0 & 0 & 0 & 0 & 0 \\ \lambda & 0 & 0 & 0 & 0 & 0 & 0 & 0 & 0 \\ 0 & \lambda & 0 & \lambda & 0 & 0 & 0 & 0 & 0 \\ 0 & 0 & \lambda & 0 & 2\lambda & 0 & 0 & 0 & 0 \\ 0 & 0 & 0 & 2\lambda & 0 & 0 & 0 & 0 & 0 \\ 0 & 0 & 0 & 0 & 2\lambda & 0 & \lambda & 0 & 0 \\ 0 & 0 & 0 & 0 & 0 & 2\lambda & 0 & 2\lambda & 0 \end{bmatrix}. \qquad (4.18)$$

The rate diagram is as shown in Figure 4.6. ∎

4.3 Examples of CTMCs: II

In this section, we consider CTMC models that are built using the Poisson processes and their properties as studied in Chapter 3.

Example 4.7. (Finite-Capacity Single Server Queue). Customers arrive at an automatic teller machine (ATM) according to a PP(λ). The space in front of the ATM can accommodate at most K customers. Thus, if there are K customers waiting at the ATM and a new customer arrives, he or she simply walks away and is lost forever. The customers form a single line and use the ATM in a first-come, first-served fashion. The processing times at the ATM for the customers are iid $Exp(\mu)$ random variables. Let $X(t)$ be the number of customers at the ATM at time t. Model $\{X(t), t \geq 0\}$ as a CTMC.

The state space of the system is $S = \{0, 1, 2, \ldots, K\}$. We shall obtain the transition rates below. In state 0, the system is empty and the only event that can occur is an arrival, which occurs after an $Exp(\lambda)$ amount of time (due to the $PP(\lambda)$ arrival process) and triggers a transition to state 1. Hence $r_{0,1} = \lambda$. In state i, $1 \leq i \leq K - 1$, there are two possible events: an arrival and a departure. An arrival occurs after an $Exp(\lambda)$ amount of time and takes the system to state $i + 1$, while the departure occurs at the next service completion time (which occurs after an $Exp(\mu)$ amount of time due to the memoryless property of the service times) and takes the system to state $i - 1$. Hence $r_{i,i+1} = \lambda$ and $r_{i,i-1} = \mu$, $1 \leq i \leq K - 1$. Finally, in state K, there is no arrival due to capacity limitation. Thus the only event that can occur is a service completion, taking the system to state $K - 1$. Hence $r_{K,K-1} = \mu$. Thus $\{X(t), t \geq 0\}$ is a CTMC with rate matrix

$$R = \begin{bmatrix} 0 & \lambda & 0 & \cdots & 0 & 0 \\ \mu & 0 & \lambda & \cdots & 0 & 0 \\ 0 & \mu & 0 & \cdots & 0 & 0 \\ \vdots & \vdots & \vdots & \ddots & \vdots & \vdots \\ 0 & 0 & 0 & \cdots & 0 & \lambda \\ 0 & 0 & 0 & \cdots & \mu & 0 \end{bmatrix}. \tag{4.19}$$

The rate diagram is as shown in Figure 4.7. We shall see in Chapter 6 that $\{X(t), t \geq 0\}$ is called an $M/M/1/K$ queue in queueing terminology. ∎

The CTMCs in Examples 4.5 and 4.7 have a similar structure: both CTMCs move from state i to state $i - 1$ or to state $i + 1$. There are no other transitions. We describe a general class of such DTMCs below.

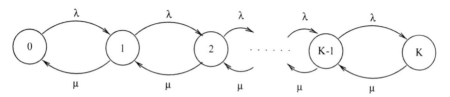

Fig. 4.7 Rate diagram of a single server queue.

Example 4.8. (Finite Birth and Death Process). A CTMC on state space $\{0, 1, 2, \ldots, K\}$ with the rate matrix

$$
R = \begin{bmatrix}
0 & \lambda_0 & 0 & \cdots & 0 & 0 \\
\mu_1 & 0 & \lambda_1 & \cdots & 0 & 0 \\
0 & \mu_2 & 0 & \cdots & 0 & 0 \\
\vdots & \vdots & \vdots & \ddots & \vdots & \vdots \\
0 & 0 & 0 & \cdots & 0 & \lambda_{K-1} \\
0 & 0 & 0 & \cdots & \mu_K & 0
\end{bmatrix}
\tag{4.20}
$$

is called a *finite birth and death process*. The transitions from i to $i + 1$ are called the births, and the λ_i's are called the birth parameters. The transitions from i to $i - 1$ are called the deaths, and the μ_i's are called the death parameters. It is convenient to define $\lambda_K = 0$ and $\mu_0 = 0$, signifying that there are no births in state K and no deaths in state 0. A birth and death process spends an $\mathrm{Exp}(\lambda_i + \mu_i)$ amount of time in state i and then jumps to state $i + 1$ with probability $\lambda_i/(\lambda_i + \mu_i)$ or to state $i - 1$ with probability $\mu_i/(\lambda_i + \mu_i)$. ∎

Birth and death processes occur quite often in applications and hence form an important class of CTMCs. For example, the CTMC model of the general machine shop in Example 4.5 is a birth and death process on $\{0, 1, \ldots, N\}$ with birth parameters

$$
\lambda_i = \min(N - i, M)\lambda, \ \ 0 \le i \le N,
$$

and death parameters

$$
\mu_i = i\mu, \ \ 0 \le i \le N.
$$

The CTMC model of the single server queue as described in Example 4.7 is a birth and death process on $\{0, 1, \ldots, K\}$ with birth parameters

$$
\lambda_i = \lambda, \ \ 0 \le i \le K - 1,
$$

and death parameters

$$
\mu_i = \mu, \ \ 1 \le i \le K.
$$

We illustrate this with more examples.

Example 4.9. (Telephone Switch). A telephone switch can handle K calls at any one time. Calls arrive according to a Poisson process with rate λ. If the switch is already serving K calls when a new call arrives, then the new call is lost. If a call is accepted, it lasts for an $\mathrm{Exp}(\mu)$ amount of time and then terminates. All call durations are independent of each other. Let $X(t)$ be the number of calls that are being handled by the switch at time t. Model $\{X(t), t \ge 0\}$ as a CTMC.

The state space of $\{X(t), t \ge 0\}$ is $\{0, 1, 2, \ldots, K\}$. In state i, $0 \le i \le K - 1$, an arrival triggers a transition to state $i + 1$ with rate λ. Hence $r_{i,i+1} = \lambda$. In state K, there are no arrivals. In state i, $1 \le i \le K$, any of the i calls may complete and trigger a transition to state $i - 1$. The transition rate is $r_{i,i-1} = i\mu$. In state 0, there are no departures. Thus $\{X(t), t \ge 0\}$ is a finite birth and death process with birth parameters

$$\lambda_i = \lambda, \ \ 0 \le i \le K - 1,$$

and death parameters

$$\mu_i = i\mu, \ \ 0 \le i \le K.$$

We shall see in Chapter 6 that $\{X(t), t \ge 0\}$ is called an $M/M/K/K$ queue in queueing terminology. ∎

Example 4.10. (Call Center). An airline phone-reservation system is called a call center and is staffed by s reservation clerks called agents. An incoming call for reservations is handled by an agent if one is available; otherwise the caller is put on hold. The system can put a maximum of H callers on hold. When an agent becomes available, the callers on hold are served in order of arrival. When all the agents are busy and there are H calls on hold, any additional callers get a busy signal and are permanently lost. Let $X(t)$ be the number of calls in the system, those handled by the agents plus any on hold, at time t. Assume the calls arrive according to a PP(λ) and the processing times of the calls are iid Exp(μ) random variables. Model $\{X(t), t \ge 0\}$ as a CTMC.

The state space is $\{0, 1, 2, \ldots, K\}$, where $K = s + H$. Suppose $X(t) = i$. For $0 \le i \le K - 1$, the transition to state $i + 1$ is caused by an arrival, which occurs at rate λ. In state K there are no arrivals. For $1 \le i \le K$, the transition to state $i - 1$ is caused by the completion of the processing of any of the $\min(i, s)$ calls that are under service. The completion rate of an individual call is μ; hence the transition rate from state i to $i - 1$ is $\min(i, s)\mu$. Thus $\{X(t), t \ge 0\}$ is a finite birth and death process with birth parameters

$$\lambda_i = \lambda, \ \ 0 \le i \le K - 1,$$

and death parameters

$$\mu_i = \min(i, s)\mu, \ \ 0 \le i \le K.$$

The rate diagram is shown in Figure 4.8. We shall see in Chapter 6 that $\{X(t), t \ge 0\}$ is called an $M/M/s/K$ queue in queueing terminology. ∎

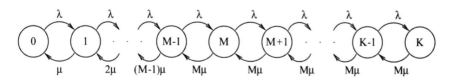

Fig. 4.8 Rate diagram of the call center.

Example 4.11. (Leaky Bucket). Leaky bucket is a traffic control mechanism used in high-speed digital telecommunication networks. It consists of two buffers: a token buffer of size M and a data buffer of size D. Tokens are generated according to a $PP(\mu)$ and are stored in the token buffer. Tokens generated while the token buffer is full are lost. Data packets arrive according to a $PP(\lambda)$. If there is a token in the token buffer when a packet arrives, it immediately removes one token from the token buffer and enters the network. If there are no tokens in the token buffer, it waits in the data buffer if the data buffer is not full. If the data buffer is full, the packet is lost. Let $Y(t)$ be the number of tokens in the token buffer at time t, and let $Z(t)$ be the number of data packets in the data buffer at time t. Model this system as a CTMC.

First note that if there are packets in the data buffer, the token buffer must be empty (otherwise the packets would have already entered the network) and vice versa. Consider a new process $\{X(t), t \geq 0\}$ defined by

$$X(t) = M - Y(t) + Z(t).$$

The state space of $\{X(t), t \geq 0\}$ is $\{0, 1, \ldots, K\}$, where $K = M + D$. Also, if $0 \leq X(t) \leq M$, we must have $Z(t) = 0$ and $Y(t) = M - X(t)$, and if $M \leq X(t) \leq K$, we must have $Y(t) = 0$ and $Z(t) = X(t) - M$. Thus we can recover $Y(t)$ and $Z(t)$ from the knowledge of $X(t)$.

Now, suppose $1 \leq X(t) \leq K - 1$. If a token arrives (this happens at rate μ), $X(t)$ decreases by 1, and if a data packet arrives (this happens at rate λ), it increases by 1. If $X(t) = 0$, then $Y(t) = M$, and tokens cannot enter the token buffer, while data packets continue to arrive. If $X(t) = K$, then $Z(t) = D$ and packets cannot enter the data buffer; however, tokens continue to enter. Thus $\{X(t), t \geq 0\}$ is a finite birth and death process with birth parameters

$$\lambda_i = \lambda, \quad 0 \leq i \leq K - 1,$$

and death parameters

$$\mu_i = \mu, \quad 1 \leq i \leq K.$$

Thus $\{X(t), t \geq 0\}$ is the same process as in the single server queue of Example 4.7!

∎

Example 4.12. (Inventory Management). A retail store manages the inventory of washing machines as follows. When the number of washing machines in stock decreases to a fixed number k, an order is placed with the manufacturer for m new washing machines. It takes a random amount of time for the order to be delivered to the retailer. If the inventory is at most k when an order is delivered (including the newly delivered order), another order for m items is placed immediately. Suppose the delivery times are iid $Exp(\lambda)$ and that the demand for the washing machines occurs according to a $PP(\mu)$. Demands that cannot be immediately satisfied are lost. Model this system as a CTMC.

Let $X(t)$ be the number of machines in stock at time t. Note that the maximum number of washing machines in stock is $K = k + m$, which happens if the order is delivered before the next demand occurs. The state space is thus $\{0, 1, \ldots, K\}$. In state 0, the demands are lost, and the stock jumps to m when the current outstanding order is delivered (which happens at rate λ). Hence we have $r_{0,m} = \lambda$. In state i, $1 \le i \le k$, one order is outstanding. The state changes to $i - 1$ if a demand occurs (which happens at rate μ) and to $i + m$ if the order is delivered. Thus we have $r_{i,i+m} = \lambda$ and $r_{i,i-1} = \mu$. Finally, if $X(t) = i$, $k + 1 \le i \le K$, there are no outstanding orders. Hence, the only transition is from i to $i - 1$, and that happens when a demand occurs. Thus, $r_{i,i-1} = \mu$. Thus $\{X(t), t \ge 0\}$ is a CTMC. The rate matrix is shown in (4.21) for the case $k = 3$, $m = 2$.

$$
R = \begin{bmatrix}
0 & 0 & \lambda & 0 & 0 & 0 \\
\mu & 0 & 0 & \lambda & 0 & 0 \\
0 & \mu & 0 & 0 & \lambda & 0 \\
0 & 0 & \mu & 0 & 0 & \lambda \\
0 & 0 & 0 & \mu & 0 & 0 \\
0 & 0 & 0 & 0 & \mu & 0
\end{bmatrix}.
\tag{4.21}
$$

The rate diagram is as shown in Figure 4.9. ∎

Example 4.13. (Manufacturing). A simple manufacturing facility consists of a single machine that can be turned on or off. If the machine is on, it produces items according to a Poisson process with rate λ. Demands for the items arrive according to a PP(μ). The machine is controlled as follows. If the number of items in stock reaches a maximum number K (the storage capacity), the machine is turned off. It is turned on again when the number of items in stock decreases to a prespecified level $k < K$. Model this system as a CTMC.

Let $X(t)$ be the number of items in stock at time t. $\{X(t), t \ge 0\}$ is not a CTMC since we don't know whether the machine is on or off if $k < X(t) < K$. Let $Y(t)$ be the state of the machine at time t, 1 if it is on and 0 if it is off. Then $\{(X(t), Y(t)), t \ge 0\}$ is a CTMC with state space

$$
S = \{(i, 1), 0 \le i < K\} \cup \{(i, 0), k < i \le K\}.
$$

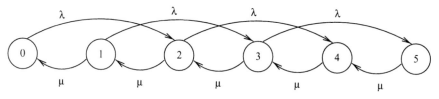

Fig. 4.9 Rate diagram of the inventory management example with $k = 3$, $r = 2$.

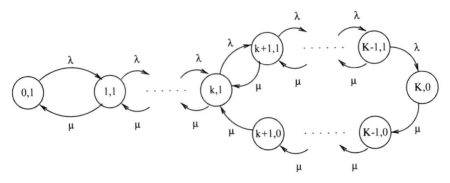

Fig. 4.10 Rate diagram of the manufacturing system.

Note that the machine is always on if the number of items is k or less. Hence we do not need the states $(i, 0), 0 \leq i \leq k$. The usual triggering event analysis yields the following transition rates:

$$r_{(i,1),(i+1,1)} = \lambda, \quad 0 \leq i < K - 1,$$
$$r_{(K-1,1),(K,0)} = \lambda,$$
$$r_{(i,1),(i-1,1)} = \mu, \quad 1 \leq i \leq K - 1,$$
$$r_{(i,0),(i-1,0)} = \mu, \quad k + 1 < i \leq K,$$
$$r_{(k+1,0),(k,1)} = \mu.$$

The rate diagram is shown in Figure 4.10. ∎

Armed with these examples, we proceed with the analysis of CTMCs in the next section.

4.4 Transient Analysis: Uniformization

Let $\{X(t), t \geq 0\}$ be a CTMC on state space $S = \{1, 2, \ldots, N\}$, and let $R = [r_{i,j}]$ be its rate matrix. In this section, we develop methods of computing the transient distribution of $\{X(t), t \geq 0\}$; that is, we compute the pmf of $X(t)$ for a given t. We assume that the probability distribution of the initial state, $X(0)$, is known. Then we have

$$P(X(t) = j) = \sum_{i=1}^{N} P(X(t) = j | X(0) = i) P(X(0) = i), \quad 1 \leq j \leq N. \quad (4.22)$$

Thus we need to compute $P(X(t) = j | X(0) = i) = p_{i,j}(t)$ in order to obtain the pmf of $X(t)$. Hence we concentrate on computing the transition matrix $P(t) = [p_{i,j}(t)]$ in this section. We first introduce the notation needed before stating the main result in Theorem 4.3.

We have seen before that the CTMC spends an $\text{Exp}(r_i)$ amount of time in state i ($r_i = \sum_{j=1}^{N} r_{i,j}$) and, if $r_i > 0$, jumps to state j with probability $p_{i,j} = r_{i,j}/r_i$. Now let r be any finite number satisfying

$$r \geq \max_{1 \leq i \leq N} \{r_i\}. \tag{4.23}$$

Define a matrix $\hat{P} = [\hat{p}_{i,j}]$ as

$$\hat{p}_{i,j} = \begin{cases} 1 - \dfrac{r_i}{r} & \text{if } i = j, \\ \dfrac{r_{i,j}}{r} & \text{if } i \neq j. \end{cases} \tag{4.24}$$

Example 4.14. Suppose R is given as

$$R = \begin{bmatrix} 0 & 2 & 3 & 0 \\ 4 & 0 & 2 & 0 \\ 0 & 2 & 0 & 2 \\ 1 & 0 & 3 & 0 \end{bmatrix}. \tag{4.25}$$

Then we have $r_1 = 5, r_2 = 6, r_3 = 4$, and $r_4 = 4$. Hence, from (4.23), we can choose $r = \max\{5, 6, 4, 4\} = 6$. Then, from (4.24), we get

$$\hat{P} = \begin{bmatrix} \frac{1}{6} & \frac{2}{6} & \frac{3}{6} & 0 \\ \frac{4}{6} & 0 & \frac{2}{6} & 0 \\ 0 & \frac{2}{6} & \frac{2}{6} & \frac{2}{6} \\ \frac{1}{6} & 0 & \frac{3}{6} & \frac{2}{6} \end{bmatrix}. \tag{4.26}$$

Note that $\hat{P}$ is stochastic. ∎

Using the notation above, the next theorem gives a method of computing $P(t)$.

Theorem 4.3. (The Matrix $P(t)$). *The transition probability matrix $P(t) = [p_{i,j}(t)]$ is given by*

$$P(t) = \sum_{k=0}^{\infty} e^{-rt} \frac{(rt)^k}{k!} \hat{P}^k. \tag{4.27}$$

We present the proof of the theorem above in Appendix 4A at the end of this chapter. Equation (4.27) provides a very stable method of computing $P(t)$. It is known as the *uniformization method* since it involves observing the CTMC at rate r that is a uniform upper bound on the transition rate r_i out of state $i \in S$. In numerical computations, we approximate $P(t)$ by using the first M terms of the infinite series in (4.27). In order to make the method numerically implementable, we need to describe how to choose M to guarantee a specified accuracy. A good rule of thumb is to choose

$$M \approx \max\{rt + 5 * \sqrt{rt}, 20\}.$$

A more precise algorithm is developed in Appendix 4B. Since the uniformization algorithm terminates more quickly (i.e., it uses less terms of the series) if the value of rt is small, it makes sense to use the smallest value of r allowed by (4.23), namely

$$r = \max_{1 \le i \le N} \{r_i\}.$$

This is what we do in the next three examples. However, in some cases it is more convenient to use an r that is larger than the one above. This is illustrated in Example 4.18.

Example 4.15. Let $\{X(t), t \ge 0\}$ be a CTMC with state space $\{1, 2, 3, 4\}$ and the rate matrix given in Example 4.14. Using the values of r and $\hat{P}$ from that example, we get

$$P(t) = \sum_{k=0}^{\infty} e^{-6t} \frac{(6t)^k}{k!} \begin{bmatrix} \frac{1}{6} & \frac{2}{6} & \frac{3}{6} & 0 \\ \frac{4}{6} & 0 & \frac{2}{6} & 0 \\ 0 & \frac{2}{6} & \frac{2}{6} & \frac{2}{6} \\ \frac{1}{6} & 0 & \frac{3}{6} & \frac{2}{6} \end{bmatrix}^k. \tag{4.28}$$

We compute $P(t)$ by using the first M terms of the infinite series in (4.28). We use $\epsilon = .00001$ in the algorithm in Appendix 4B and display the value of $P(t)$ and the value of M:

$$P(.5) = \begin{bmatrix} .2506 & .2170 & .3867 & .1458 \\ .2531 & .2384 & .3744 & .1341 \\ .1691 & .1936 & .4203 & .2170 \\ .1580 & .1574 & .3983 & .2862 \end{bmatrix}, \quad M = 13,$$

$$P(1) = \begin{bmatrix} .2062 & .2039 & .3987 & .1912 \\ .2083 & .2053 & .3979 & .1885 \\ .1968 & .1984 & .4010 & .2039 \\ .1920 & .1940 & .4015 & .2125 \end{bmatrix}, \quad M = 19,$$

$$P(5) = \begin{bmatrix} .2000 & .2000 & .4000 & .2000 \\ .2000 & .2000 & .4000 & .2000 \\ .2000 & .2000 & .4000 & .2000 \\ .2000 & .2000 & .4000 & .2000 \end{bmatrix}, \quad M = 56.$$

It can be numerically verified that $P(1) = P(.5)P(.5)$, thus verifying the Chapman–Kolmogorov equations (4.8). Some rows do not sum to 1 due to rounding errors. ∎

Example 4.16. (General Machine Shop). Consider the machine shop of Example 4.5. Suppose the shop has four machines and two repair persons. Suppose the lifetimes of the machines are exponential random variables with mean 3 days, while the repair times are exponential random variables with mean 2 hours. The shop operates 24 hours a day. Suppose all the machines are operating at 8:00 a.m. Monday. What is the expected number of working machines at 5:00 p.m. the same Monday?

Using the notation of Example 4.5, we see that the repair rate is $\lambda = .5$ per hour and the failure rate is $\mu = 1/72$ per hour. (Note that we must use consistent units.) Let $X(t)$ be the number of working machines at time t. Then, from Example 4.5, we see that $\{X(t), t \geq 0\}$ is a CTMC on $\{0, 1, 2, 3, 4\}$ with the rate matrix

$$R = \begin{bmatrix} 0 & 1 & 0 & 0 & 0 \\ \frac{1}{72} & 0 & 1 & 0 & 0 \\ 0 & \frac{2}{72} & 0 & 1 & 0 \\ 0 & 0 & \frac{3}{72} & 0 & \frac{1}{2} \\ 0 & 0 & 0 & \frac{4}{72} & 0 \end{bmatrix}. \tag{4.29}$$

We can take 8:00 a.m. as time 0 (hours). Thus we are given $X(0) = 4$. We are asked to compute the expected number of working machines at 5:00 p.m.; i.e., after 9 hours. Thus we need $E(X(9))$. From (4.23), we get $r = \frac{74}{72}$, and from (4.24), we get

$$\hat{P} = \begin{bmatrix} \frac{2}{74} & \frac{72}{74} & 0 & 0 & 0 \\ \frac{1}{74} & \frac{1}{74} & \frac{72}{74} & 0 & 0 \\ 0 & \frac{2}{74} & 0 & \frac{72}{74} & 0 \\ 0 & 0 & \frac{3}{74} & \frac{35}{74} & \frac{36}{74} \\ 0 & 0 & 0 & \frac{4}{74} & \frac{70}{74} \end{bmatrix}. \tag{4.30}$$

Using an error tolerance of $\epsilon = .00001$ and $t = 9$ hours in the uniformization algorithm of Appendix 4B, we get

$$P(9) = \begin{bmatrix} .0002 & .0020 & .0136 & .1547 & .8295 \\ .0000 & .0006 & .0080 & .1306 & .8607 \\ .0000 & .0002 & .0057 & .1154 & .8787 \\ .0000 & .0002 & .0048 & .1070 & .8880 \\ .0000 & .0001 & .0041 & .0987 & .8971 \end{bmatrix}. \tag{4.31}$$

The algorithm used the first 26 terms of the infinite series to compute the matrix above. Now we are given that $X(0) = 4$. Hence the pmf of $X(9)$ is given by the last row of the matrix above. Hence the expected number of working machines at 5:00 p.m. is given by

$$E(X(9)) = 0 * .0000 + 1 * .0001 + 2 * .0041 + 3 * .0987 + 4 * .8971 = 3.8928.$$

Note that the probability that all machines are working at 5:00 p.m. is .8971; however, this is not the probability that there were no failures from 8:00 a.m. to 5:00 p.m. There may have been one or more failures and the corresponding repairs. ∎

Example 4.17. (Airplane Reliability). Consider the four-engine jet airplane of Example 4.6. Suppose each engine can give trouble-free service for 200 hours on average before it needs to be shut down due to any problem. Assume that all four engines are operational at the beginning of a 6-hour flight. What is the probability that this flight lands safely?

We assume that the only trouble on the flight is from the engine failures. Let $\{X(t), t \geq 0\}$ be the CTMC as in Example 4.6. The rate matrix is as given in (4.18) with $\lambda = 1/200 = .005$ per hour. The flight lands safely if the state of the CTMC stays in the set $\{5, 6, 8, 9\}$ for the duration $[0, 6]$. Since this CTMC cannot return to this set of states once it leaves it, the required probability is given by the probability that $X(6)$ is in $\{5, 6, 8, 9\}$ given that $X(0) = 9$, the initial state. Using the uniformization algorithm of Appendix 4B, we get as the probability of a safe landing

$$p_{9,5}(6) + p_{9,6}(6) + p_{9,8}(6) + p_{9,9}(6) = .0033 + .0540 + .0540 + .8869 = .9983.$$

Thus the probability of a crash is .0017. This is far too large for any commercial airline! ∎

Example 4.18. (Two-State CTMC). Consider the two-state CTMC of Example 4.3 with state space $\{0, 1\}$ and rate matrix

$$R = \begin{bmatrix} 0 & \lambda \\ \mu & 0 \end{bmatrix}. \tag{4.32}$$

Here $r_0 = \lambda$ and $r_1 = \mu$. Hence (4.23) would suggest $r = \max\{\lambda, \mu\}$. However, it is more convenient to choose

$$r = \lambda + \mu.$$

Then the $\hat{P}$ matrix of (4.24) is given by

$$\hat{P} = \begin{bmatrix} \dfrac{\mu}{\lambda + \mu} & \dfrac{\lambda}{\lambda + \mu} \\ \dfrac{\mu}{\lambda + \mu} & \dfrac{\lambda}{\lambda + \mu} \end{bmatrix}. \tag{4.33}$$

It can be readily verified that

$$
\hat{P}^k = \begin{bmatrix} \dfrac{\mu}{\lambda+\mu} & \dfrac{\lambda}{\lambda+\mu} \\ \dfrac{\mu}{\lambda+\mu} & \dfrac{\lambda}{\lambda+\mu} \end{bmatrix}, \quad k \ge 1.
\tag{4.34}
$$

Substituting in (4.27), we get

$$
P(t) = \sum_{k=0}^{\infty} e^{-rt} \frac{(rt)^k}{k!} \hat{P}^k
$$

$$
= e^{-rt} \cdot \begin{bmatrix} 1 & 0 \\ 0 & 1 \end{bmatrix} + \sum_{k=1}^{\infty} e^{-rt} \frac{(rt)^k}{k!} \begin{bmatrix} \dfrac{\mu}{\lambda+\mu} & \dfrac{\lambda}{\lambda+\mu} \\ \dfrac{\mu}{\lambda+\mu} & \dfrac{\lambda}{\lambda+\mu} \end{bmatrix}
$$

$$
= e^{-rt} \cdot \begin{bmatrix} 1 & 0 \\ 0 & 1 \end{bmatrix} + \begin{bmatrix} \dfrac{\mu}{\lambda+\mu} & \dfrac{\lambda}{\lambda+\mu} \\ \dfrac{\mu}{\lambda+\mu} & \dfrac{\lambda}{\lambda+\mu} \end{bmatrix} \sum_{k=1}^{\infty} e^{-rt} \frac{(rt)^k}{k!}
$$

$$
= e^{-rt} \cdot \begin{bmatrix} 1 & 0 \\ 0 & 1 \end{bmatrix} + \begin{bmatrix} \dfrac{\mu}{\lambda+\mu} & \dfrac{\lambda}{\lambda+\mu} \\ \dfrac{\mu}{\lambda+\mu} & \dfrac{\lambda}{\lambda+\mu} \end{bmatrix} (1 - e^{-rt})
$$

$$
= \begin{bmatrix} \dfrac{\mu}{\lambda+\mu} & \dfrac{\lambda}{\lambda+\mu} \\ \dfrac{\mu}{\lambda+\mu} & \dfrac{\lambda}{\lambda+\mu} \end{bmatrix} + \begin{bmatrix} \dfrac{\lambda}{\lambda+\mu} & -\dfrac{\lambda}{\lambda+\mu} \\ -\dfrac{\mu}{\lambda+\mu} & \dfrac{\mu}{\lambda+\mu} \end{bmatrix} e^{-rt}.
\tag{4.35}
$$

From this we can read

$$
p_{1,1}(t) = \frac{\lambda}{\lambda+\mu} + \frac{\mu}{\lambda+\mu} e^{-(\lambda+\mu)t}.
$$

Thus the probability that a machine is up at time t given that it was up at time 0 is given by the expression above. Note that the machine could have gone through many failures and repairs up to time t. The probability above includes the sum total of all such possibilities. ∎

4.5 Occupancy Times

In this section, we shall do the short-term analysis of a CTMC $\{X(t), t \ge 0\}$; i.e., we study the behavior of the CTMC over a finite interval $[0, T]$. We concentrate on the *occupancy time* of a given state; i.e., the expected length of time the system spends in that state during a given interval of time. Here we develop a method for computing such quantities.

Let $X(t)$ be the state of a system at time t, and assume that $\{X(t), t \geq 0\}$ is a CTMC with state space $S = \{1, 2, \ldots, N\}$ and rate matrix R. Let $m_{i,j}(T)$ be the expected amount of time the CTMC spends in state j during the interval $[0, T]$, starting in state i. The quantity $m_{i,j}(T)$ is called the occupancy time of state j until time T starting from state i. The following theorem shows a way of computing $m_{i,j}(T)$.

Theorem 4.4. (The Occupancy Times). *Let $P(t) = [p_{i,j}(t)]$ be the transition probability matrix of $\{X(t), t \geq 0\}$. Then*

$$m_{i,j}(T) = \int_0^T p_{i,j}(t)dt, \quad 1 \leq i, j \leq N. \tag{4.36}$$

Proof. Let i and j be fixed, and define

$$Y_j(t) = \begin{cases} 1 & \text{if } X(t) = j, \\ 0 & \text{otherwise.} \end{cases}$$

The total amount of time spent in state j by the CTMC during $[0, T]$ is then given by

$$\int_0^T Y_j(t)dt.$$

Hence we get

$$m_{i,j}(T) = \mathsf{E}\left(\int_0^T Y_j(t)dt \,\middle|\, X(0) = i \right)$$

$$= \int_0^T \mathsf{E}(Y_j(t)|X(0) = i)dt$$

(the integral and E can be interchanged here)

$$= \int_0^T \mathsf{P}(Y_j(t) = 1|X(0) = i)dt$$

$$= \int_0^T \mathsf{P}(X(t) = j|X(0) = i)dt$$

$$= \int_0^T p_{i,j}(t)dt$$

as desired. ■

Example 4.19. (Two-State Machine). Consider the two-state machine of Example 4.3. Suppose the expected time until failure of the machine is 10 days, while the expected repair time is 1 day. Suppose the machine is working at the beginning of January. Compute the expected total uptime of the machine in the month of January.

Let $\{X(t), t \geq 0\}$ be as in Example 4.3. Using the data given here in time units of days, we get $\mu = .1$ days^{-1} and $\lambda = 1$ days^{-1}. Since there are 31 days in January, the expected uptime in January is given by $m_{1,1}(31)$. From Example 4.18, we have

$$
\begin{aligned}
p_{1,1}(t) &= \frac{\lambda}{\lambda + \mu} + \frac{\mu}{\lambda + \mu} e^{-(\lambda + \mu)t} \\
&= \frac{10}{11} + \frac{1}{11} e^{-1.1t}.
\end{aligned}
\tag{4.37}
$$

Substituting in (4.36), we get

$$
\begin{aligned}
m_{1,1}(31) &= \int_0^{31} p_{1,1}(t) dt \\
&= \int_0^{31} \left(\frac{10}{11} + \frac{1}{11} e^{-1.1t} \right) dt \\
&= \frac{10}{11} \cdot 31 + \frac{1}{11} \cdot \frac{1}{1.1} (1 - e^{-34.1}) \\
&= 28.26 \text{ days.}
\end{aligned}
$$

Thus the expected uptime during January is 28.26 days. Hence the expected downtime must be $31 - 28.26 = 2.74$ days. The reader should compute $m_{1,0}(31)$ and verify this independently. ∎

Note that Theorem 4.4 can be used directly if $p_{i,j}(t)$ is known as a function of t in closed form, as was the case in the example above. However, $p_{i,j}(t)$ is rarely known in closed form and mostly has to be computed using the infinite series representation of (4.27). How do we compute $m_{i,j}(T)$ in such cases?

First we compute the uniformization constant r from (4.23) and the $\hat{P}$ matrix from (4.24). Next we construct an $N \times N$ *occupancy matrix*,

$$
M(t) = [m_{i,j}(t)].
\tag{4.38}
$$

The next theorem gives an infinite series expression for $M(T)$, analogous to Theorem 4.3.

Theorem 4.5. (The Matrix $M(T)$). *Let Y be a* P(rT) *random variable. Then*

$$
M(T) = \frac{1}{r} \sum_{k=0}^{\infty} P(Y > k) \hat{P}^k, \quad T \geq 0.
\tag{4.39}
$$

Proof. From (4.27), we get

$$
p_{i,j}(t) = \sum_{k=0}^{\infty} e^{-rt} \frac{(rt)^k}{k!} [\hat{P}^k]_{i,j}.
$$

We also have

$$\int_0^T e^{-rt} \frac{(rt)^k}{k!} dt = \frac{1}{r} \left(1 - \sum_{l=0}^{k} e^{-rT} \frac{(rT)^l}{l!} \right)$$

$$= \frac{1}{r} P(Y > k).$$

Substituting in (4.36), we get

$$m_{i,j}(T) = \int_0^T p_{i,j}(t) dt$$

$$= \int_0^T \sum_{k=0}^{\infty} e^{-rt} \frac{(rt)^k}{k!} [\hat{P}^k]_{i,j} dt$$

$$= \sum_{k=0}^{\infty} \left(\int_0^T e^{-rt} \frac{(rt)^k}{k!} dt \right) [\hat{P}^k]_{i,j}$$

$$= \sum_{k=0}^{\infty} \frac{1}{r} P(Y > k) [\hat{P}^k]_{i,j}.$$

Putting the equation above in matrix form yields (4.39). ∎

The theorem above provides a very stable method of computing the matrix $M(T)$. An algorithm to compute $M(T)$ based on the theorem above is developed in Appendix 4C. We illustrate the theorem above with several examples below.

Example 4.20. (Single Server Queue). Consider the ATM queue of Example 4.7. Suppose the customers arrive at a rate of 10 per hour and take on average 4 minutes at the machine to complete their transactions. Suppose that there is space for at most five customers in front of the ATM and that it is idle at 8:00 a.m. What is the expected amount of time the machine is idle during the next hour?

The arrival process is Poisson with rate $\lambda = 10$ per hour. The processing times are iid exponential with mean $4/60$ hours; i.e., with parameter $\mu = 15$ per hour. The capacity is $K = 5$. Let $X(t)$ be the number of customers in the queue at time t. With these parameters, $\{X(t), t \geq 0\}$ is a CTMC with state space $\{0, 1, 2, 3, 4, 5\}$ and rate matrix given by

$$R = \begin{bmatrix} 0 & 10 & 0 & 0 & 0 & 0 \\ 15 & 0 & 10 & 0 & 0 & 0 \\ 0 & 15 & 0 & 10 & 0 & 0 \\ 0 & 0 & 15 & 0 & 10 & 0 \\ 0 & 0 & 0 & 15 & 0 & 10 \\ 0 & 0 & 0 & 0 & 15 & 0 \end{bmatrix}. \tag{4.40}$$

We are given $X(0) = 0$. The expected idle time is thus given by $m_{0,0}(1)$. Using the uniformization algorithm of Appendix 4C with $\epsilon = .00001$, we get

$$
M(1) = \begin{bmatrix}
.4451 & .2548 & .1442 & .0814 & .0465 & .0280 \\
.3821 & .2793 & .1604 & .0920 & .0535 & .0326 \\
.3246 & .2407 & .2010 & .1187 & .0710 & .0441 \\
.2746 & .2070 & .1780 & .1695 & .1046 & .0663 \\
.2356 & .1806 & .1598 & .1569 & .1624 & .1047 \\
.2124 & .1648 & .1489 & .1492 & .1571 & .1677
\end{bmatrix}.
$$

The algorithm uses the first 44 terms of the infinite series to compute the result above. Thus $m_{0,0}(1) = .4451$ hours; i.e., if the ATM is idle at the beginning, it is idle on average $.4451 * 60 = 26.71$ minutes of the first hour. Similarly, the queue is filled to capacity for $m_{0,5} = .0280$ hours $= 1.68$ minutes of the first hour. ∎

Example 4.21. (Manufacturing). Consider the manufacturing operation described in Example 4.13. Suppose the system operates 24 hours a day, the demands occur at the rate of five per hour, and the average time to manufacture one item is 10 minutes. The machine is turned on whenever the stock of manufactured items is reduced to two, and it stays on until the stock rises to four, at which point it is turned off. It stays off until the stock is reduced to two, and so on. Suppose the stock is at four (and the machine is off) at the beginning. Compute the expected amount of time during which the machine is on during the next 24 hours.

Let $X(t)$ be the number of items in stock at time t, and let $Y(t)$ be the state of the machine at time t, 1 if it is on and 0 if it is off. Then $\{(X(t), Y(t)), t \geq 0\}$ is a CTMC with state space

$$
S = \{1 = (0, 1), 2 = (1, 1), 3 = (2, 1), 4 = (3, 1), 5 = (4, 0), 6 = (3, 0)\}.
$$

The demand rate is $\mu = 5$ per hour and the production rate is $\lambda = 6$ per hour. Following the analysis of Example 4.13, we get as the rate matrix of the CTMC

$$
R = \begin{bmatrix}
0 & 6 & 0 & 0 & 0 & 0 \\
5 & 0 & 6 & 0 & 0 & 0 \\
0 & 5 & 0 & 6 & 0 & 0 \\
0 & 0 & 5 & 0 & 6 & 0 \\
0 & 0 & 0 & 0 & 0 & 5 \\
0 & 0 & 5 & 0 & 0 & 0
\end{bmatrix}. \tag{4.41}
$$

We are given that the initial state of the system is $(X(0), Y(0)) = (4, 0)$; i.e., the initial state is 5. The machine is on whenever the system is in state 1, 2, 3, or 4. Hence the total expected on time during 24 hours is given by $m_{5,1}(24) + m_{5,2}(24) + m_{5,3}(24) + m_{5,4}(24)$. We use the uniformization algorithm of Appendix 4C with $T = 24$ and $\epsilon = .00001$. The algorithm uses the first 331 terms of the infinite series to yield

$$M(24) = \begin{bmatrix} 4.0004 & 4.6322 & 5.4284 & 2.9496 & 3.5097 & 3.4798 \\ 3.8602 & 4.6639 & 5.4664 & 2.9703 & 3.5345 & 3.5047 \\ 3.7697 & 4.5553 & 5.5361 & 3.0084 & 3.5802 & 3.5503 \\ 3.7207 & 4.4965 & 5.4656 & 3.0608 & 3.6431 & 3.6132 \\ 3.7063 & 4.4793 & 5.4448 & 2.9586 & 3.7204 & 3.6906 \\ 3.7380 & 4.5173 & 5.4905 & 2.9835 & 3.5503 & 3.7204 \end{bmatrix}.$$

Hence the total expected on time is given by

$$3.7063 + 4.4793 + 5.4448 + 2.9586 = 16.5890 \text{ hours.}$$

If the system had no stock in the beginning, the total expected uptime would be

$$4.0004 + 4.6322 + 5.4284 + 2.9496 = 17.0106 \text{ hours.} \blacksquare$$

4.6 Limiting Behavior

In the study of the limiting behavior of DTMCs in Section 2.5, we studied three quantities: the limiting distribution, the stationary distribution, and the occupancy distribution. In this section, we study similar quantities for the CTMCs. Let $\{X(t), t \geq 0\}$ be a CTMC with state space $S = \{1, 2, \ldots, N\}$ and rate matrix R. In Section 4.4, we saw how to compute $\mathsf{P}(X(t) = j)$ for $1 \leq j \leq N$. In this section, we study

$$p_j = \lim_{t \to \infty} \mathsf{P}(X(t) = j). \tag{4.42}$$

If the limit above exists, we call it the *limiting probability* of being in state j. If it exists for all $1 \leq j \leq N$, we call

$$p = [p_1, p_2, \ldots, p_N]$$

the *limiting distribution* of the CTMC. If the limiting distribution exists, we ask if it is unique and how to compute it if it is unique.

Here we recapitulate some important facts from Appendix 4A. The reader does not need to follow the appendix if he or she is willing to accept these facts as true. Recall the definition of the uniformization constant from (4.23). We opt to choose strict inequality here as follows:

$$r > \max_{1 \leq j \leq N} \{r_j\}.$$

Construct $\hat{P}$ from R by using (4.24). We have shown in Theorem 4.13 that $\hat{P}$ is a stochastic matrix. Let $\{\hat{X}_n, n \geq 0\}$ be a DTMC on state space $\{1, 2, \ldots, N\}$ with transition probability matrix $\hat{P}$. Let $N(t)$ be a $P(rt)$, and assume that it is independent of $\{\hat{X}_n, n \geq 0\}$. Then we saw in Appendix 4A that

$$P(X(t) = j) = P(\hat{X}_{N(t)} = j).$$

Hence the limiting distribution of $X(t)$, if it exists, is identical to the limiting distribution of $\hat{X}_n$. We use this fact below.

Suppose $\{\hat{X}_n, n \geq 0\}$ is irreducible; i.e., it can go from any state i to any state j in a finite number of steps. However, $\hat{X}_n = i$ for some n if and only if $X(t) = i$ for some t. This implies that the CTMC $\{X(t), t \geq 0\}$ can go from any state i to any state j in a finite amount of time. This justifies the following definition.

Definition 4.2. (Irreducible CTMC). The CTMC $\{X(t), t \geq 0\}$ is said to be irreducible if the corresponding DTMC $\{\hat{X}_n, n \geq 0\}$ is irreducible.

Using this definition, we get the next theorem.

Theorem 4.6. (Limiting Distribution). *An irreducible CTMC $\{X(t), t \geq 0\}$ with rate matrix $R = [r_{i,j}]$ has a unique limiting distribution $p = [p_1, p_2, \ldots, p_N]$. It is given by the solution to*

$$p_j r_j = \sum_{i=1}^{N} p_i r_{i,j}, \quad 1 \leq j \leq N, \tag{4.43}$$

$$\sum_{i=1}^{N} p_i = 1. \tag{4.44}$$

Proof. The CTMC $\{X(t), t \geq 0\}$ is irreducible; hence the DTMC $\{\hat{X}_n, n \geq 0\}$ is irreducible. The choice of the uniformization constant r implies that $\hat{p}_{j,j} = 1 - r_j/r > 0$ for all j. Thus the DTMC $\{\hat{X}_n, n \geq 0\}$ is aperiodic and irreducible. Hence, from Theorem 2.10, it has a limiting distribution

$$\lim_{n \to \infty} P(\hat{X}_n = j) = \pi_j, \quad 1 \leq j \leq N.$$

The limiting distribution satisfies (2.35) and (2.36) as follows:

$$\pi_j = \sum_{i=1}^{N} \pi_i \hat{p}_{i,j}$$

and

$$\sum_{j=1}^{N} \pi_j = 1.$$

Now, $N(t)$ goes to ∞ as t goes to ∞. Hence

$$
\begin{aligned}
p_j &= \lim_{t \to \infty} P(X(t) = j) \\
&= \lim_{t \to \infty} P(\hat{X}_{N(t)} = j) \\
&= \lim_{n \to \infty} P(\hat{X}_n = j) \\
&= \pi_j.
\end{aligned}
\tag{4.45}
$$

Hence, the limiting distribution p exists and is unique, and is equal to the limiting distribution of the DTMC! Substituting $p_j = \pi_j$ in the balance equations satisfied by π, we get

$$
p_j = \sum_{i=1}^{N} p_i \hat{p}_{i,j}
\tag{4.46}
$$

and

$$
\sum_{j=1}^{N} p_j = 1.
$$

The last equation produces (4.44). Substituting in (4.46) for $\hat{p}_{i,j}$ from (4.24), we get

$$
\begin{aligned}
p_j &= \sum_{i=1}^{N} p_i \hat{p}_{i,j} \\
&= p_j \hat{p}_{j,j} + \sum_{i \neq j} p_i \hat{p}_{i,j} \\
&= p_j \left(1 - \frac{r_j}{r} \right) + \sum_{i \neq j} p_i \frac{r_{i,j}}{r}.
\end{aligned}
$$

Canceling p_j from both sides and rearranging, we get

$$
p_j \frac{r_j}{r} = \sum_{i \neq j} p_i \frac{r_{i,j}}{r}.
$$

Canceling the common factor r and recalling that $r_{i,i} = 0$, we get (4.43). ∎

We can think of $p_j r_j$ as the rate at which the CTMC leaves state j and $p_i r_{i,j}$ as the rate at which the CTMC enters state j from state i. Thus, (4.43) says that the limiting probabilities are such that the rate of entering state j from all other states is the same as the rate of leaving state j. Hence it is called the *balance equation* or *rate-in–rate-out* equation.

The balance equations can easily be written down from the rate diagram as follows. Imagine that there is a flow from node i to node j at rate $p_i r_{i,j}$ on every arc (i, j) in the rate diagram. Then the balance equations simply say that the total flow into a node from all other nodes in the rate diagram equals the total flow out of that node to all other nodes.

In Chapter 2, we saw that the limiting distribution of a DTMC is also its stationary distribution. Is the same true for CTMCs as well? The answer is given by the next theorem.

Theorem 4.7. (Stationary Distribution). *Let* $\{X(t), t \geq 0\}$ *be an irreducible CTMC with limiting distribution* p. *Then it is also the stationary distribution of the CTMC; i.e.,*

$$P(X(0) = j) = p_j \ \text{for} \ 1 \leq j \leq N \Rightarrow$$
$$P(X(t) = j) = p_j \ \text{for} \ 1 \leq j \leq N, t \geq 0.$$

Proof. From the proof of Theorem 4.6, $p_j = \pi_j$ for all $1 \leq j \leq N$. Suppose $P(X(0) = j) = P(\hat{X}_0 = j) = p_j$ for $1 \leq j \leq N$. Then, from Corollary 2.3,

$$P(\hat{X}_k = j) = \pi_j = p_j, \quad 1 \leq j \leq N, k \geq 0,$$

and

$$P(X(t) = j) = \sum_{i=1}^{N} P(X(t) = j \mid X(0) = i) P(X(0) = i)$$

$$= \sum_{i=1}^{N} P(X(0) = i) \sum_{k=0}^{\infty} e^{-rt} \frac{(rt)^k}{k!} [\hat{P}^k]_{i,j}$$

$$= \sum_{k=0}^{\infty} e^{-rt} \frac{(rt)^k}{k!} \sum_{i=1}^{N} P(X(0) = i) P(\hat{X}_k = j \mid \hat{X}_0 = i)$$

$$= \sum_{k=0}^{\infty} e^{-rt} \frac{(rt)^k}{k!} \sum_{i=1}^{N} P(\hat{X}_k = j \mid \hat{X}_0 = i) P(\hat{X}_0 = i)$$

$$= \sum_{k=0}^{\infty} e^{-rt} \frac{(rt)^k}{k!} P(\hat{X}_k = j)$$

$$= \sum_{k=0}^{\infty} e^{-rt} \frac{(rt)^k}{k!} p_j$$

$$= p_j$$

as desired. ∎

The next question is about the occupancy distribution. In DTMCs, the limiting distribution is also the occupancy distribution. Does this hold true for the CTMCs? The answer is yes if we define the occupancy of state i to be the expected long-run fraction of the time the CTMC spends in state i. The main result is given in the next theorem.

Theorem 4.8. (Occupancy Distribution). *Let $m_{i,j}(T)$ be the expected total time spent in state j up to time T by an irreducible CTMC starting in state i. Then*

$$\lim_{T \to \infty} \frac{m_{i,j}(T)}{T} = p_j,$$

where p_j is the limiting probability that the CTMC is in state j.

Proof. From Theorem 4.5, we see that

$$M(T) = \frac{1}{r} \sum_{k=0}^{\infty} P(Y > k) \hat{P}^k, \quad T \geq 0,$$

where $M(T) = [m_{i,j}(T)]$ and Y is a $P(rT)$ random variable. We have

$$P(Y > k) = \sum_{m=k+1}^{\infty} a_m,$$

where

$$a_m = e^{-rT} \frac{(rT)^m}{m!}.$$

Substituting, we get

$$M(T) = \frac{1}{r} \sum_{k=0}^{\infty} \sum_{m=k+1}^{\infty} a_m \hat{P}^k$$

$$= \frac{1}{r} \sum_{m=1}^{\infty} a_m \sum_{k=0}^{m-1} \hat{P}^k$$

$$= \frac{1}{r} \sum_{m=1}^{\infty} e^{-rT} \frac{(rT)^{m-1}}{(m-1)!} \frac{rT}{m} \sum_{k=0}^{m-1} \hat{P}^k.$$

Hence

$$\frac{M(T)}{T} = \sum_{m=1}^{\infty} e^{-rT} \frac{(rT)^{m-1}}{(m-1)!} \frac{1}{m} \sum_{k=0}^{m-1} \hat{P}^k$$

$$= \sum_{m=0}^{\infty} e^{-rT} \frac{(rT)^m}{m!} \frac{1}{m+1} \sum_{k=0}^{m} \hat{P}^k$$

$$= E\left(\frac{1}{N(T)+1} \sum_{k=0}^{N(T)} \hat{P}^k \right),$$

where $N(T)$ is the number of events in a PP(r) up to time T. From Theorems 4.7 and 4.8, we see that

$$\lim_{n \to \infty} \frac{1}{n+1} \sum_{k=0}^{n} [\hat{P}^k]_{i,j} = \pi_j = p_j.$$

Also, as $T \to \infty$, $N(T) \to \infty$. Using these two observations, we get

$$\lim_{T \to \infty} \frac{m_{i,j}(T)}{T} = \lim_{T \to \infty} \mathsf{E}\left(\frac{1}{N(T)+1} \sum_{k=0}^{N(T)} [\hat{P}^k]_{i,j} \right)$$

$$= \lim_{n \to \infty} \frac{1}{n+1} \sum_{k=0}^{n} [\hat{P}^k]_{i,j}$$

$$= p_j$$

as desired. ∎

Thus, an irreducible CTMC has a unique limiting distribution, which is also its stationary distribution and its occupancy distribution. It can be computed by solving the balance equation (4.43) along with the normalizing equation (4.44). We illustrate the concepts above by means of several examples below.

Example 4.22. (Two-State Machine). Consider the two-state machine as described in Example 4.3. Compute the limiting distribution of the state of the machine.

The two-state CTMC of Example 4.3 is irreducible. Hence it has a unique limiting distribution $[p_0, p_1]$. The two balance equations are

$$\lambda p_0 = \mu p_1,$$

$$\mu p_1 = \lambda p_0.$$

Note that they are the same! The normalizing equation is

$$p_0 + p_1 = 1.$$

The solution is given by

$$p_0 = \frac{\mu}{\lambda + \mu}, \quad p_1 = \frac{\lambda}{\lambda + \mu}.$$

Note that this is also the stationary distribution and the occupancy distribution of the machine. (Check this using the $P(t)$ matrix from Example 4.18.)

Suppose the expected time until failure of the machine is 10 days, while the expected repair time is 1 day. Thus $\lambda = 1$ and $\mu = .1$. Hence

$$p_0 = \frac{1}{11}, \quad p_1 = \frac{10}{11}.$$

Thus, in the long run, the machine spends 10 out of 11 days in working condition. This makes intuitive sense in view of the fact that the machine stays down for 1 day, followed by being up for 10 days on average! ∎

Example 4.23. (Finite Birth and Death Process). Let $\{X(t), t \geq 0\}$ be a birth and death process as described in Example 4.8. Assume that all the birth rates $\{\lambda_i, 0 \leq i < K\}$ and death rates $\{\mu_i, 1 \leq i \leq K\}$ are positive. Then the CTMC is irreducible and hence has a unique limiting distribution. We compute it below.

The balance equations are as follows:

$$\lambda_0 p_0 = \mu_1 p_1,$$
$$(\lambda_1 + \mu_1)p_1 = \lambda_0 p_0 + \mu_2 p_2,$$
$$(\lambda_2 + \mu_2)p_2 = \lambda_1 p_1 + \mu_3 p_3,$$
$$\vdots \quad \vdots$$
$$(\lambda_{K-1} + \mu_{K-1})p_{K-1} = \lambda_{K-2}p_{K-2} + \mu_K p_K,$$
$$\mu_K p_K = \lambda_{K-1}p_{K-1}.$$

From the first equation above, we get

$$p_1 = \frac{\lambda_0}{\mu_1}p_0.$$

Adding the first two balance equations, we get

$$\lambda_1 p_1 = \mu_2 p_2,$$

which yields

$$p_2 = \frac{\lambda_1}{\mu_2}p_1 = \frac{\lambda_0\lambda_1}{\mu_1\mu_2}p_0.$$

Proceeding this way (adding the first three balance equations, etc.), we get, in general,

$$p_i = \rho_i p_0, \quad 0 \leq i \leq K, \tag{4.47}$$

where $\rho_0 = 1$ and

$$\rho_i = \frac{\lambda_0\lambda_1 \cdots \lambda_{i-1}}{\mu_1\mu_2 \cdots \mu_i}, \quad 1 \leq i \leq K. \tag{4.48}$$

The only unknown left is p_0, which can be computed by using the normalization equation as follows:

$$p_0 + p_1 + \cdots + p_{K-1} + p_K = (\rho_0 + \rho_1 + \cdots + \rho_{K-1} + \rho_K)p_0 = 1.$$

This yields

$$p_0 = \frac{1}{\sum_{j=0}^{K} \rho_j}. \tag{4.49}$$

Combining this with (4.47), we get

$$p_i = \frac{\rho_i}{\sum_{j=0}^{K} \rho_j}, \quad 0 \le i \le K. \tag{4.50}$$

This also gives the stationary distribution and the occupancy distribution of the birth and death processes. ∎

Example 4.24. (The General Machine Shop). Consider the machine shop of Example 4.16, with four machines and two repair persons. Compute the long-run probability that all the machines are up. Also compute the long-run fraction of the time that both repair persons are busy.

Let $X(t)$ be the number of working machines at time t. Then $\{X(t), t \ge 0\}$ is a birth and death process on $\{0, 1, 2, 3, 4\}$ with birth rates $\lambda_0 = 1, \lambda_1 = 1, \lambda_2 = 1, \lambda_3 = \frac{1}{2}$ and death rates $\mu_1 = \frac{1}{72}, \mu_2 = \frac{2}{72}, \mu_3 = \frac{3}{72}, \mu_4 = \frac{4}{72}$. Hence we get

$$\rho_0 = 1, \rho_1 = 72, \rho_2 = 2592, \rho_3 = 62,208, \rho_4 = 559,872.$$

Thus

$$\sum_{i=0}^{4} \rho_i = 624,745.$$

Hence, using (4.50), we get

$$p_0 = \frac{1}{624,745} = 1.6001 \times 10^{-6},$$

$$p_1 = \frac{72}{624,745} = 1.1525 \times 10^{-4},$$

$$p_2 = \frac{2592}{624,745} = 4.149 \times 10^{-3},$$

$$p_3 = \frac{62,208}{624,745} = 9.957 \times 10^{-2},$$

$$p_4 = \frac{559,872}{624,745} = 8.962 \times 10^{-1}.$$

Thus, in the long run, all the machines are functioning 89.6% of the time. Both repair persons are busy whenever the system is in state 0, 1, or 2. Hence the long-run fraction of the time that both repair persons are busy is given by $p_0 + p_1 + p_2 = .00427$. ∎

Example 4.25. (Finite-Capacity Single Server Queue). Consider the finite-capacity single server queue of Example 4.7. Let $X(t)$ be the number of customers in the queue at time t. Then we have seen that $\{X(t), t \ge 0\}$ is a birth and death process on state space $\{0, 1, \ldots, K\}$ with birth rates

$$\lambda_i = \lambda, \quad 0 \le i < K,$$

and death rates

$$\mu_i = \mu, \quad 1 \le i \le K.$$

Now define

$$\rho = \frac{\lambda}{\mu}.$$

Using the results of Example 4.23, we get

$$\rho_i = \rho^i, \quad 0 \le i \le K,$$

and hence, assuming $\rho \ne 1$,

$$\sum_{i=0}^{K} \rho_i = \sum_{i=0}^{K} \rho^i = \frac{1 - \rho^{K+1}}{1 - \rho}.$$

Substituting in (4.50), we get

$$p_i = \frac{1 - \rho}{1 - \rho^{K+1}} \rho^i, \quad 0 \le i \le K. \tag{4.51}$$

This gives the limiting distribution of the number of customers in the single server queue. We leave it to the reader to verify that, if $\rho = 1$, the limiting distribution is given by

$$p_i = \frac{1}{K + 1}, \quad 0 \le i \le K. \quad \blacksquare \tag{4.52}$$

Example 4.26. (Leaky Bucket). Consider the leaky bucket of Example 4.11. Suppose the token buffer size is $M = 10$ and the data buffer size is $D = 14$. The tokens are generated at the rate $\mu = 1$ per millisecond, and data packets arrive at the rate $\lambda = 1$ per millisecond. What fraction of the time is the data buffer empty? What fraction of the time is it full?

Let $Y(t)$ be the number of tokens in the token buffer at time t, and let $Z(t)$ be the number of data packets in the data buffer at time t. In Example 4.11, we saw that the process $\{X(t), t \ge 0\}$ defined by

$$X(t) = M - Y(t) + Z(t)$$

behaves like a single server queue with capacity $K = M + D = 24$, arrival rate $\lambda = 1$, and service rate $\mu = 1$. Hence $\rho = \lambda/\mu = 1$. Thus, from (4.52), we get

$$p_i = \frac{1}{25}, \quad 0 \le i \le 24.$$

Now, from the analysis in Example 4.11, the data buffer is empty whenever the $\{X(t), t \ge 0\}$ process is in states 0 through $M = 10$. Hence the long-run fraction of the time the data buffer is empty is given by

$$\sum_{i=0}^{10} p_i = \frac{11}{25} = .44.$$

We can interpret this to mean that 44% of the packets enter the network with no delay at the leaky bucket. Similarly, the data buffer is full whenever the $\{X(t), t \geq 0\}$ process is in state $M + D = 24$. Hence the long-run fraction of the time the data buffer is full is given by

$$p_{24} = \frac{1}{25} = .04.$$

This means that 4% of the packets will be dropped due to buffer overflow. ∎

Example 4.27. (Manufacturing). Consider the manufacturing setup of Example 4.21. Compute the long-run fraction of the time the machine is off.

Let $X(t)$ and $Y(t)$ be as in Example 4.21. We see that $\{(X(t), Y(t)), t \geq 0\}$ is an irreducible CTMC with state space $\{1, 2, \ldots, 6\}$ and rate matrix as given in (4.41). Hence it has a unique limiting distribution that is also its occupancy distribution. The machine is off in states $5 = (4, 0)$ and $6 = (3, 0)$. Thus the long-run fraction of the time that the machine is off is given by $p_5 + p_6$. The balance equations and the normalizing equation can be solved numerically to obtain

$$p_1 = .1585, \quad p_2 = .1902, \quad p_3 = .2282;$$
$$p_4 = .1245, \quad p_5 = .1493, \quad p_6 = .1493.$$

Hence the required answer is $.1493 + .1493 = .2986$. Thus the machine is turned off about 30% of the time! ∎

4.7 Cost Models

Following the development in Chapter 2, we now study cost models associated with CTMCs. Recall Example 1.4(a), where we were interested in the expected total cost of operating a machine for T units of time and also in the long-run cost per unit time of operating the machine. In this section, we shall develop methods of computing such quantities.

We assume the following cost model. Let $\{X(t), t \geq 0\}$ be a CTMC with state space $\{1, 2, \ldots, N\}$ and rate matrix R. Whenever the CTMC is in state i, it incurs costs continuously at rate $c(i)$, $1 \leq i \leq N$.

4.7.1 Expected Total Cost

In this subsection, we study the ETC, the *expected total cost* up to a finite time T, called the horizon. Note that the cost rate at time t is $c(X(t))$. Hence the total cost up to time T is given by

$$\int_0^T c(X(t))dt.$$

The expected total cost up to T, starting from state i, is given by

$$g(i, T) = \mathsf{E}\left(\int_0^T c(X(t))dt \,\middle|\, X(0) = i \right), \quad 1 \le i \le N. \qquad (4.53)$$

The next theorem gives a method of computing these costs. First, it is convenient to introduce the following vector notation:

$$c = \begin{bmatrix} c(1) \\ c(2) \\ \vdots \\ c(N-1) \\ c(N) \end{bmatrix},$$

and

$$g(T) = \begin{bmatrix} g(1, T) \\ g(2, T) \\ \vdots \\ g(N-1, T) \\ g(N, T) \end{bmatrix}.$$

Theorem 4.9. (Expected Total Cost). *Let $M(T) = [m_{i,j}(T)]$ be the occupancy matrix as given in Theorem 4.5. Then*

$$g(T) = M(T)c. \qquad (4.54)$$

Proof. We have

$$g(i, T) = \mathsf{E}\left(\int_0^T c(X(t))dt \,\middle|\, X(0) = i \right)$$

$$= \int_0^T \mathsf{E}(c(X(t))|X(0) = i)dt$$

$$= \int_0^T \sum_{j=1}^N c(j)\mathsf{P}(X(t) = j|X(0) = i)dt$$

$$= \int_0^T \sum_{j=1}^N c(j)p_{i,j}(t)dt$$

$$= \sum_{j=1}^{N} c(j) \int_{0}^{T} p_{i,j}(t) dt$$

$$= \sum_{j=1}^{N} c(j) m_{i,j}(T),$$

where the last equation follows from Theorem 4.4. In matrix form, this yields Equation (4.54). ∎

We illustrate this with several examples.

Example 4.28. (Two-State Machine). Consider the two-state machine as described in Example 4.3. Suppose the machine produces revenue at rate A per day when it is up and the repair costs B per day of downtime. What is the expected net income from the machine up to time T (days) if the machine started in the up state?

Let $\{X(t), t \geq 0\}$ be a CTMC on $\{0, 1\}$ as described in Example 4.3. (Assume that the parameters λ and μ are per day.) Since we are interested in revenues, we set $c(0)$, the revenue rate in state 0, to be $-B$ per day of downtime and $c(1) = A$ per day of uptime. From Example 4.18, we get

$$p_{1,1}(t) = \frac{\lambda}{\lambda + \mu} + \frac{\mu}{\lambda + \mu} e^{-(\lambda+\mu)t}$$

and

$$p_{1,0}(t) = \frac{\mu}{\lambda + \mu} - \frac{\mu}{\lambda + \mu} e^{-(\lambda+\mu)t}.$$

Using Theorem 4.4, we get

$$m_{1,1}(T) = \int_{0}^{T} p_{1,1}(t) \, dt$$

$$= \int_{0}^{T} \left(\frac{\lambda}{\lambda + \mu} + \frac{\mu}{\lambda + \mu} e^{-(\lambda+\mu)t} \right) dt$$

$$= \frac{\lambda T}{\lambda + \mu} + \frac{\mu}{(\lambda + \mu)^2} (1 - e^{-(\lambda+\mu)T}).$$

Similarly,

$$m_{1,0}(T) = \frac{\mu T}{\lambda + \mu} - \frac{\mu}{(\lambda + \mu)^2} (1 - e^{-(\lambda+\mu)T}).$$

The net income up to T can then be obtained from Theorem 4.9 as

$$g(1, T) = m_{1,0}(T)c(0) + m_{1,1}(T)c(1)$$

$$= \frac{(\lambda A - \mu B)T}{\lambda + \mu} + \frac{\mu(B + A)}{(\lambda + \mu)^2} (1 - e^{-(\lambda+\mu)T}).$$

As a numerical example, consider the data from Example 4.19. Assume $A = \$2400$ per day and $B = \$480$ per day. The net revenue in the month of January $(T = 31)$ is given by

$$g(1, 31) = -m_{1,0}(31)B + m_{1,1}(31)A.$$

In Example 4.19, we have computed $m_{1,0}(31) = 2.7355$ days and $m_{1,1}(31) = 28.2645$ days. Substituting in the equation above, we get

$$g(1, 31) = -2.7355 \cdot 480 + 28.2645 \cdot 2400 = \$66,521.60. \ \blacksquare$$

Example 4.29. (General Machine Shop). Consider the general machine shop of Example 4.16. Each machine produces revenue of \$50 per hour of uptime, and the downtime costs \$15 per hour per machine. The repair time costs an additional \$10 per hour per machine. Compute the net expected revenue during the first day, assuming all the machines are up in the beginning.

Let $X(t)$ be the number of up machines at time t. Then $\{X(t), t \geq 0\}$ is a CTMC on $\{0, 1, 2, 3, 4\}$ with the rate matrix given in (4.29). The revenue rates in various states are as follows:

$$c(0) = 0 * 50 - 4 * 15 - 2 * 10 = -80,$$
$$c(1) = 1 * 50 - 3 * 15 - 2 * 10 = -15,$$
$$c(2) = 2 * 50 - 2 * 15 - 2 * 10 = 50,$$
$$c(3) = 3 * 50 - 1 * 15 - 1 * 10 = 125,$$
$$c(4) = 4 * 50 - 0 * 15 - 0 * 10 = 200.$$

The net expected revenue vector is given by

$$g(24) = M(24) * c$$

$$= \begin{bmatrix} 1.0143 & 1.0318 & 1.1496 & 3.6934 & 17.1108 \\ 0.0143 & 1.0319 & 1.1538 & 3.7931 & 18.0069 \\ 0.0004 & 0.0320 & 1.1580 & 3.8940 & 18.9155 \\ 0.0001 & 0.0044 & 0.1623 & 3.9964 & 19.8369 \\ 0.0000 & 0.0023 & 0.0876 & 2.2041 & 21.7060 \end{bmatrix} \begin{bmatrix} -80 \\ -15 \\ 50 \\ 125 \\ 200 \end{bmatrix}$$

$$= \begin{bmatrix} 3844.69 \\ 4116.57 \\ 4327.23 \\ 4474.96 \\ 4621.05 \end{bmatrix}.$$

The required answer is given by $g(4, 24) = \$4621.05. \ \blacksquare$

4.7.2 Long-Run Cost Rates

In Example 4.28, suppose the downtime cost B is given. We want to know how much the revenue rate during uptime should be so that it is economically profitable to operate the machine. If we go by total cost, the answer will depend on the planning horizon T and also the initial state of the machine. An alternative is to compute the long-run net revenue per unit time for this machine and insist that this be positive for profitability. This answer will not depend on T, and, as we shall see, not even on the initial state of the machine. Hence, computing such long-run cost rates or revenue rates is very useful. We shall show how to compute these quantities in this subsection.

First, define the long-run cost rate as

$$g(i) = \lim_{T \to \infty} \frac{g(i, T)}{T}. \tag{4.55}$$

The next theorem shows an easy way to compute $g(i)$.

Theorem 4.10. (Long-Run Cost Rate). *Suppose $\{X(t), t \geq 0\}$ is an irreducible CTMC with limiting distribution $p = [p_1, p_2, \ldots, p_N]$. Then*

$$g = g(i) = \sum_{j=1}^{N} p_j c(j), \quad 1 \leq i \leq N.$$

Proof. We have

$$
\begin{aligned}
g(i) &= \lim_{T \to \infty} \frac{g(i, T)}{T} \\
&= \lim_{T \to \infty} \frac{\sum_{j=1}^{N} m_{i,j}(T) c(j)}{T} \\
&\quad \text{(from Theorem 4.9)} \\
&= \sum_{j=1}^{N} c(j) \lim_{T \to \infty} \frac{m_{i,j}(T)}{T} \\
&= \sum_{j=1}^{N} c(j) p_j,
\end{aligned}
$$

where the last equality follows from Theorem 4.8. This proves the theorem. ∎

Note that the theorem is intuitively obvious: the CTMC incurs a cost at rate $c(j)$ per unit time it spends in state j. It spends p_j fraction of the time in state j, regardless of the starting state. Hence it must incur costs at rate $\sum_{j=1}^{N} p_j c(j)$ in the long run. We illustrate this result with several examples.

Example 4.30. (Two-State Machine). Consider the two-state machine of Example 4.28. Suppose the downtime cost of the machine is B per unit time. What is the minimum revenue rate A during the uptime needed to break even in the long run?

From Example 4.22, the machine is up for $p_1 = \lambda/(\lambda + \mu)$ fraction of the time in the long run and down for $p_0 = \mu/(\lambda + \mu)$ fraction of the time in the long run. Using $c(0) = -B$ and $c(1) = A$ in Theorem 4.10, we obtain as the long-run net revenue per unit time

$$g = \frac{\lambda A - \mu B}{\lambda + \mu}.$$

For breakeven, we must have

$$g \geq 0;$$

i.e.,

$$\frac{A}{\mu} \geq \frac{B}{\lambda}.$$

Note that this condition makes sense: A/μ is the expected revenue during one uptime, and B/λ is the expected cost of one downtime. If the former is greater than the latter, the machine is profitable to operate; otherwise it is not! ■

Example 4.31. (Telephone Switch). Consider the CTMC model of the telephone switch described in Example 4.9. The switch has the capacity to handle at most six calls simultaneously. The call arrival rate is four calls per minute, and the arrival process is Poisson. The average duration of a call is 2 minutes, and the call durations are iid exponential random variables. Compute the expected revenue per minute if each caller is charged 10 cents per minute of the call duration. What is the rate at which revenue is lost due to the switch being full?

Let $X(t)$ be the number of calls carried by the switch at time t. Then $\{X(t), t \geq 0\}$ is a birth and death process with birth rates

$$\lambda_i = 4, \quad 0 \leq i \leq 5,$$

and death rates

$$\mu_i = i/2, \quad 1 \leq i \leq 6.$$

The revenue rate in state i is $c(i) = 10i$ cents per minute. Hence the long-run rate of revenue is

$$g = \sum_{j=0}^{6} p_j c(j) = 10 \sum_{j=0}^{6} j p_j.$$

Note that the last sum is just the expected number of calls in the switch in steady state. Using the results of Example 4.23, we get

$$\rho_i = \frac{8^i}{i!}, \quad 0 \leq i \leq 6,$$

and

$$p_i = \frac{\rho_i}{\sum_{j=0}^{6} \rho_j}, \quad 0 \leq i \leq 6.$$

Hence we get

$$g = 10\frac{\sum_{i=0}^{6} i\rho_i}{\sum_{j=0}^{6} \rho_j} = 48.8198.$$

Hence the switch earns revenue at a rate of 48.82 cents per minute.

Note that calls arrive at the rate of four per minute, and the average revenue from each call is 20 cents (duration is 2 minutes). Hence, if the switch had ample capacity so that no call was rejected, the revenue would be $4 * 20 = 80$ cents per minute. But the actual revenue is only 48.82 cents per minute. Hence revenue is lost at the rate $80 - 48.82 = 31.18$ cents per minute. ∎

4.8 First-Passage Times

In this section, we study the *first-passage times* in CTMCs, following the development in the DTMCs in Section 2.7. Let $\{X(t), t \geq 0\}$ be a CTMC with state space $\{1, \ldots, N\}$ and rate matrix R. The first-passage time into state N is defined to be

$$T = \min\{t \geq 0 : X(t) = N\}. \tag{4.56}$$

As in Section 2.7, we study the expected value $\mathsf{E}(T)$. Let

$$m_i = \mathsf{E}(T | X(0) = i). \tag{4.57}$$

Note that $m_N = 0$. The next theorem gives a method of computing $m_i, 1 \leq i \leq N - 1$.

Theorem 4.11. (First-Passage Times). $\{m_i, 1 \leq i \leq N - 1\}$ *satisfy*

$$r_i m_i = 1 + \sum_{j=1}^{N-1} r_{i,j} m_j, \quad 1 \leq i \leq N - 1. \tag{4.58}$$

Proof. We shall prove the theorem by conditioning on the time of the first transition and the state of the CTMC after the first transition. Let $X(0) = i, 1 \leq i \leq N - 1$, and let $X(Y) = j$, where Y is the sojourn time in the initial state. Then, $T = Y$ if $X(Y) = N$, and $T = Y +$ the first passage time into state N from state j if $j \neq N$. Using this observation, we get

$$\mathsf{E}(T | X(0) = i, X(Y) = j) = \mathsf{E}(Y | X(0) = i) + m_j, \quad 1 \leq i, j \leq N - 1,$$

and

$$\mathsf{E}(T | X(0) = i, X(Y) = N) = \mathsf{E}(Y | X(0) = i), \quad 1 \leq i \leq N - 1.$$

Now, Y is an $Exp(r_i)$ random variable if $X(0) = i$. Hence

$$E(Y \mid X(0) = i) = \frac{1}{r_i}.$$

From the properties of CTMCs, we also have

$$P(X(Y) = j \mid X(0) = i) = \frac{r_{i,j}}{r_i},$$

where $r_{i,i} = 0$ and $r_i = \sum_{j=1}^{N} r_{i,j}$. Using these facts, we get, for $1 \le i \le N - 1$,

$$
\begin{aligned}
m_i &= E(T \mid X(0) = i) \\
&= \sum_{j=1}^{N} E(T \mid X(0) = i, X(Y) = j) P(X(Y) = j \mid X(0) = i) \\
&= \sum_{j=1}^{N-1} \left(\frac{1}{r_i} + m_j \right) \frac{r_{i,j}}{r_i} + \frac{1}{r_i} \frac{r_{i,N}}{r_i} \\
&= \frac{1}{r_i} + \sum_{j=1}^{N-1} \frac{r_{i,j}}{r_i} m_j.
\end{aligned}
$$

Multiplying the last equation by r_i yields (4.58). ∎

We illustrate the theorem above with several examples.

Example 4.32. Let $\{X(t), t \ge 0\}$ be the CTMC in Example 4.14. Compute the expected time the CTMC takes to reach state 4 starting from state 1.
   Equations (4.58) are

$$
\begin{aligned}
5m_1 &= 1 + 2m_2 + 3m_3, \\
6m_2 &= 1 + 4m_1 + 2m_3, \\
4m_3 &= 1 + 2m_2.
\end{aligned}
$$

Solving simultaneously, we get

$$m_1 = 1.2727, m_2 = 1.3182, m_3 = .9091.$$

Thus it takes on average 1.2727 units of time to go from state 1 to state 4. ∎

Example 4.33. (Single Server Queue). Consider the single server queue of Example 4.20. Compute the expected time until the queue becomes empty, given that it has one customer at time 0.
   Let $X(t)$ be the number of customers in the queue at time t. With the parameters given in Example 4.20, $\{X(t), t \ge 0\}$ is a CTMC with state space $\{0, 1, 2, 3, 4, 5\}$

and rate matrix given by (4.40). Let T be the first-passage time to state 0, and let $m_i = \mathsf{E}(T \mid X(0) = i)$. Equations (4.58) for this case are

$$m_0 = 0,$$
$$25m_i = 1 + 10m_{i+1} + 15m_{i-1}, \quad 1 \le i \le 4,$$
$$15m_5 = 1 + 15m_4.$$

Solving numerically, we get

$$m_1 = .1737, m_2 = .3342, m_3 = .4749, m_4 = .5860, m_5 = .6527.$$

Thus the expected time until the queue becomes empty (starting with one person) is .1737 hours, or 10.42 minutes. ■

As in Section 2.7, we now study the expected time to reach a set of states, A. Analogously to (2.62), define

$$T = \min\{t \ge 0 : X(t) \in A\}. \tag{4.59}$$

Theorem 4.11 can be easily extended to the case of the first-passage time defined above. Let $m_i(A)$ be the expected time to reach the set A starting from state $i \notin A$. Note that $m_i(A) = 0$ for $i \in A$. Following the same argument as in the proof of Theorem 4.11, we can show that

$$r_i m_i(A) = 1 + \sum_{j \notin A} r_{i,j} m_j(A), \quad i \notin A. \tag{4.60}$$

We illustrate this with an example below.

Example 4.34. (Airplane Reliability). Consider the four-engine airplane of Example 4.17. Suppose that in a testing experiment the airplane takes off with four properly functioning engines and keeps on flying until it crashes. Compute the expected time of the crash.

Let $\{X(t), t \ge 0\}$ be the CTMC of Example 4.6 with $\lambda = .005$ per hour. The airplane crashes as soon as the CTMC visits the set $\{1, 2, 3, 4, 7\}$. Using this as A in (4.60), we get $m_i = 0$ for $i \in A$ and

$$.02m_9 = 1 + .01m_8 + .01m_6,$$
$$.015m_8 = 1 + .01m_5,$$
$$.015m_6 = 1 + .01m_5,$$
$$.01m_5 = 1.$$

Solving backward recursively, we get

$$m_5 = 100, m_6 = \frac{400}{3}, m_8 = \frac{400}{3}, m_9 = \frac{550}{3}.$$

Thus the expected time to failure is $m_9 = 550/3 = 183$ hours and 20 minutes. ■

4.9   Case Study: Cars-R-Us Used Car Dealer

This case is inspired by a paper by Y. Amihud and H. Mendelson (1980).

Cars-R-Us is a small used car dealer that buys and sells used cars. It generally sells a car at a 10% markup over its purchase price. It deals primarily in five models, labeled 1, 2, 3, 4, and 5. Its parking lot can hold at most 40 cars at a time. Table 4.1 shows the purchase price, the weekly supply of used cars to the dealer at this price, the selling price, and weekly demand for the used cars at this price for the five different models.

The dealer currently follows a static stocking policy of reserving eight spaces for each model. Thus the dealer won't buy a car of model 1 if there are already eight cars of that model on his lot. The dealer's son feels that they should allocate spaces to models based on their supply, demand, and profitability. For example, he feels that the number of spaces allocated to a model should equal the smaller of the weekly supply and demand, and the spaces should be allocated to the most expensive (and hence more profitable) models first. This leads him to suggest 12 spaces for model 5, 7 for model 4, 14 for model 3, 7 for model 2, and none for model 1. However, he also knows that this does not sound right and that further analysis is needed to find the optimal stocking policy. Our job is to help him do this.

We shall use CTMCs to model the random evolution of the number of cars on the lot. To facilitate this, we assume that the arrival processes of the car sellers and car buyers are independent Poisson processes with rates given in Table 4.1. We shall concentrate on static policies. A static policy is described by a vector $[K_1, \ K_2, \ K_3, \ K_4, \ K_5]$ of integers that add up to 40, the capacity of the parking lot at the dealership. Under this policy, the dealer stocks at most K_i cars of model i, $1 \leq i \leq 5$. The father's policy is given by the vector [8, 8, 8, 8, 8], while the son's policy is given by the vector [0, 7, 14, 7, 12].

Let us study the first car model. We assume that sellers arrive according to a Poisson process with rate $\lambda_1 = 20$ per week and buyers arrive according to a Poisson process with rate $\mu_1 = 16$ per week. We stock at most K_1 cars of this type. Let $X(t)$ be the number of model 1 cars on the lot at time t. The stock goes up if the stock level is less than K_1 and a seller arrives, and it goes down if the stock level is positive and a buyer arrives. It is clear that $\{X(t), t \geq 0\}$ is identical to the finite-capacity single server queue of Example 4.7 with capacity $K = K_1$, arrival process Poisson with rate $\lambda = \lambda_1 = 20$, and iid exponential service times with parameter $\mu = \mu_1 = 16$.

		Purchase	Supply	Selling	Demand
Table 4.1 Data for Cars-R-Us used car dealer.	Model	Price ($)	Cars/Week	Price ($)	Cars/Week
	1	8000	20	8800	16
	2	10800	12	11880	10
	3	15500	14	17050	15
	4	18000	8	19800	7
	5	22000	15	24200	12

The aim is, of course, to maximize the net profit. We need to take into account one significant cost of running the business: the cost of carrying the inventory. The dealership pays for the inventory out of a line of credit for which it has to pay interest. It also pays insurance premiums based on the value of the inventory. All these expenses can be approximated as follows. Each week the dealer incurs a cost equal to 1% of the average value of the inventory. Thus, if the average value of the inventory is $100,000, the dealer pays $1000 per week as the inventory carrying cost.

With this we are ready to derive the objective function. Let p_j be the limiting distribution of the $\{X(t), t \geq 0\}$ process. It is given by Equation (4.51) of Example 4.25. Thus, the expected number of cars of model 1 on the lot in steady state is given by

$$L = \sum_{j=0}^{K_1} j p_j.$$

Since each car has cost the dealer $8000, the average value of the inventory is $8000L. Hence the inventory carrying cost per week is $.01 * 8000 * L = 80L$. The dealer sells cars at rate 16 in states 1 through K_1 and no cars are sold in state 0. Hence the weekly rate of car sales is given by $16(1 - p_0)$. The dealer purchases cars at rate 20 in states 0 through $K_1 - 1$, and no cars are purchased in state K_1. Hence the weekly rate of car purchases is given by $20(1 - p_{K_1})$. Clearly these two rates must be identical. Each car generates a profit of $.1*8000 = 800$ dollars. Hence the profit rate in steady state is given by $800 * 16 * (1 - p_0) = 800 * 20 * (1 - p_{K_1})$. Thus the net profit rate in dollars per week is given by

$$g_1(K_1) = 12800(1 - p_0) - 80L.$$

Note that both L and p_0 are functions of K_1. Hence g_1 is a function of K_1. Figure 4.11 shows the graph of g_1 as a function of K_1. Note that it is a concave function of K_1 and is maximized at $K_1 = 10$, implying that it is not optimal to stock more than 10 cars of model 1, even if there is space for them. Of course, how many the dealer should actually stock will depend on the profit functions for the other models. These can be obtained in a similar fashion.

Now, the weekly profit from following a static stocking policy $K = [K_1, K_2, K_3, K_4, K_5]$ is given by

$$g(K) = \sum_{i=1}^{5} g_i(K_i).$$

Thus, the optimal static stocking policy can be obtained by solving the following maximization problem:

$$\text{Maximize } g(K)$$

$$\text{Subject to } \sum_{i=1}^{5} K_i \leq 40$$

$$\text{and } K_i \geq 0 \text{ integers.}$$

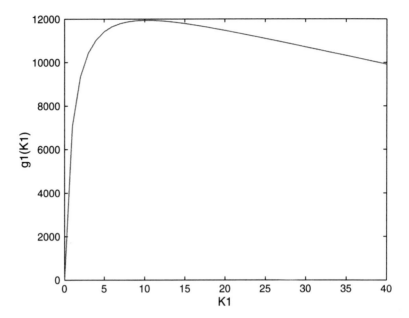

Fig. 4.11 Plot of g_1 as a function of K_1.

Now, in general, solving optimization problems with integral constraints is a very difficult proposition. However, in this case we are in luck. Since the individual net profit functions are concave, we can solve the problem using a simple greedy algorithm. Basically such an algorithm allocates each successive spot in the parking lot to the model that produces the highest incremental profit and continues this way until all spots are allocated or all increments in profit are negative. The details of the algorithms are given below.

Allocation Algorithm:

Step 1: Initialize $K = [0, 0, 0, 0, 0]$.
Step 2: Compute
$$\Delta_i = g_i(K_i + 1) - g_i(K_i), \quad 1 \le i \le 5.$$
Let
$$r = \operatorname{argmax}\{\Delta_i, \ 1 \le i \le 5\}.$$
Step 3: If $\sum_{i=1}^{5} K_i < 40$ and $\Delta_r > 0$, set
$$K_r = K_r + 1$$
and go back to Step 2.
Step 4: If $\sum_{i=1}^{5} K_i = 40$ or $\Delta_r \le 0$, stop. K is the optimal allocation vector.

Table 4.2 Results for
Cars-R-Us used car dealer.

Policy	Allocation Vector	Expected Weekly Profit
Son	[0, 7, 14, 7, 12]	64,618
Father	[8, 8, 8, 8, 8]	76,085
Optimal	[7, 7, 11, 7, 8]	76,318

Using this algorithm, we get as the optimal allocation vector [7, 7, 11, 7, 8]. Table 4.2 shows the profit rates for the three policies: son, father, and optimal.

Thus it seems the son's policy does not do too well. The father's policy seems pretty close to optimal. The weekly profits under the optimal policy are $233 more than under the father's policy and $11,986 more than under the son's policy. Assuming 50 weeks of operation per year, this translates into annual profit increases of $11,700 and $584,980, respectively.

At this point, it is worth discussing what else could be done. For example, does the policy have to be static? Could we come up with a policy that tells whether or not to accept a seller's car depending upon the current mix of cars on the lot? This is a much more complicated problem. Another question we should raise with the dealership is: are the prices under the dealer's control? In our analysis, we have assumed that they are fixed (possibly due to competition). However, if they are under the dealer's control, that will create another interesting policy question, even under static stocking policies. Typically the demand and supply will depend upon the prices, although getting reliable data for this price dependence is rather difficult. The dealer can presumably set the buying and selling prices for a given model depending upon how many cars of that model are on the lot. This creates a highly complex, but doable, optimization problem. We will not get into the details here.

We end this section with the Matlab functions that were used to compute the results given above.

```
*********************************
function SProfit = carsrussinglemodel(lam,mu,bp,sp,Kmax,h)
%analysis of a single model policy in the Cars-R-Us case study
%Usage A = carsrussinglemodel(lam,mu,bp,sp,Kmax,h)
%lam = arrival rate of sellers to the dealer per week
%mu = arrival rate of buyers at the dealer per week
%bp = the dealer buys a car from a seller at this price
%sp = the dealer sells a car to a buyer at this price
%Kmax = maximum number of used cars the dealer holds in inventory
%h = holding cost parameter as a fraction of the value of the inventory
%Output: A(k) = weekly expected profit if at most k−1 cars are stocked
SProfit = zeros(1,Kmax+1)
%SProfit(k) = net weekly profit when the user stocks at most k − 1 cars
rho = lam/mu
for K = 1:Kmax
%finite capacity queue calculations
p = zeros(1,K+1)
```

```
%p(j) = limiting probability that there are j − 1 cars on the lot
p(1) = 1
for j = 2:K+1
p(j) = p(j−1)*rho
end
p = p/sum(p)
Inv = [0:K]*p'; %average inventory on the lot
SProfit(K+1) = mu*sp*(1−p(1)) − bp*lam*(1−p(K+1)) − h*bp*Inv
end

**********************************

function [Optalloc, Optprofit] = carsrusoptimalpolicy(l,m,b,s,Kmax,h)
%computes the optimal allocation
%and optimal profit in the Cars-R-Us case study
 %Usage: [A, B] = carsrusoptimalpolicy(l,m,b,s,Kmax,h)
%l(i) = arrival rate of sellers of model i to the dealer per week
%m(i) = arrival rate of buyers of model i at the dealer per week
%b(i) = the dealer buys a model i from a seller at price b(i)
%s(i) = the dealer sells a model i to a buyer at price s(i)
%Kmax = maximum number of used cars the dealer holds in inventory
%h = holding cost parameter as a fraction of the value of the inventory
%Output: A(i) = spaces allocated to model i
%B = expected weekly profit from the optimal allocation
M = length(l); % number of models
Profit = zeros(M,Kmax+1)
%Profit(i,k) = net weekly profit for model i when the user stocks at
%most k−1 cars of that model
for i = 1:M
Pri = carsrusfixedpolicy(l(i),m(i),b(i),s(i),Kmax,h)
%profit vector from model i
Profit(i,:) = Pri'
end
%greedy algorithm
alloc = zeros(M,1)
%alloc(i) = number of spots allocated to model i
for k = 1:Kmax
for i = 1:M
diff(i) = Profit(i,alloc(i)+2)−Profit(i,alloc(i)+1)
end
 [maxd,I] = max(diff)
if maxd > 0
alloc(I) = alloc(I)+1
end
end for i = 1:M
maxp(i) = Profit(i,alloc(i)+1)
```

%profit from model i by stocking alloc(i) cars
end
Optalloc = alloc; %optimal allocation vector
Optprofit = sum(maxp) %optimal weekly profit

Appendix 4A: Proof Of Theorem 4.3

In this appendix, we give the proof of Theorem 4.3, which is restated below.

Theorem 4.12. (Theorem 4.3). *The transition probability matrix $P(t) = [p_{i,j}(t)]$ is given by*

$$P(t) = \sum_{k=0}^{\infty} e^{-rt} \frac{(rt)^k}{k!} \hat{P}^k. \tag{4.61}$$

The proof is developed in stages below. We start with the important properties of the $\hat{P}$ matrix.

Theorem 4.13. (Properties of $\hat{P}$). *Let R be a rate matrix of a CTMC. The $\hat{P}$ matrix as defined in (4.24) is a stochastic matrix; i.e., it is a one-step transition matrix of a DTMC.*

Proof. **(1)** If $i \neq j$, then $\hat{p}_{i,j} = r_{i,j}/r \geq 0$. For $i = j$, we have

$$\hat{p}_{i,i} = 1 - \frac{r_i}{r} \geq 0$$

since r is bounded below by the largest r_i.
(2) For $1 \leq i \leq N$, we have

$$\sum_{j=1}^{N} \hat{p}_{i,j} = \hat{p}_{i,i} + \sum_{j \neq i} \hat{p}_{i,j}$$

$$= 1 - \frac{r_i}{r} + \sum_{j \neq i} \frac{r_{i,j}}{r}$$

$$= 1 - \frac{r_i}{r} + \sum_{j=1}^{N} \frac{r_{i,j}}{r}$$

$$= 1 - \frac{r_i}{r} + \frac{r_i}{r} = 1.$$

(We have used the fact that $r_{i,i} = 0$.) This proves the theorem. ■

Now suppose we observe the CTMC $\{X(t), t \geq 0\}$ at times $0 = S_0 < S_1 < S_2 < S_3 < \cdots$, etc., and let $\hat{X}_n = X(S_n)$; i.e., $\hat{X}_n$ is the nth observation. The sequence $\{S_n, n \geq 0\}$ is defined recursively as follows. We start with $S_0 = 0$

and $\hat{X}_0 = X(0)$ as the initial state. Let $n \geq 0$ be a fixed integer. Suppose $\hat{X}_n = i$. Let T_{n+1} be the remaining sojourn time of the CTMC in state i. Let Y_{n+1} be an $\text{Exp}(r - r_i)$ random variable that is independent of T_{n+1} and the history of $\{X(t), t \geq 0\}$ up to time S_n. The next observation time S_{n+1} is defined as

$$S_{n+1} = S_n + \min\{T_{n+1}, Y_{n+1}\},$$

and the next observation is given by

$$\hat{X}_{n+1} = X(S_{n+1}).$$

The next theorem describes the structure of the observation process $\{\hat{X}_n, n \geq 0\}$.

Theorem 4.14. (The Embedded DTMC). *The stochastic process $\{\hat{X}_n, n \geq 0\}$ is a DTMC with one-step transition probability matrix $\hat{P}$.*

Proof. The Markov property of $\{\hat{X}_n, n \geq 0\}$ follows from the Markov property of $\{X(t), t \geq 0\}$ and the construction of the $\{Y_n, n \geq 1\}$ variables. Next we compute the transition probabilities. Suppose $\hat{X}_n = i$. If $T_{n+1} > Y_{n+1}$, the next observation occurs at time $S_n + Y_{n+1}$. However, the state of the CTMC at this time is the same as at time S_n. Hence, $\hat{X}_{n+1} = i$. (Conversely, $\hat{X}_{n+1} = i$ implies that $T_{n+1} > Y_{n+1}$.) This yields

$$P(\hat{X}_{n+1} = i | \hat{X}_n = i) = P(T_{n+1} > Y_{n+1} | X(S_n) = i)$$
$$= \frac{r - r_i}{r - r_i + r_i} = 1 - \frac{r_i}{r},$$

where the second equality follows from (3.16). Next, if $T_{n+1} \leq Y_{n+1}$, the next observation occurs at time $S_n + T_{n+1}$. Now, the state of the CTMC at this time is j with probability $r_{i,j}/r$. This yields, for $i \neq j$,

$$P(\hat{X}_{n+1} = j | \hat{X}_n = i) = P(X(S_{n+1}) = j | X(S_n) = i)$$
$$= P(X(S_n + T_{n+1}) = j, T_{n+1} < Y_{n+1} | X(S_n) = i)$$
$$= \frac{r_{i,j}}{r_i} \cdot \frac{r_i}{r_i + r - r_i} = \frac{r_{i,j}}{r}.$$

The theorem follows from this. ∎

Now let $N(t)$ be the number of observations up to time t. The next theorem describes the structure of the counting process $\{N(t), t \geq 0\}$.

Theorem 4.15. (The Observation Process). *$\{N(t), t \geq 0\}$ is a Poisson process with rate r and is independent of $\{\hat{X}_n, n \geq 0\}$.*

Proof. Given $\hat{X}_n = i$, we see that $T_{n+1} \sim \text{Exp}(r_i)$ and $Y_{n+1} \sim \text{Exp}(r - r_i)$ are independent random variables. Hence, using (3.14), we see that $\min\{T_{n+1}, Y_{n+1}\}$ is an exponential random variable with parameter $r_i + r - r_i = r$. But this is independent of $\hat{X}_n$! Hence the intervals between consecutive observations are iid Exp(r) random variables. Hence, by the results in Section 3.3, we see that $\{N(t), t \geq 0\}$ is PP(r). This also establishes the independence of $\{N(t), t \geq 0\}$ and $\{\hat{X}_n, n \geq 0\}$. ∎

It is easy to see that

$$X(t) = \hat{X}_{N(t)}, t \geq 0. \tag{4.62}$$

Using Theorems 4.14, 4.15 and the equation above we now prove Theorem 4.3.

Proof of Theorem 4.3. We have

$$
\begin{aligned}
p_{i,j}(t) &= P(X(t) = j \mid X(0) = i) \\
&= \sum_{k=0}^{\infty} P(X(t) = j \mid X(0) = i, N(t) = k) P(N(t) = k \mid X(0) = i) \\
&= \sum_{k=0}^{\infty} P(\hat{X}_{N(t)} = j \mid \hat{X}_0 = i, N(t) = k) P(N(t) = k \mid \hat{X}_0 = i)
\end{aligned}
$$

(using (4.62))

$$= \sum_{k=0}^{\infty} P(\hat{X}_k = j \mid \hat{X}_0 = i) e^{-rt} \frac{(rt)^k}{k!}$$

(due to the independence of $\{\hat{X}_n, n \geq 0\}$ and $\{N(t), t \geq 0\}$, and Theorem 3.6)

$$= \sum_{k=0}^{\infty} [\hat{P}^k]_{i,j} e^{-rt} \frac{(rt)^k}{k!}.$$

Here we have used Theorem 2.2 to write

$$P(\hat{X}_k = j \mid \hat{X}_0 = i) = [\hat{P}^k]_{i,j}.$$

Putting the equation above in matrix form, we get (4.27). ∎

Appendix 4B: Uniformization Algorithm to Compute $P(t)$

Here we develop an algorithm to compute $P(t)$ based on the results of Theorem 4.3. We first develop bounds on the error resulting from approximating the infinite series in (4.27) by a finite series of the first M terms.

Theorem 4.16. (Error Bounds for $P(t)$). *For a fixed $t \geq 0$, let*

$$P^M(t) = [p_{i,j}^M(t)] = \sum_{k=0}^{M} e^{-rt} \frac{(rt)^k}{k!} \hat{P}^k. \tag{4.63}$$

Then

$$|p_{i,j}(t) - p_{i,j}^M(t)| \le \sum_{k=M+1}^{\infty} e^{-rt} \frac{(rt)^k}{k!} \tag{4.64}$$

for all $1 \le i, j \le N$.

Proof. We have

$$|p_{i,j}(t) - p_{i,j}^M(t)| = \left| \sum_{k=0}^{\infty} e^{-rt} \frac{(rt)^k}{k!} [\hat{P}^k]_{i,j} - \sum_{k=0}^{M} e^{-rt} \frac{(rt)^k}{k!} [\hat{P}^k]_{i,j} \right|$$

$$= \left| \sum_{k=M+1}^{\infty} e^{-rt} \frac{(rt)^k}{k!} [\hat{P}^k]_{i,j} \right|$$

$$\le \sum_{k=M+1}^{\infty} e^{-rt} \frac{(rt)^k}{k!},$$

where the last inequality follows because $[\hat{P}^k]_{i,j} \le 1$ for all i, j, k. ∎

The error bound in (4.64) can be used as follows. Suppose we want to compute $P(t)$ within a tolerance of ϵ. Choose an M such that

$$\sum_{k=M+1}^{\infty} e^{-rt} \frac{(rt)^k}{k!} \le \epsilon.$$

Such an M can always be chosen. However, it depends upon t, r, and ϵ. For this M, Theorem 4.16 guarantees that the approximation $P^M(t)$ is within ϵ of $P(t)$. This yields the following algorithm.

The Uniformization Algorithm for P(t). This algorithm computes the transition probability matrix $P(t)$ (within a given numerical accuracy ϵ) for a CTMC with a rate matrix R:

1. Given $R, t, 0 < \epsilon < 1$.
2. Compute r by using equality in (4.23).
3. Compute $\hat{P}$ by using (4.24).
4. $A = \hat{P}; B = e^{-rt} I; c = e^{-rt}; sum = c; k = 1$.
5. While $sum < 1 - \epsilon$, do:
   $c = c * (rt)/k;$
   $B = B + cA;$
   $A = A\hat{P};$
   $sum = sum + c;$
   $k = k + 1;$
   end.
6. B is within ϵ of $P(t)$.

Appendix 4C: Uniformization Algorithm to Compute $M(T)$

Here we develop an algorithm to compute $M(T)$ based on the representation in Theorem 4.5. First, the following theorem gives the accuracy of truncating the infinite series in (4.39) at $k = K$ to approximate $M(T)$.

Theorem 4.17. (Error Bounds for $M(T)$).

$$\left| m_{i,j}(T) - \sum_{k=0}^{K} \frac{1}{r} P(Y > k)[\hat{P}^k]_{i,j} \right| \leq \frac{1}{r} \sum_{k=K+1}^{\infty} P(Y > k) \qquad (4.65)$$

for all $1 \leq i, j \leq N$.

Proof. The proof follows along the same lines as that of Theorem 4.16. ■

We have

$$rT = E(Y) = \sum_{k=0}^{\infty} P(Y > k).$$

Hence

$$\frac{1}{r} \sum_{k=K+1}^{\infty} P(Y > k) = \frac{1}{r} \left(\sum_{k=0}^{\infty} P(Y > k) - \sum_{k=0}^{K} P(Y > k) \right)$$

$$= \frac{1}{r} \left(rT - \sum_{k=0}^{K} P(Y > k) \right)$$

$$= T - \frac{1}{r} \left(\sum_{k=0}^{K} P(Y > k) \right).$$

Thus, for a given error tolerance $\epsilon > 0$, if we choose K such that

$$\frac{1}{r} \sum_{k=0}^{K} P(Y > k) > T - \epsilon,$$

then the error in truncating the infinite series at $k = K$ is bounded above by ϵ. This observation is used in the following algorithm.

The Uniformization Algorithm for $M(T)$. This algorithm computes the matrix $M(T)$ (within a given numerical accuracy ϵ) for a CTMC with a rate matrix R:

1. Given $R, T, 0 < \epsilon < 1$.
2. Compute r by using equality in (4.23).
3. Compute $\hat{P}$ by using (4.24).
4. $A = \hat{P}; k = 0$.

5. $yek(= P(Y = k)) = e^{-rT}; ygk(= P(Y > k)) = 1 - yek;$
   $sum(= \sum_{r=0}^{k} P(Y > r)) = ygk.$
6. $B = ygk * I.$
7. While $sum/r < T - \epsilon$, do:
   $k = k + 1;$
   $yek = yek * (rT)/k; ygk = ygk - yek;$
   $B = B + ygk * A;$
   $A = A\hat{P};$
   $sum = sum + ygk;$
   end.
8. B/r is within ϵ of $M(T)$ and uses the first $k + 1$ terms of the infinite series in (4.39).

4.10 Problems

CONCEPTUAL PROBLEMS

4.1. A weight of L tons is held up by K cables that share the load equally. When one of the cables breaks, the remaining unbroken cables share the entire load equally. When the last cable breaks, we have a failure. The failure rate of a cable subject to M tons of load is λM per year. The lifetimes of the K cables are independent of each other. Let $X(t)$ be the number of cables that are still unbroken at time t. Show that $\{X(t), t \geq 0\}$ is a CTMC, and find its rate matrix.

4.2. Customers arrive at a bank according to a Poisson process with rate λ per hour. The bank lobby has enough space for ten customers. When the lobby is full, an incoming customer goes to another branch and is lost. The bank manager assigns one teller to customer service as long as the number of customers in the lobby is three or less, she assigns two tellers if the number is more than three but less than eight, and otherwise she assigns three tellers. The service times of the customers are iid $Exp(\mu)$ random variables. Let $X(t)$ be the number of customers in the bank at time t. Model $\{X(t), t \geq 0\}$ as a birth and death process, and derive the birth and death rates.

4.3. A computer has five processing units (PUs). The lifetimes of the PUs are iid $Exp(\mu)$ random variables. When a PU fails, the computer tries to isolate it automatically and reconfigure the system using the remaining PUs. However, this process succeeds with probability c, called the *coverage factor*. If the reconfiguring succeeds, the system continues with one less PU. If the process fails, the entire system crashes. Assume that the reconfiguring process is instantaneous and that once the system crashes it stays down forever. Let $X(t)$ be 0 if the system is down at time t; otherwise it equals the number of working PUs at time t. Model $\{X(t), t \geq 0\}$ as a CTMC, and show its rate matrix.

4.4. A service station has three servers, indexed 1, 2, and 3. When a customer arrives, he is assigned to the idle server with the lowest index. If all servers are busy, the customer goes away. The service times at server i are iid $\text{Exp}(\mu_i)$ random variables. (Different μ's represent different speeds of the servers.) The customers arrive according to a PP(λ). Model this system as a CTMC.

4.5. A single-server service station serves two types of customers. Customers of type i, $i = 1, 2$, arrive according to a PP(λ_i), independently of each other. The station has space to handle at most K customers. The service times are iid $\text{Exp}(\mu)$ for both types of customers. The admission policy is as follows. If, at the time of an arrival, the total number of customers in the system is M or less (here $M < K$ is a fixed integer), the arriving customer is allowed to join the queue; otherwise he is allowed to join only if he is of type 1. This creates preferential treatment for type 1 customers. Let $X(t)$ be the number of customers (of both types) in the system at time t. Show that $\{X(t), t \geq 0\}$ is a birth and death process. Derive its birth and death rates.

4.6. A system consisting of two components is subject to a series of shocks that arrive according to a PP(λ). A shock can cause the failure of component 1 alone with probability p, component 2 alone with probability q, both components with probability r, or have no effect with probability $1 - p - q - r$. No repairs are possible. The system fails when both the components fail. Model the state of the system as a CTMC.

4.7. A machine shop consists of two borers and two lathes. The lifetimes of the borers are $\text{Exp}(\mu_b)$ random variables, and those of the lathes are $\text{Exp}(\mu_l)$ random variables. The machine shop has two repair persons: Al and Bob. Al can repair both lathes and borers, while Bob can only repair lathes. Repair times for the borers are $\text{Exp}(\lambda_b)$ and for the lathes $\text{Exp}(\lambda_l)$, regardless of who repairs the machines. Borers have priority in repairs. Repairs can be preempted. Making appropriate independence assumptions, model this machine shop as a CTMC.

4.8. An automobile part needs three machining operations performed in a given sequence. These operations are performed by three machines. The part is fed to the first machine, where the machining operation takes an $\text{Exp}(\mu_1)$ amount of time. After the operation is complete, the part moves to machine 2, where the machining operation takes $\text{Exp}(\mu_2)$ amount of time. It then moves to machine 3, where the operation takes $\text{Exp}(\mu_3)$ amount of time. There is no storage room between the two machines, and hence if machine 2 is working, the part from machine 1 cannot be removed even if the operation at machine 1 is complete. We say that machine 1 is blocked in such a case. There is an ample supply of unprocessed parts available so that machine 1 can always process a new part when a completed part moves to machine 2. Model this system as a CTMC. (*Hint*: Note that machine 1 may be working or blocked, machine 2 may be working, blocked, or idle, and machine 3 may be working or idle.)

4.9. A single-server queueing system has a finite capacity K. The customers arrive according to a PP(λ) and demand iid Exp(μ) service times. The server is subject to failures and repairs as follows. The server serves for an Exp(θ) amount of time and then fails. Once it fails, all the customers in the system go away instantaneously. The server takes an Exp(α) time to get repaired. While the server is down, no new customers are allowed to enter the system. If the system is full, the arriving customers depart immediately without getting served. Model this system as a CTMC.

4.10. Consider the five-processor system described in Conceptual Problem 4.3. Suppose that the system maintenance starts as soon as the system crashes. The amount of time needed to repair the system is an Exp(λ) random variable, at the end of which all five processors are functioning. Model this system as a CTMC.

4.11. Consider the two-component system of Conceptual Problem 4.6. Furthermore, suppose that a single repair person starts the system repair as soon as the system fails. The repair time of each component is an Exp(α) random variable. The system is turned on when both components are repaired. Assume that shocks have no effect on the components unless the system is on. Model this as a CTMC.

4.12. Do Conceptual Problem 4.11, but now assume that the repair on a component can start as soon as it fails, even though the system is still up, and that the component is put back in service as soon as it is repaired. The shocks have no effect on a component under repair. The repair person gives priority to component 1 over component 2 for repairs.

4.13. Consider the model of the telephone switch in Example 4.9. Show that the limiting distribution of the number of calls in the switch is given by

$$p_j = \frac{\rho^j / j!}{\sum_{i=0}^{K} \rho^i / i!}, \quad 0 \le j \le K,$$

where $\rho = \lambda/\mu$. (*Hint*: Use the results of Example 4.23.)

COMPUTATIONAL PROBLEMS

4.1. Compute the transition probability matrix $P(t)$ at $t = 0.20$ for a CTMC on $S = \{1, 2, 3, 4, 5\}$ with the rate matrix

$$R = \begin{bmatrix} 0 & 4 & 4 & 0 & 0 \\ 5 & 0 & 5 & 5 & 0 \\ 5 & 5 & 0 & 4 & 4 \\ 0 & 5 & 5 & 0 & 4 \\ 0 & 0 & 5 & 5 & 0 \end{bmatrix}.$$

4.2. Compute the transition probability matrix $P(t)$ at $t = 0.10$ for a CTMC on $S = \{1, 2, 3, 4, 5, 6\}$ with the rate matrix

$$R = \begin{bmatrix} 0 & 6 & 0 & 0 & 0 & 0 \\ 0 & 0 & 6 & 0 & 0 & 0 \\ 0 & 0 & 0 & 6 & 0 & 0 \\ 0 & 0 & 0 & 0 & 6 & 0 \\ 0 & 0 & 0 & 0 & 0 & 6 \\ 6 & 0 & 0 & 0 & 0 & 0 \end{bmatrix}.$$

4.3. Compute the transition probability matrix $P(t)$ at $t = 0.10$ for a CTMC on $S = \{1, 2, 3, 4, 5, 6\}$ with the rate matrix

$$R = \begin{bmatrix} 0 & 6 & 6 & 8 & 0 & 0 \\ 0 & 0 & 6 & 8 & 0 & 0 \\ 0 & 0 & 0 & 6 & 0 & 0 \\ 0 & 6 & 8 & 0 & 0 & 0 \\ 8 & 0 & 0 & 0 & 0 & 6 \\ 6 & 0 & 0 & 0 & 6 & 0 \end{bmatrix}.$$

4.4. Consider the model in Conceptual Problem 4.1. Suppose the individual cables have a failure rate of .2 per year per ton of load. We need to build a system to support 18 tons of load, to be equally shared by the cables. If we use three cables, what is the probability that the system will last for more than 2 years?

4.5. In Computational Problem 3.8, how many cables are needed to ensure with probability more than .999 that the system will last for more than 2 years?

4.6. Consider Conceptual Problem 4.3. Suppose the mean lifetime of a processor is 2 years and the coverage factor is .94. What is the probability that the five-processor system functions for 5 years without fail? Assume all processors are operating at time 0.

4.7. Consider Conceptual Problem 4.4. Suppose the mean service time at the third server is 8 minutes. Server 2 is twice as fast as server 3, and server 1 is twice as fast as server 2. The customers arrive at a rate of 20 per hour. Compute the probability that all three servers are busy at time $t = 20$ minutes, assuming that the system is empty at time 0.

4.8. Consider the system of Conceptual Problem 4.6. The shocks arrive on average once a day. About half of the shocks cause no damage, one out of ten damages component 1 alone, about one out of five damages component 2 alone, and the rest damage both components. Compute the probability that the system survives for at least 4 weeks if both components are working initially.

4.9. Consider the call center of Example 4.10. Suppose the arrival rate is 60 calls per hour and each call takes an average of 6 minutes to handle. If there are eight reservation agents and the system can put four callers on hold at one time, what is the probability that the system is full after 1 hour, assuming it is idle at time 0?

4.10. Consider the single-server model of Conceptual Problem 4.9. Suppose the server is working at time 0. Compute the probability that the server is working at time t.

4.11. Compute the occupancy matrix $M(t)$ for the CTMC in Computational Problem 4.1 using the t values given there.

4.12. Compute the occupancy matrix $M(t)$ for the CTMC in Computational Problem 4.2 using the t values given there.

4.13. Compute the occupancy matrix $M(t)$ for the CTMC in Computational Problem 4.3 using the t values given there.

4.14. Consider the model of Conceptual Problem 4.2. Suppose the arrival rate is 20 customers per hour and the average service time is 4 minutes. Compute the expected amount of time three tellers are working during an 8-hour day, assuming there are no customers in the bank at the beginning of the day.

4.15. Consider Computational Problem 4.7. Compute the expected amount of time the three servers are busy during the first hour.

4.16. Consider the three-machine production system described in Conceptual Problem 4.8. Suppose the mean processing times at the three machines are 1 minute, 1.2 minutes, and 1 minute, respectively. Compute the expected amount of time machine 1 is blocked during the first hour, assuming all machines are working at time 0.

4.17. Consider the telephone switch of Example 4.31. Compute the expected amount of time the switch is full during 1 day, assuming it is idle at the beginning of the day.

4.18. Consider the call center of Example 4.10. Suppose the arrival rate is 60 calls per hour and each call takes an average of 6 minutes to handle. If there are eight reservation agents, what is the expected time that all of them are busy during an 8-hour shift, assuming none is busy at the beginning? The system can put four callers on hold at one time.

4.19. In Computational Problem 4.16, compute the expected amount of time machine 3 is working during the first hour, assuming all machines are working at time 0.

4.20. Classify the CTMCs in Computational Problems 4.1, 4.2, and 4.3 as irreducible or reducible.

4.21. Compute the limiting distribution of the CTMC from Computational Problem 4.1.

4.22. Compute the limiting distribution of the CTMC from Computational Problem 4.2.

4.23. Compute the limiting distribution of the state of the system of Computational Problem 4.7.

4.24. Compute the limiting distribution of the state of the system of Conceptual Problem 4.2 assuming the arrival rate is 20 customers per hour and the average service time is 6 minutes.

4.25. Compute the long-run fraction of the time that the last machine is working in the three-machine production system of Computational Problem 4.16.

4.26. Consider the call center described in Example 4.10 with the data given in Computational Problem 4.9. Compute:

1. The long-run fraction of the time that all the agents are busy.
2. The long-run fraction of the time that the call center has to turn away calls.
3. The expected number of busy agents in steady state.

4.27. Consider the telephone switch of Example 4.31. Compute the expected number of calls handled by the switch in steady state.

4.28. Consider the five-processor system described in Conceptual Problem 4.3 with parameters given in Computational Problem 4.8. Suppose that the system maintenance starts as soon as the system crashes. The amount of time needed to repair the system is exponentially distributed with mean 5 days, at the end of which all five processors are functioning. Compute the limiting probability that the system is under repair.

4.29. Consider the two-component system of Conceptual Problem 4.11 with the data given in Computational Problem 4.8. The repair time of each component is an exponential random variable with mean 2 days. Compute the long-run fraction of the time the system is down.

4.30. Do Computational Problem 4.29, but now assume the system behaves as described in Conceptual Problem 4.12.

4.31. Suppose the CTMC in Computational Problem 4.1 incurs costs at the rates $c(i) = 2 * i + 1, 1 \le i \le 5$. Compute the ETC incurred over the time interval $[0, 10]$ if the initial state is 2.

4.32. Compute the long-run cost rate for the CTMC in Computational Problem 4.31.

4.33. Suppose the CTMC in Computational Problem 4.2 incurs costs at the rates $c(i) = 4 * i^2 + 1, 1 \le i \le 6$. Compute the ETC incurred over the time interval $[0, 20]$ if the initial state is 6.

4.34. Compute the long-run cost rate for the CTMC in Computational Problem 4.33.

4.35. Suppose the CTMC in Computational Problem 4.3 incurs costs at the rates $c(i) = 2 * i^2 - 3 * i + 1, 1 \leq i \leq 6$. Compute the ETC incurred over the time interval $[0, 15]$ if the initial state is 4.

4.36. Consider the call center of Example 4.10 with data from Computational Problem 4.9. Suppose each busy agent costs $20 per hour and each customer in the system produces revenue of $25 per hour spent in the system. Compute the long-run net revenue per unit time of operating the system.

4.37. Consider Conceptual Problem 4.4 with data from Computational Problem 4.7. Suppose the costs of the three servers are $40, $20, and $10 per hour, respectively. (The server costs money only when it is busy.) Each customer can be charged c per hour that they spend in the system. What is the smallest value of c that can make the system profitable in the long run?

4.38. Consider the three-machine production system described in Conceptual Problem 4.8 with data from Computational Problem 4.16. Each machine costs $40 per hour while it is working on a component. Each machine adds value at a rate of $75 per hour to the part that it works on. The value added, or the cost of operation, is zero when the machine is idle or blocked. Compute the net contribution of the three machines per unit time in the long run.

4.39. Consider the telephone switch of Example 4.31. Suppose the switch is idle at time 0. Compute the expected total revenue in the first hour if each admitted call is charged 10 cents per minute.

4.40. Consider the five-processor system described in Computational Problem 4.28. Suppose each working processor produces revenue at a rate of $100 per hour, while repairs cost $200 per hour. What is the expected revenue produced during the first year if all five processors are working in the beginning?

4.41. The system in Computational Problem 4.40 is available for lease for $2 million per year. Is it worth leasing?

4.42. Consider the two-component system in Computational Problem 4.29. The system produces revenue at a rate of $400 per day per working component. The repair costs $200 per day. Compute the expected total revenue in the first week of a newly installed system.

4.43. Compute the long-run revenue rate of the system in Computational Problem 4.42.

4.44. Consider the single server queue of Example 4.7 with capacity 10. Suppose the customers arrive at a rate of five per hour. We have a choice of using one of two servers: server 1 serves a customer in 8 minutes, while server 2 takes 10 minutes. Each admitted customer pays $10 for service. Server 1 can be hired for $20 per hour, while server 2 is available for $15 per hour. Which server should be hired to maximize the net revenue per hour in the long run? (*Hint*: Argue that server 1 brings in revenue at a rate of $75 per hour spent serving a customer, while server 2 brings in revenue at a rate of $60 per hour when it is busy. Idle servers bring in no revenue.)

4.45. Compute the expected time to go from state 1 to state 5 in the CTMC of Computational Problem 4.1.

4.46. Compute the expected time to go from state 2 to state 4 in the CTMC of Computational Problem 4.2.

4.47. Compute the expected time to go from state 6 to state 4 in the CTMC of Computational Problem 4.3.

4.48. Consider the system described in Computational Problem 4.4. Compute the expected time of failure.

4.49. In the system described in Computational Problem 4.4, how many cables should be used in order to get the expected lifetime of the system to be at least 2 years?

4.50. Compute the expected lifetime of the system described in Computational Problem 4.8.

4.51. Compute the expected lifetime of the five-processor system described in Computational Problem 4.10.

Case Study Problems. You may use the Matlab programs of Section 4.9 to answer the following problems.

4.52. The manager of the dealership has yet another suggestion: allocate spaces in proportion to the profit margin. What allocation vector K does this imply? How do the weekly profits from this policy compare with those from the optimal policy?

4.53. The dealership is considering increasing the markup from the current 10% to 15% while leaving the purchase price unchanged. The dealer estimates that this will decrease the demand by 20% from the current levels. How will this policy affect the profits, assuming the dealer uses optimal space allocation in both cases?

4.54. The dealer wants to simplify the business by stocking only four models rather than five. Assuming the removal of a model does not affect the supply or demand for the rest, which model should the dealer eliminate? Does this improve the profits?

4.55. The dealership is considering increasing the markup from the current 10% to 15% while leaving the selling price unchanged. The dealer estimates that this will decrease the supply by 20% from current levels. How will this policy affect the profits, assuming the dealer uses optimal space allocation in both cases?

4.56. The dealer's son has another idea: stock a sixth model. Its purchase price is $30,000 and its selling price is $33,000. The supply rate is 8 per week, and the demand rate is 7 per week. Should the dealer stock this model? What is the new optimal allocation strategy?

4.57. The dealer has an opportunity to extend his parking lot. How many new spaces should he add?

Chapter 5
Generalized Markov Models

5.1 Introduction

The main focus of this book is to study systems that evolve randomly in time. We encountered several applications in Chapter 2 where the system is observed at time $n = 0, 1, 2, 3, \ldots$. In such cases, we define X_n as the state of the system at time n and study the discrete-time stochastic process $\{X_n, n \geq 0\}$. In Chapter 2, we studied the systems that have the Markov property at each time $n = 0, 1, 2, 3, \ldots$; i.e., the future of the system from any time n onward depends on its history up to time n only through the state of the system at time n. We found this property to be immensely helpful in studying the behavior of these systems.

We also encountered several applications in Chapter 4 where the system is observed continuously at time $t \geq 0$ and hence we need to study the continuous-time stochastic process $\{X(t), t \geq 0\}$. In Chapter 4, we studied the systems that have the Markov property at all times $t \geq 0$; i.e., the future of the system from any time t onward depends on its history up to time t only through the state of the system at time t. Again, we found that the Markov property made it possible to study many aspects of such systems.

This creates a natural question: what can be done if the system does not have the Markov property at all the integer points n or all the real points t? We address this question in this chapter. Essentially, we shall study *generalized Markov models*; i.e., continuous-time stochastic processes $\{X(t), t \geq 0\}$ that have the Markov property at a set of (possibly random) time points $0 = S_0 \leq S_1 \leq S_2 \leq S_3, \ldots$, etc. In other words, the future of the system from time S_n onward depends on the history of the system up to time S_n only via the state of the system at time S_n. This relaxation comes with a price. The transient and short-term analysis of these systems becomes much harder, and hence we shall not attempt it in this book. (We refer the reader to more advanced texts for more information on this.) Here, we shall concentrate the long-term analysis and the long-run cost models.

We begin this development with a simple process, called the renewal process, in the next section.

V.G. Kulkarni, *Introduction to Modeling and Analysis of Stochastic Systems*,
Springer Texts in Statistics, DOI 10.1007/978-1-4419-1772-0_5,
© Springer Science+Business Media, LLC 2011

5.2 Renewal Processes

Suppose we are observing a series of events occurring randomly over time. For example, the events may be accidents taking place at a particular traffic intersection, the births in a maternity ward, the failures of a repairable system, the earthquakes in a given state, the arrivals of customers at a service station, transactions at a bank, the transmission of packets in a data communication network, the completion of jobs by a computer system, etc.

Let S_n be the time when the nth event occurs. We assume that the events are indexed in increasing time of their occurrence; i.e., $S_n \leq S_{n+1}$. If the nth and $(n + 1)$st events occur at the same time, $S_n = S_{n+1}$. We shall also assume that $S_0 = 0$. Define

$$T_n = S_n - S_{n-1}, n \geq 1.$$

Thus T_n, called the *inter-event time*, is the (random) time interval between the occurrence of the $(n - 1)$st and the nth event.

Now let $N(t)$ be the total number of events observed during $(0, t]$. Note that the event at time 0 is not counted in $N(t)$, but any event at time t is counted. The process $\{N(t), t \geq 0\}$ is called a *counting process* generated by $\{T_n, n \geq 1\}$. A typical sample path of a counting process is shown in Figure 5.1. The figure also illustrates the relationship between the random variables S_n, T_n, and $N(t)$.

Next we define a *renewal process* as a special case of a counting process.

Definition 5.1. (Renewal Process). A counting process $\{N(t), t \geq 0\}$ generated by $\{T_n, n \geq 1\}$ is called a renewal process if $\{T_n, n \geq 1\}$ is a sequence of nonnegative iid random variables.

We present several examples of renewal processes below.

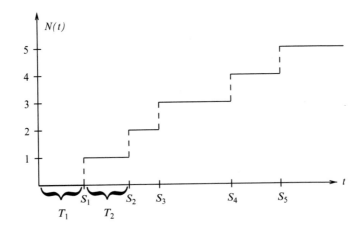

Fig. 5.1 A typical sample path of a counting process.

Example 5.1. (Poisson Process). A Poisson process with rate λ as defined in Section 3.3 is a renewal process because, by its definition, the inter-event times $\{T_n, n \geq 1\}$ are iid $Exp(\lambda)$ random variables. ∎

Example 5.2. (Battery Replacement). Mr. Smith replaces the battery in his car as soon as it dies. We ignore the time required to replace the battery since it is small compared with the lifetime of the battery. Let $N(t)$ be the number of batteries replaced during the first t years of the life of the car, not counting the one that the car came with. Then $\{N(t), t \geq 0\}$ is a renewal process if the lifetimes of the successive batteries are iid.

Now, it is highly inconvenient for Mr. Smith to have the car battery die on him at random. To avoid this, he replaces the battery once it becomes 3 years old, even if it hasn't failed yet. Of course, if the battery fails before it is 3 years old, he has to replace it anyway. By following this policy, he tries to reduce the unscheduled replacements, which are more unpleasant than the planned replacements. Now let $N(t)$ be the number of batteries replaced up to time t, planned or unplanned. Is $\{N(t), t \geq 0\}$ a renewal process?

By definition, $\{N(t), t \geq 0\}$ is a renewal process if the inter-replacement times $\{T_n, n \geq 1\}$ are iid nonnegative random variables. Let L_n be the lifetime of the nth battery (in years), counting the original battery as battery number 1. Suppose $\{L_n, n \geq 1\}$ are iid nonnegative random variables. Mr. Smith will do the first replacement, at time $T_1 = 3$ if $L_1 > 3$ or else he has to do it at time $T_1 = L_1$. Thus $T_1 = \min\{L_1, 3\}$. Similarly, the time between the first and second replacements is $T_2 = \min\{L_2, 3\}$ and, in general, $T_n = \min\{L_n, 3\}$. Since $\{L_n, n \geq 1\}$ are iid nonnegative random variables, so are the inter-replacement times $\{T_n, n \geq 1\}$. Hence $\{N(t), t \geq 0\}$ is a renewal process. ∎

Example 5.3. (Renewal Processes in DTMCs). Let X_n be the state of a system at time n, and suppose $\{X_n, n \geq 0\}$ is a DTMC on state space $\{1, 2, \ldots, N\}$. Suppose the DTMC starts in a fixed state $i, 1 \leq i \leq N$. Define S_n as the nth time the process visits state i. Mathematically, S_n's can be recursively defined as follows: $S_0 = 0$ and

$$S_n = \min\{k > S_{n-1} : X_k = i\}, \quad n \geq 1.$$

In this case, $N(t)$ counts the number of visits to state i during $(0, t]$. Note that although a DTMC is a discrete-time process, the $\{N(t), t \geq 0\}$ process is a continuous-time process. Is $\{N(t), t \geq 0\}$ a renewal process?

Let $T_n = S_n - S_{n-1}$ be the time between the $(n - 1)$st and nth visits to state i. The DTMC is in state i at time 0 and returns to state i at time T_1. The next visit to state i occurs after T_2 transitions. Due to the Markov property and time homogeneity, the random variable T_2 is independent of T_1 and has the same distribution. By continuing this way, we see that $\{T_n, n \geq 1\}$ is a sequence of iid nonnegative random variables. Hence $\{N(t), t \geq 0\}$ is a renewal process. ∎

Example 5.4. (Renewal Processes in CTMCs). Let $\{X(t), t \geq 0\}$ be a CTMC on state space $\{1, 2, \ldots, N\}$. Suppose the CTMC starts in state $i, 1 \leq i \leq N$. Define

S_n as the nth time the process enters state i. In this case, $N(t)$ counts the number of visits to state i during $(0, t]$. Is $\{N(t), t \geq 0\}$ a renewal process?

The answer is "yes" using the same argument as in the previous example since the CTMC has the Markov property at all times $t \geq 0$. ∎

Next we study the *long-run renewal rate*; i.e., we study

$$\lim_{t \to \infty} \frac{N(t)}{t}.$$

We first need to define what we mean by the limit above since $N(t)/t$ is a random variable for each t. We do this in the following general definition.

Definition 5.2. (Convergence with Probability One). Let Ω be a sample space and $Z_n : \Omega \to (-\infty, \infty)$ $(n \geq 1)$ be a sequence of random variables defined on it. The random variables Z_n are said to converge with probability 1 to a constant c as n tends to ∞ if

$$P\left(\omega \in \Omega : \lim_{n \to \infty} Z_n(\omega) = c\right) = 1.$$

In words, convergence with probability 1 implies that the set of ω's for which the convergence holds has probability 1. There may be some ω's for which the convergence fails; however, the probability of all such ω's is 0. The "convergence with probability 1" is also known as "strong convergence" or "almost sure convergence," although we will not use these terms in this book. One of the most important "convergence with probability 1" results is stated without proof in the next theorem.

Theorem 5.1. (Strong Law of Large Numbers (SLLN)). *Let $\{T_n, n \geq 1\}$ be a sequence of iid random variables with common mean τ, and let*

$$S_n = T_1 + T_2 + \cdots + T_n.$$

Then

$$\lim_{n \to \infty} \frac{S_n}{n} = \tau \tag{5.1}$$

with probability 1.

With these preliminaries, we are ready to study the long-run renewal rate. First consider a special case where the inter-event times are all exactly equal to τ; i.e., the nth event takes place at time $n\tau$. Then we see that the renewal process $\{N(t), t \geq 0\}$ is a deterministic process in this case, and the long-run rate at which the events take place is $1/\tau$; i.e.,

$$\lim_{t \to \infty} \frac{N(t)}{t} = \frac{1}{\tau}.$$

Intuitively, we would expect that the limit above should continue to hold when the inter-event times are iid random variables with mean τ. This intuition is made precise in the next theorem.

Theorem 5.2. (Long-Run Renewal Rate). *Let $\{N(t), t \geq 0\}$ be a renewal process generated by a sequence of nonnegative iid random variables $\{T_n, n \geq 1\}$ with common mean $0 < \tau < \infty$. Then*

$$\lim_{t \to \infty} \frac{N(t)}{t} = \frac{1}{\tau} \tag{5.2}$$

with probability 1.

Proof. Let $S_0 = 0$ and

$$S_n = T_1 + T_2 + \cdots + T_n, \quad n \geq 1.$$

Thus the nth event takes place at time S_n. Since $N(t)$ is the number of events that occur up to time t, it can be seen that the last event at or before t occurs at time $S_{N(t)}$ and the first event after t occurs at time $S_{N(t)+1}$. Thus

$$S_{N(t)} \leq t < S_{N(t)+1}.$$

Dividing by $N(t)$, we get

$$\frac{S_{N(t)}}{N(t)} \leq \frac{t}{N(t)} < \frac{S_{N(t)+1}}{N(t)}. \tag{5.3}$$

Now, since $\tau < \infty$,

$$\lim_{t \to \infty} N(t) = \infty$$

with probability 1. Hence,

$$\lim_{t \to \infty} \frac{S_{N(t)}}{N(t)} = \lim_{n \to \infty} \frac{S_n}{n}$$

$$= \tau \tag{5.4}$$

with probability 1 from the SLLN. Similarly,

$$\lim_{t \to \infty} \frac{S_{N(t)+1}}{N(t)} = \lim_{t \to \infty} \frac{S_{N(t)+1}}{N(t)+1} \frac{N(t)+1}{N(t)}$$

$$= \lim_{n \to \infty} \frac{S_{n+1}}{n+1} \lim_{n \to \infty} \frac{n+1}{n}$$

$$= \tau. \tag{5.5}$$

Taking limits in (5.3), we get

$$\lim_{t \to \infty} \frac{S_{N(t)}}{N(t)} \leq \lim_{t \to \infty} \frac{t}{N(t)} \leq \lim_{t \to \infty} \frac{S_{N(t)+1}}{N(t)}.$$

Substituting from (5.4) and (5.5) in the above, we get

$$\tau \le \lim_{t \to \infty} \frac{t}{N(t)} \le \tau.$$

Hence, taking reciprocals and assuming $0 < \tau < \infty$, we get

$$\lim_{t \to \infty} \frac{N(t)}{t} = \frac{1}{\tau}. \quad \blacksquare$$

Remark 5.1. If $\tau = 0$ or $\tau = \infty$, we can show that (5.2) remains valid if we interpret $1/0 = \infty$ and $1/\infty = 0$. $\blacksquare$

Thus the long-run renewal rate equals the reciprocal of the mean inter-event time, as intuitively expected. This seemingly simple result proves to be tremendously powerful in the long-term analysis of generalized Markov models. We first illustrate it with several examples.

Example 5.5. (Poisson Process). Let $\{N(t), t \ge 0\}$ be a PP(λ). Then the inter-event times are iid Exp(λ) with mean $\tau = 1/\lambda$. Hence, from Theorem 5.2,

$$\lim_{t \to \infty} \frac{N(t)}{t} = \frac{1}{\tau} = \lambda.$$

Thus the long-run rate of events in a Poisson process is λ, as we have seen before. We have also seen in Section 3.3 that for a fixed t, $N(t)$ is a Poisson random variable with parameter λt. Hence

$$\lim_{t \to \infty} \frac{E(N(t))}{t} = \frac{\lambda t}{t} = \lambda.$$

Thus the expected number of events per unit time in a PP(λ) is also λ! Note that this does not follow from Theorem 5.2 but happens to be true. $\blacksquare$

Example 5.6. (Battery Replacement). Consider the battery replacement problem of Example 5.2. Suppose the lifetimes of successive batteries are iid random variables that are uniformly distributed over $(1, 5)$ years. Suppose Mr. Smith replaces batteries only upon failure. Compute the long-run rate of replacement.

The mean of a U$(1, 5)$ random variable is 3. Hence the mean inter-replacement time is 3 years. Using Theorem 5.2, we see that the rate of replacement is $1/3 = .333$ per year. $\blacksquare$

Example 5.7. (Battery Replacement (Continued)). Suppose Mr. Smith follows the policy of replacing the batteries upon failure or upon becoming 3 years old. Compute the long-run rate of replacement under this policy.

Let L_i be the lifetime of the ith battery. The ith inter-replacement time is then $T_i = \min\{L_i, 3\}$, as seen in Example 5.2. Hence,

$$\tau = E(T_i)$$
$$= E(\min\{L_i, 3\})$$

$$= \int_1^3 x f_{L_i}(x)dx + \int_3^5 3 f_{L_i}(x)dx$$

$$= \frac{1}{4}\left(\int_1^3 x dx + \int_3^5 3 dx\right)$$

$$= \frac{1}{4}\left(\frac{8}{2} + 3 * 2\right)$$

$$= 2.5 \text{ years.}$$

Hence, from Theorem 5.2, the long-run rate of replacement is $1/2.5 = .4$ under this policy. This is clearly more than $1/3$, the rate of replacement under the "replace upon failure" policy of Example 5.6. Why should Mr. Smith accept a higher rate of replacement? Presumably to reduce the rate of unplanned replacements. So, let us compute the rate of unplanned replacements under the new policy.

Let $N_1(t)$ be the number of unplanned replacements up to time t. $\{N_1(t), t \geq 0\}$ is a renewal process (see Conceptual Problem 5.2). Hence, we can use Theorem 5.2 if we can compute τ, the expected value of T_1, the time of the first unplanned replacement. Now, if $L_1 < 3$, the first replacement is unplanned and hence $T_1 = L_1$. Otherwise the first replacement is a planned one and takes place at time 3. The expected time until the unplanned replacement from then on is still τ. Using the fact that L_1 is $U(1, 5)$, we get

$$\tau = E(T_1)$$
$$= E(T_1|L_1 < 3)P(L_1 < 3) + E(T_1|L_1 \geq 3)P(L_1 \geq 3)$$
$$= E(L_1|L_1 < 3)P(L_1 < 3) + (3 + \tau)P(L_1 \geq 3)$$
$$= E(\min(L_1, 3)) + \tau P(L_1 \geq 3)$$
$$= 2.5 + .5\tau.$$

Solving for τ, we get

$$\tau = 5.$$

Hence the unplanned replacements occur at a rate of $1/5 = .2$ per year in the long run. This is less than $1/3$, the rate of unplanned replacements under the "replace upon failure" policy. Thus, if unplanned replacements are really inconvenient, it may be preferable to use this modified policy. We need to wait until the next section to quantify how "inconvenient" the unplanned replacements have to be in order to justify the higher total rate of replacements. ∎

Example 5.8. (Two-State Machine). Consider the CTMC model of a machine subject to failures and repairs as described in Example 4.3. Suppose the machine is working at time 0. Compute the long-run rate of repair completions for this machine.

Let $X(t)$ be the state of the machine at time t, 1 if it is up and 0 if it is down. $\{X(t), t \geq 0\}$ is a CTMC, as explained in Example 4.3. Then, $N(t)$, the number of repair completions up to time t, is the number of times the CTMC, enters state 1 up to time t. Since $X(0)$ is given to be 1, we see from Example 5.4 that $\{N(t), t \geq 0\}$

is a renewal process. The expected time until the first repair completion, τ, is the expected time in state 1 plus the expected time in state 0, which is $1/\mu + 1/\lambda$. Hence, using Theorem 5.2, the long-run rate of repair completions is given by

$$\frac{1}{\tau} = \frac{\lambda\mu}{\lambda + \mu}. \quad \blacksquare$$

Example 5.9. (Manpower Model). Consider the manpower planning model of Paper Pushers Inc., as described in Example 2.6. Compute the long-run rate of turnover (the rate at which an existing employee is replaced by a new one) for the company.

Note that the company has 100 employees, each of whom moves from grade to grade according to a DTMC. When the ith employee quits, he is replaced by a new employee in grade 1 and is assigned employee ID i. Suppose we keep track of the turnover of the employees with ID i, $1 \leq i \leq 100$. Let $N_i(t)$ be the number of new employees that have joined with employee ID i up to time t, not counting the initial employee. Since each new employee starts in grade 1 and moves from grade to grade according to a DTMC, the lengths of their tenures with the company are iid random variables. Hence $\{N_i(t), t \geq 0\}$ is a renewal process. From Example 2.31, the expected length of stay of an employee is 1.4 years. Hence, using Theorem 5.2, we get

$$\lim_{t\to\infty} \frac{N_i(t)}{t} = \frac{1}{1.4} \text{ per year.}$$

Since there are 100 similar slots, the turnover for the company is given by

$$\lim_{t\to\infty} \sum_{i=1}^{100} \frac{N_i(t)}{t} = \frac{100}{1.4} = 71.43 \text{ per year.}$$

This is very high and suggests that some steps are needed to bring it down. $\blacksquare$

5.3 Cumulative Processes

Consider a system that incurs costs and earns rewards as time goes on. Thinking of rewards as negative costs, let $C(t)$ be the total net cost (i.e., cost $-$ reward) incurred by a system over the interval $(0, t]$. Since $C(t)$ may increase and decrease with time and take both positive and negative values, we think of $\{C(t), t \geq 0\}$ as a continuous-time stochastic process with state space $(-\infty, \infty)$.

Now suppose we divide the time into consecutive intervals, called *cycles*, with T_n being the (possibly random, but nonnegative) length of the nth cycle. Define $S_0 = 0$ and

$$S_n = T_1 + T_2 + \cdots + T_n, \quad n \geq 1.$$

Thus the nth interval is $(S_{n-1}, S_n]$. Define

$$C_n = C(S_n) - C(S_{n-1}), \quad n \geq 1.$$

Then C_n is the net cost (positive or negative) incurred over the nth cycle.

Definition 5.3. (Cumulative Processes). The stochastic process $\{C(t), t \geq 0\}$ is called a cumulative process if $\{(T_n, C_n), n \geq 1\}$ is a sequence of iid bivariate random variables.

Thus the total cost process is a cumulative process if the successive interval lengths are iid and the costs incurred over these intervals are iid. However, the cost over the nth interval may depend on its length. Indeed, it is this possibility of dependence that gives the cumulative processes their power! The *cumulative processes* are also known as *renewal reward processes*. These processes are useful in the long-term analysis of systems when we want to compute the long-term cost rates, the long-run fraction of time spent in different states, etc. We illustrate this with a few examples.

Example 5.10. (Battery Replacement). Consider the battery replacement problem described in Example 5.7. Now suppose the cost of replacing a battery is \$75 if it is a planned replacement and \$125 if it is an unplanned replacement. Thus the cost of unexpected delays, the towing involved, and disruption to the normal life of Mr. Smith from the sudden battery failure is figured to be \$50! Let $C(t)$ be the total battery replacement cost up to time t. Is $\{C(t), t \geq 0\}$ a cumulative process?

Let L_n be the lifetime of the nth battery and $T_n = \min\{L_n, 3\}$ be the nth inter-replacement time. The nth replacement takes place at time S_n and costs $\$C_n$. Think of $(S_{n-1}, S_n]$ as the nth interval. Then C_n is the cost incurred over the nth interval. Now, we have

$$C_n = \begin{cases} 125 & \text{if } L_n < 3, \\ 75 & \text{if } L_n \geq 3. \end{cases}$$

Note that C_n explicitly depends on L_n and hence on T_n. However, $\{(T_n, C_n), n \geq 1\}$ is a sequence of iid bivariate random variables. Hence, $\{C(t), t \geq 0\}$ is a cumulative process. ∎

We have always talked about $C(t)$ as the total cost up to time t. However, t need not represent time, as shown by the next example.

Example 5.11. (Prorated Warranty). A tire company issues the following 50,000 mile prorated warranty on its tires that sell for \$95 per-tire. Suppose the new tire fails when it has L miles on it. If $L > 50{,}000$, the customer has to buy a new tire for the full price. If $L \leq 50{,}000$, the customer gets a new tire for $\$95 * L/50{,}000$. Assume that the customer continues to buy the same brand of tire after each failure. Let $C(t)$ be the total cost to the customer over the first t miles. Is $\{C(t), t \geq 0\}$ a cumulative process?

Let T_n be the life (in miles) of the nth tire purchased by the customer. Let C_n be the cost of the nth tire. The pro rata warranty implies that

$$C_n = \$95(\min\{50{,}000, T_n\})/50{,}000.$$

Thus the customer incurs a cost C_n after $S_n = T_1 + T_2 + \cdots + T_n$ miles. Think of T_n as the nth interval and C_n as the cost incurred over it. Now suppose that successive tires have iid lifetimes. Then $\{(T_n, C_n), n \geq 1\}$ is a sequence of iid bivariate random variables. Hence, $\{C(t), t \geq 0\}$ is a cumulative process. ∎

In the examples so far, the cost was incurred in a lump sum fashion at the end of an interval. In the next example, costs and rewards are incurred continuously over time.

Example 5.12. (Two-State Machine). Consider the CTMC model of a machine subject to failures and repairs as described in Example 4.3. Suppose the machine is working at time 0. Suppose the machine produces revenue at rate $\$A$ per unit time when it is up and costs $\$B$ per unit time for a repair person to repair it. Let $C(t)$ be the net total cost up to time t. Show that $\{C(t), t \geq 0\}$ is a cumulative process.

Let $X(t)$ be the state of the machine at time t, 1 if it is up and 0 if it is down. Let U_n be the nth uptime and D_n be the nth downtime. Then $\{U_n, n \geq 1\}$ is a sequence of iid $\text{Exp}(\lambda)$ random variables and is independent of $\{D_n, n \geq 1\}$, a sequence of iid $\text{Exp}(\mu)$ random variables. Since the machine goes through an up–down cycle repeatedly, it seems natural to take the nth interval to be $T_n = U_n + D_n$. The cost over the nth interval is

$$C_n = BD_n - AU_n.$$

Now, the independence of the successive up and down times implies that $\{(T_n, C_n), n \geq 1\}$ is a sequence of iid bivariate random variables, although C_n does depend upon T_n. Hence $\{C(t), t \geq 0\}$ is a cumulative process. ∎

Example 5.13. (Compound Poisson Process). Suppose customers arrive at a restaurant in batches. Let $N(t)$ be the total number of batches that arrive during $(0, t]$ and $C(t)$ the total number of customers that arrive during $(0, t]$. Suppose $\{C(t), t \geq 0\}$ is a compound Poisson process as defined in Section 3.6. We shall show that $\{C(t), t \geq 0\}$ is a cumulative process.

Let C_n be the size of the nth batch and T_n be the inter-batch arrival time. Since $\{C(t), t \geq 0\}$ is a CPP, we know that $\{N(t), t \geq 0\}$ is a PP and $\{C_n, n \geq 0\}$ is a sequence of iid random variables that is independent of the batch arrival process. Furthermore, from (3.28), we see that

$$C(t) = \sum_{n=1}^{N(t)} C_n, \quad t \geq 0.$$

From the definition of the CPP, we see that $\{(T_n, C_n), n \geq 1\}$ is a sequence of iid bivariate random variables. (In fact, C_n is also independent of T_n!) Hence it follows that the CPP $\{C(t), t \geq 0\}$ is a cumulative process. ∎

In general, it is very difficult to compute $\mathsf{E}(C(t))$ for a general cumulative process. One case where this calculation is easy is the compound Poisson process as worked out in Equation 3.29 of Section 3.6. Hence we shall restrict our attention to the long-term analysis. In our long-term analysis, we are interested in the long-run cost rate,

$$\lim_{t \to \infty} \frac{C(t)}{t}.$$

The next theorem gives the main result.

Theorem 5.3. (Long-Run Cost Rate). *Let $\{C(t), t \geq 0\}$ be a cumulative process with the corresponding sequence of iid bivariate random variables $\{(T_n, C_n), n \geq 1\}$. Then*

$$\lim_{t \to \infty} \frac{C(t)}{t} = \frac{E(C_1)}{E(T_1)} \tag{5.6}$$

with probability 1.

Proof. We shall prove the theorem under the assumption that the cost C_n over the nth cycle is incurred as a lump sum cost at the end of the cycle. We refer the reader to more advanced texts for the proof of the general case.

Let $N(t)$ be the number of cycles that are completed by time t. Then $\{N(t), t \geq 0\}$ is a renewal process generated by $\{T_n, n \geq 1\}$. The total cost up to time t can be written as

$$C(t) = \begin{cases} 0 & \text{if } N(t) = 0, \\ \sum_{n=1}^{N(t)} C_n & \text{if } N(t) > 0. \end{cases}$$

Assuming $N(t) > 0$, we have

$$\frac{C(t)}{t} = \frac{\sum_{n=1}^{N(t)} C_n}{t}$$

$$= \frac{\sum_{n=1}^{N(t)} C_n}{N(t)} \cdot \frac{N(t)}{t}.$$

Taking limits as $t \to \infty$ and using Theorem 5.2, we get

$$\lim_{t \to \infty} \frac{C(t)}{t} = \lim_{t \to \infty} \frac{\sum_{n=1}^{N(t)} C_n}{N(t)} \cdot \lim_{t \to \infty} \frac{N(t)}{t}$$

$$= \lim_{k \to \infty} \frac{\sum_{n=1}^{k} C_n}{k} \cdot \frac{1}{E(T_1)}$$

(follows from Theorem 5.2)

$$= E(C_1) \cdot \frac{1}{E(T_1)},$$

where the last step follows from Theorem 5.1. This yields the theorem. ∎

We show the application of the theorem above with a few examples.

Example 5.14. (Battery Replacement). Consider the battery replacement problem of Example 5.10. Compute the long-run cost per year of following the preventive replacement policy.

Let $C(t)$ be the total cost up to time t. We have seen in Example 5.10 that $\{C(t), t \geq 0\}$ is a cumulative process with the iid bivariate sequence $\{(T_n, C_n), n \geq 1\}$ given there. Hence the long-run cost rate can be computed from Theorem 5.3. We have already computed $E(T_1) = 2.5$ in Example 5.7. Next we compute

$$E(C_1) = 75 * P(L_1 > 3) + 125 * P(L_1 \leq 3)$$
$$= 75 * .5 + 125 * .5$$
$$= 100.$$

Hence the long-run cost rate is

$$\lim_{t \to \infty} \frac{C(t)}{t} = \frac{E(C_1)}{E(T_1)} = \frac{100}{2.5} = 40$$

dollars per year. What is the cost rate of the "replace upon failure" policy? In this case, the total cost is a cumulative process with $\{(T_n, C_n), n \geq 1\}$ with $T_n = L_n$ and $C_n = 125$ (since all replacements are unplanned) for all $n \geq 1$. Hence, from Theorem 5.3,

$$\lim_{t \to \infty} \frac{C(t)}{t} = \frac{125}{3} = 41.67$$

dollars per year. Thus the preventive maintenance policy is actually more economical than the "replace upon failure!" ∎

Example 5.15. (Prorated Warranty). Consider the prorated warranty of Example 5.11. The lifetimes of the tires are iid random variables with common pdf

$$f(x) = 2 * 10^{-10}x \text{ for } 0 \leq x \leq 100,000 \text{ miles.}$$

Suppose the customer has the option of buying the tire without warranty for $90. Based upon the long-run cost per mile, should the customer get the warranty? (We assume that the customer either always gets the warranty or never gets the warranty.)

First consider the case of no warranty. The nth cycle length is T_n, the lifetime of the nth tire. The cost over the nth cycle is $90. Clearly, $\{(T_n, C_n), n \geq 1\}$ is a sequence of iid bivariate random variables. Hence the total cost is a cumulative process, and the long-run cost rate is given by Theorem 5.3. We have

$$E(T_1) = \int_0^{100,000} 2 * 10^{-10}x^2 dx$$
$$= 2 * 10^{-10} \frac{x^3}{3} \Big|_0^{100,000}$$
$$= \tfrac{2}{3} \cdot 10^5.$$

Hence the long-run cost rate of no warranty is

$$\frac{E(C_1)}{E(T_1)} = 135 * 10^{-5}$$

dollars per mile, or $1.35 per 1000 miles.

Next consider the policy of always buying the tires under warranty. The nth cycle here is as in the no warranty case. However, the cost C_n is given by

$$C_n = 95\frac{\min\{50{,}000, T_n\}}{50{,}000}.$$

Hence

$$E(C_1) = \int_0^{100{,}000} 2 * 10^{-10} * 95\frac{\min\{50{,}000, x\}}{50{,}000}x\,dx$$

$$= \int_0^{50{,}000} 2 * 10^{-10} * \frac{95}{50{,}000}x^2\,dx + \int_{50{,}000}^{100{,}000} 95 * 2 * 10^{-10}x\,dx$$

$$= \frac{95}{6} + \frac{95 * 3}{4}$$

$$= 87.08333.$$

Hence the long-run cost under warranty is

$$\frac{87.08333}{.666 * 10^5} = 130.625 * 10^{-5}$$

dollars per mile, or $1.31 per 1000 miles. This is less than the cost of not buying the warranty. Hence the customer should buy the warranty. ∎

Example 5.16. (Two-State Machine). Consider the two-state machine described in Example 5.12. Compute the long-run net cost per unit time.

The machine produces revenue at rate $A per unit time when it is up and it costs $B per unit time to repair it. $C(t)$ is the net total cost up to time t. We have already shown that $\{C(t), t \geq 0\}$ is a cumulative process. Hence, from Theorem 5.3, we get

$$\lim_{t \to \infty} \frac{C(t)}{t} = \frac{E(C_1)}{E(T_1)}$$

$$= \frac{E(BD_1 - AU_1)}{E(U_1 + D_1)},$$

where U_1 is the first uptime and D_1 is the first downtime of the machine. We have

$$E(D_1) = \frac{1}{\lambda}, \quad E(U_1) = \frac{1}{\mu}.$$

Substituting, we get as the long-run cost rate

$$\lim_{t \to \infty} \frac{C(t)}{t} = \frac{B/\lambda - A/\mu}{1/\lambda + 1/\mu}$$

$$= B \frac{\mu}{\lambda + \mu} - A \frac{\lambda}{\lambda + \mu}. \tag{5.7}$$

This is consistent with the result in Example 4.30, where we have computed the long-run revenue rate using the CTMC methods. ∎

Example 5.17. (A General Two-State Machine). In Examples 5.8, 5.12, and 5.16, we studied a two-state machine whose up and down times are independent exponential random variables. Here we consider a machine whose up and down times are generally distributed. Let U_n be the nth uptime and D_n be the nth downtime. We assume that $\{(U_n, D_n), n \geq 1\}$ are iid bivariate random variables. Thus the nth downtime may be dependent on the nth uptime. Compute the long-run fraction of the time that the machine is up, assuming that the machine is up initially.

Let $C(t)$ be the total time the machine is up during $(0, t]$. Then $\{C(t), t \geq 0\}$ is a cumulative process with nth cycle length $T_n = U_n + D_n$ and nth "cost" $C_n = U_n$. Hence, using Theorem 5.3, we get

$$\lim_{t \to \infty} \frac{C(t)}{t} = \frac{\mathsf{E}(C_1)}{\mathsf{E}(T_1)} = \frac{\mathsf{E}(U_1)}{\mathsf{E}(U_1) + \mathsf{E}(D_1)}.$$

The result above is intuitive; however, it is surprising that it is not influenced by the dependence between U_1 and D_1.

As a numerical example, suppose the uptimes are uniformly distributed over $[4, 6]$ weeks and that the nth downtime is exactly 20% of the nth uptime; i.e., $D_n = .2U_n$. In this case, the long-run fraction of the time the machine is up is given by

$$\frac{\mathsf{E}(U_1)}{\mathsf{E}(U_1) + \mathsf{E}(D_1)} = \frac{5}{5 + .2 * 5} = \frac{5}{6}. \ \blacksquare$$

Example 5.18. (Compound Poisson Process). Let $\{C(t), t \geq 0\}$ be a compound Poisson process as defined in Example 5.13. We have seen that this is a cumulative process with

$$\mathsf{E}(C_1) = \tau, \quad \mathsf{E}(T_1) = 1/\lambda.$$

From Theorem 3.11 we get

$$\mathsf{E}(C(t)) = \tau \lambda t, \ t \geq 0.$$

Hence

$$\frac{\mathsf{E}(C(t))}{t} = \frac{\tau \lambda t}{t} = \frac{\tau}{1/\lambda} = \frac{\mathsf{E}(C_1)}{\mathsf{E}(T_1)}.$$

Thus Theorem 5.3 holds. Indeed, the expected cost rate over any time interval equals the long-run cost rate for the CPP! ∎

Armed with Theorem 5.3, we are ready to attack the non-Markovian world!

5.4 Semi-Markov Processes: Examples

Consider a system with state space $\{1, 2, \ldots, N\}$. Suppose the system enters the initial state X_0 at time $S_0 = 0$. It stays there for a nonnegative random amount of time and jumps to another state X_1 (which could be the same as X_0) at time S_1. It stays in the new state for another nonnegative random amount of time and then jumps to the next state X_2 (which could be the same as X_1) at time S_2 and continues this way forever. Thus S_n is the time of the nth transition and X_n is the nth state visited by the system. Let $X(t)$ be the state of the system at time t. Then $X(S_n) = X_n$ for $n \geq 0$.

Definition 5.4. (Semi-Markov Process (SMP)). The stochastic process $\{X(t), t \geq 0\}$ described above is called a semi-Markov process if it has the Markov property at every transition epoch S_n; i.e., the evolution of the process from time $t = S_n$ onward depends on the history of the process up to time S_n only via X_n.

In other words, if $\{X(t), t \geq 0\}$ is an SMP, the process $\{X(t + S_n), t \geq 0\}$, given the entire history $\{X(t), 0 \leq t \leq S_n\}$ and $X(S_n) = i$, is independent of $\{X(t), 0 \leq t < S_n\}$ and is probabilistically identical to $\{X(t), t \geq 0\}$ given $X(0) = i$. The definition of an SMP also explains why such processes are called *semi*-Markov: they have the Markov property only at transition epochs and not at all times. Notice that the future from time S_n cannot depend upon n either. In that sense, the definition above forces the SMP to be "time homogeneous"!

Now, note that the Markov property of the SMP $\{X(t), t \geq 0\}$ at each transition epoch $S_n, n \geq 0$, implies the Markov property of $\{X_n, n \geq 0\}$ at every $n \geq 0$. Thus $\{X_n, n \geq 0\}$ is a time-homogeneous DTMC with state space $\{1, 2, \ldots, N\}$ and is called the *embedded DTMC* of the SMP. Let $P = [p_{i,j}]$ be its transition probability matrix,

$$p_{i,j} = \mathsf{P}(X_{n+1} = j | X_n = i), \quad 1 \leq i, j \leq N.$$

Note that $p_{i,i}$ may be positive, implying that an SMP can jump from state i to itself in one step. Next define

$$w_i = \mathsf{E}(S_1 | X_0 = i), \quad 1 \leq i \leq N, \tag{5.8}$$

to be the mean sojourn time in state i. Since the SMP has the Markov property at each transition epoch, we see that the expected sojourn time in state i is w_i on any visit to state i. Let

$$w = [w_1, w_2, \ldots, w_N]$$

be the vector of expected sojourn times.

With the transition matrix P and sojourn time vector w, we can visualize the evolution of the SMP as follows. The SMP starts in state i. It stays there for w_i amount of time on average and then jumps to state j with probability $p_{i,j}$. It then stays in state j for w_j amount of time on average and jumps to state k with probability $p_{j,k}$, and so on. A typical sample path of an SMP is shown in Figure 5.2.

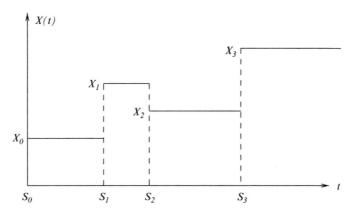

Fig. 5.2   A typical sample path of a semi-Markov process.

Recall that a DTMC is described by its one-step transition probability matrix and a CTMC by its rate matrix. The natural question is: are P and w sufficient to describe the SMP? The answer depends upon what questions we plan to ask about the SMP. As stated earlier, the transient and occupancy-time analyses of SMPs is beyond the scope of this book and we shall concentrate mainly on the long-term analysis. We shall show in the next section that P and w are sufficient for this purpose. They are not sufficient if we are interested in transient and occupancy-time analyses.

Next we discuss several examples of SMPs. In each case, we describe the SMP by giving the transition probability matrix P of the embedded DTMC and the vector w of the expected sojourn times.

Example 5.19. (Two-State Machine). Consider the two-state machine described in Example 5.17. The machine alternates between "up (1)" and "down (0)" states, with U_n being the nth uptime and D_n being the nth downtime. In Example 5.17, we had assumed that $\{(U_n, D_n), n \geq 1\}$ is a sequence of iid bivariate random variables. Here we make a further assumption that U_n is also independent of D_n. Let the mean uptime be $\mathsf{E}(U)$ and the mean downtime be $\mathsf{E}(D)$. Suppose $X(t)$ is the state of the machine at time t. Is $\{X(t), t \geq 0\}$ an SMP?

Suppose the machine starts in state $X_0 = 1$. Then it stays there for U_1 amount of time and then moves to state $X_1 = 0$. It stays there for D_1 amount of time, independent of history, and then moves back to state $X_2 = 1$ and proceeds this way forever. This shows that $\{X(t), t \geq 0\}$ is an SMP with state space $\{0, 1\}$. The embedded DTMC has a transition probability matrix given by

$$P = \begin{bmatrix} 0 & 1 \\ 1 & 0 \end{bmatrix}. \tag{5.9}$$

Also

$$w_0 = \mathsf{E}(S_1 | X_0 = 0) = \mathsf{E}(D),$$
$$w_1 = \mathsf{E}(S_1 | X_0 = 1) = \mathsf{E}(U). \quad \blacksquare$$

Example 5.20. (A CTMC Is an SMP). Let $\{X(t), t \geq 0\}$ be a CTMC with state space $\{1, 2, \ldots, N\}$ and rate matrix R. Show that $\{X(t), t \geq 0\}$ is an SMP, and compute the matrix P and vector w.

Since a CTMC has the Markov property at all times, it certainly has it at the transition epochs. Hence a CTMC is an SMP. Suppose the CTMC starts in state i. From the properties of the CTMC, we know that S_1, the time the CTMC spends in state i before it jumps, is an $Exp(r_i)$ random variable, where

$$r_i = \sum_{j=1}^{N} r_{i,j}.$$

Hence

$$w_i = \frac{1}{r_i}.$$

The next state X_1 is j with probability $r_{i,j}/r_i$, independently of S_1. (Recall that $r_{i,i} = 0$.) Thus we have

$$p_{i,j} = \mathsf{P}(X_1 = j \mid X_0 = i) = \frac{r_{i,j}}{r_i}. \tag{5.10}$$

Note that $p_{i,i} = 0$ in this case, signifying that a CTMC does not jump from state i to itself in one step. ∎

Example 5.21. (Series System). Consider a series system of N components. The system fails as soon as any one of the components fails. Suppose the lifetime of the ith $(1 \leq i \leq N)$ component is an $Exp(v_i)$ random variable and that the lifetimes of the components are independent. When a component fails, the entire system is shut down and the failed component is repaired. The mean repair time of the ith component is τ_i. The repair times are mutually independent as well. When a component is under repair, no other component can fail since the system is down. Model this system as an SMP.

We first decide the state space of the system. We say that the system is in state 0 if it is functioning (i.e., all components are up) and is in state i if the system is down and the ith component is under repair. Let $X(t)$ be the state of the system at time t. The state space of $\{X(t), t \geq 0\}$ is $\{0, 1, 2, \ldots, N\}$. The memoryless property of the exponential random variables and the independence assumptions about the lifetimes and repair times imply that the system has the Markov property every time it changes state. Hence it is an SMP. Next we derive its P matrix and w vector.

Suppose the system starts in state 0. Then it stays there until one of the N components fails, and if the ith component is first to fail it moves to state i. Hence, using the properties of independent exponential random variables (see Section 3.1), we get

$$p_{0,i} = \mathsf{P}(X_1 = i \mid X_0 = 0) = \frac{v_i}{v}, \quad 1 \leq i \leq N,$$

where

$$v = \sum_{i=1}^{N} v_i.$$

If the system starts in state i, then it stays there until the repair of the ith component is complete and then moves to state 0. Hence we get

$$p_{i,0} = P(X_1 = 0 | X_0 = i) = 1, \ \ 1 \le i \le N.$$

Thus the transition probability matrix of the embedded DTMC $\{X_n, n \ge 0\}$ is given by

$$P = \begin{bmatrix} 0 & v_1/v & v_2/v & \cdots & v_N/v \\ 1 & 0 & 0 & \cdots & 0 \\ 1 & 0 & 0 & \cdots & 0 \\ \vdots & \vdots & \vdots & \ddots & \vdots \\ 1 & 0 & 0 & \cdots & 0 \end{bmatrix}. \tag{5.11}$$

Also, the mean sojourn time in state 0 is given by

$$w_0 = \frac{1}{v},$$

and the mean sojourn time in state i is given by

$$w_i = \tau_i, \ \ 1 \le i \le N. \ \blacksquare$$

Example 5.22. (Machine Maintenance). Consider the following maintenance policy for a machine. If the machine fails before it reaches age v (a fixed positive number), it is sent for repair; otherwise it is replaced upon failure with no attempt to repair it. If the machine is sent for repair and the repair takes less than u (another fixed positive number) amount of time, the machine becomes as good as new and is put back into use. If the repair takes longer than u, the repair attempt is abandoned and the machine is replaced by a new one. The successive lifetimes of the machine are iid with common cdf $A(\cdot)$ and mean a, the repair times are iid with common cdf $B(\cdot)$, and the replacement times are iid with common cdf $C(\cdot)$ and mean c. Let $X(t)$ be the state of the machine at time t (1 if it is up, 2 if it is under repair, and 3 if it is under replacement). Is $\{X(t), t \ge 0\}$ an SMP, and if it is, what are its P matrix and w vector?

The $\{X(t), t \ge 0\}$ process is an SMP since it has the Markov property at every transition epoch due to the independence assumptions about the lifetimes, repair times, and replacement times. We compute the P matrix of the embedded DTMC next.

Suppose a new machine is put into use at time 0; i.e., $X(0) = X_0 = 1$. Then S_1, the amount of time it spends in state 1, is the lifetime of the machine. If $S_1 \le v$, then the system moves to state $X_1 = 2$; otherwise it moves to state $X_1 = 3$. Using this, we get

$$p_{1,2} = P(X_1 = 2 | X_0 = 1) = P(S_1 \leq v) = A(v)$$

and

$$p_{1,3} = P(S_1 > v) = 1 - A(v).$$

A similar argument in the remaining states yields the following transition probability matrix for the embedded DTMC $\{X_n, n \geq 0\}$:

$$P = \begin{bmatrix} 0 & A(v) & 1 - A(v) \\ B(u) & 0 & 1 - B(u) \\ 1 & 0 & 0 \end{bmatrix}. \tag{5.12}$$

Also, we get the following expected sojourn times:

$$w_1 = \text{expected lifetime} = a,$$

$$w_2 = E(\min(u, \text{repair time})) = \int_0^u (1 - B(x))dx = b(u) \text{ (say)},$$

$$w_3 = \text{expected replacement time} = c. \blacksquare$$

The examples above should suffice to show the usefulness of SMPs. We study the long-term analysis (this includes the limiting behavior and the average cost models) in the next section.

5.5 Semi-Markov Processes: Long-Term Analysis

We start this section with a study of the first-passage times in SMPs leading to the long-term analysis of SMPs.

5.5.1 Mean Inter-visit Times

Let $\{X(t), t \geq 0\}$ be an SMP on state space $\{1, 2, \ldots, N\}$ with transition matrix P and sojourn time vector w. In this subsection, we shall compute the expected time between two consecutive visits, called the *inter-visit time*, to state j. This expected value will be useful in the next two subsections. First we need some further notation.

Let Y_j be the first time the SMP *enters* state j,

$$Y_j = \min\{S_n : n > 0, X_n = j\}.$$

(Note the strict inequality $n > 0$.) Also let

$$m_{i,j} = E(Y_j | X_0 = i), \quad 1 \leq i, j \leq N.$$

The $m_{i,j}, i \neq j$, is the mean first-passage time from state i to state j. The mean inter-visit time of state j is given by $m_{j,j}$. The next theorem gives the equations satisfied by the mean first-passage times.

Theorem 5.4. (Mean First-Passage Times). *The mean first-passage times, $m_{i,j}s$, satisfy*

$$m_{i,j} = w_i + \sum_{k=1, k \neq j}^{N} p_{i,k} m_{k,j}, \quad 1 \leq i, j \leq N. \tag{5.13}$$

Proof. We derive the result by conditioning on X_1. Suppose $X_0 = i$. If $X_1 = j$, then $Y_j = S_1$, and hence

$$\mathsf{E}(Y_j | X_0 = i, X_1 = j) = \mathsf{E}(S_1 | X_0 = i, X_1 = j).$$

On the other hand, if $X_1 = k \neq j$, then the SMP enters state k at time S_1 and the mean time to enter state j from then on is $m_{k,j}$ due to the Markov property and time homogeneity. Hence, for $k \neq j$,

$$\mathsf{E}(Y_j | X_0 = i, X_1 = k) = \mathsf{E}(S_1 | X_0 = i, X_1 = k) + m_{k,j}.$$

Combining these two equations, we get

$$
\begin{aligned}
m_{i,j} &= \mathsf{E}(Y_j | X_0 = i) \\
&= \sum_{k=1}^{N} \mathsf{E}(Y_j | X_0 = i, X_1 = k) \mathsf{P}(X_1 = k | X_0 = i) \\
&= \mathsf{E}(Y_j | X_0 = i, X_1 = j) \mathsf{P}(X_1 = j | X_0 = i) \\
&\quad + \sum_{k=1, k \neq j}^{N} \mathsf{E}(Y_j | X_0 = i, X_1 = k) \mathsf{P}(X_1 = k | X_0 = i) \\
&= \mathsf{E}(S_1 | X_0 = i, X_1 = j) \mathsf{P}(X_1 = j | X_0 = i) \\
&\quad + \sum_{k=1, k \neq j}^{N} [\mathsf{E}(S_1 | X_0 = i, X_1 = k) + m_{k,j}] \mathsf{P}(X_1 = k | X_0 = i) \\
&= \sum_{k=1}^{N} \mathsf{E}(S_1 | X_0 = i, X_1 = k) \mathsf{P}(X_1 = k | X_0 = i) \\
&\quad + \sum_{k=1, k \neq j}^{N} m_{k,j} \mathsf{P}(X_1 = k | X_0 = i) \\
&= \mathsf{E}(S_1 | X_0 = i) + \sum_{k=1, k \neq j}^{N} m_{k,j} \mathsf{P}(X_1 = k | X_0 = i).
\end{aligned}
$$

We have

$$P(X_1 = k | X_0 = i) = p_{i,k},$$

and from (5.8) we get

$$E(S_1 | X_0 = i) = w_i.$$

Using these results in the last equation, we get (5.13). ∎

Our main interest is in $m_{j,j}$, the mean inter-visit time of state j. Using the theorem above to compute these quantities is rather inefficient since we need to solve different sets of N equations in N unknowns to obtain $m_{j,j}$ for different j's. The next theorem gives a more efficient method of computing all $m_{j,j}$'s by solving a single set of N equations in N unknowns under some restrictions.

Theorem 5.5. (Mean Inter-visit Times). *Assume that the embedded DTMC* $\{X_n, n \geq 0\}$ *is irreducible, and let* $\pi = [\pi_1, \pi_2, \ldots, \pi_N]$ *be a nonzero solution to the balance equations*

$$\pi_j = \sum_{i=1}^{N} \pi_i p_{i,j}, \quad 1 \leq j \leq N.$$

Then

$$m_{j,j} = \frac{\sum_{i=1}^{N} \pi_i w_i}{\pi_j}, \quad 1 \leq j \leq N. \tag{5.14}$$

Proof. It follows from the results of Section 2.5 that there is a (nonzero) solution to the balance equations. Now, multiply both sides of (5.13) by π_i and sum over all i. We get

$$\sum_{i=1}^{N} \pi_i m_{i,j} = \sum_{i=1}^{N} \pi_i w_i + \sum_{i=1}^{N} \pi_i \sum_{k=1, k \neq j}^{N} p_{i,k} m_{k,j}$$

$$= \sum_{i=1}^{N} \pi_i w_i + \sum_{k=1, k \neq j}^{N} m_{k,j} \sum_{i=1}^{N} \pi_i p_{i,k}$$

$$= \sum_{i=1}^{N} \pi_i w_i + \sum_{k=1, k \neq j}^{N} m_{k,j} \pi_k$$

(using the balance equations)

$$= \sum_{i=1}^{N} \pi_i w_i + \sum_{i=1, i \neq j}^{N} \pi_i m_{i,j}$$

(replacing k by i in the last sum)

$$= \sum_{i=1}^{N} \pi_i w_i + \sum_{i=1}^{N} \pi_i m_{i,j} - \pi_j m_{j,j}.$$

Canceling the second sum on the right-hand side with the sum on the left-hand side, we get (5.14). ∎

Note that we do not need the embedded DTMC to be aperiodic. We also do not need a normalized solution to the balance equations; any solution will do. The condition of irreducibility in the theorem ensures that none of the π_i's will be zero.

Example 5.23. (Two-State Machine). Compute the mean time between two repair completions in the case of the two-state machine described in Example 5.19.

The embedded DTMC $\{X_n, n \geq 0\}$ has the transition probability matrix given in (5.9). This DTMC is irreducible, and a solution to the balance equation is

$$\pi_0 = 1; \quad \pi_1 = 1.$$

The mean sojourn times $[w_0, w_1]$ are given in Example 5.19. Hence, using Theorem 5.5, the mean time between two consecutive repair completions is given by

$$m_{1,1} = \frac{\pi_0 w_0 + \pi_1 w_1}{\pi_1} = E(U) + E(D).$$

This is what we expect. ∎

Example 5.24. (Machine Maintenance). Consider the machine maintenance policy of Example 5.22. Compute the expected time between two consecutive repair commencements.

Let $\{X(t), t \geq 0\}$ be the SMP model developed in Example 5.22. The embedded DTMC has the probability transition matrix given in (5.12). The DTMC is irreducible. The balance equations are

$$\begin{aligned}
\pi_1 &= \pi_2 B(u) + \pi_3, \\
\pi_2 &= \pi_1 A(v), \\
\pi_3 &= (1 - A(v))\pi_1 + (1 - B(u))\pi_2.
\end{aligned}$$

One possible solution is

$$\pi_1 = 1, \quad \pi_2 = A(v), \quad \pi_3 = 1 - A(v)B(u).$$

The mean sojourn times $[w_1, w_2, w_3]$ are given in Example 5.22. The time between two consecutive repair commencements is the same as the inter-visit time of state 2. Substituting in (5.14), we get

$$\begin{aligned}
m_{2,2} &= \frac{\pi_1 w_1 + \pi_2 w_2 + \pi_3 w_3}{\pi_2} \\
&= \frac{a + A(v)b(u) + c(1 - A(v)B(u))}{A(v)}.
\end{aligned} \tag{5.15}$$

As a numerical example, suppose the lifetimes are uniformly distributed over [10, 40] days, the repair times are exponentially distributed with mean 2 days, and the replacement times are exactly 1 day. Suppose the cutoff numbers are $v = 30$ days and $u = 3$ days. The relevant quantities are

$$A(30) = \mathsf{P}(\text{lifetime} \le 30) = .6667,$$
$$B(3) = \mathsf{P}(\text{repair time} \le 3) = 1 - e^{-3/2} = .7769,$$
$$a = \text{expected lifetime} = 25,$$
$$b(3) = \mathsf{E}(\min(3, \text{repair time})) = \tfrac{1}{.5}(1 - e^{-.5*3}) = 1.5537,$$
$$c = \text{expected replacement time} = 1.$$

Substituting in (5.15), we get

$$
\begin{aligned}
m_{2,2} &= \frac{25 + (.6667) * (1.5537) + (1)(1 - (.6667)(.7769))}{.6667} \\
&= \frac{26.5179}{.6667} \\
&= 39.7769 \text{ days. } \blacksquare
\end{aligned}
$$

With Theorem 5.5, we are ready to study the limiting behavior of the SMP in the next subsection.

5.5.2 Occupancy Distributions

In this subsection, we study the limiting behavior of the SMPs. In our study of the limiting behavior of DTMCs and CTMCs, we encountered three related quantities:

1. the limiting distribution;
2. the stationary distribution; and
3. the occupancy distribution.

In the case of the SMPs, we shall concentrate only on the last quantity. Study of the first two quantities requires more advanced techniques and is outside the scope of this book.

Let $M_j(t)$ be the total (random) amount of time the SMP spends in state j during the interval $[0, t]$. We are interested in computing the long-run fraction of the time the SMP spends in state j; i.e., we want to compute

$$\lim_{t \to \infty} \frac{M_j(t)}{t}.$$

If the limit above exists (with probability 1), we denote it by p_j and we call the vector $[p_1, p_2, \ldots, p_N]$ the *occupancy distribution* of the SMP. The next theorem shows when the occupancy distribution exists and how to compute it.

Theorem 5.6. (Occupancy Distribution). *Suppose* $\{X(t), t \geq 0\}$ *is an SMP with an irreducible embedded DTMC* $\{X_n, n \geq 0\}$ *with transition probability matrix* P. *The occupancy distribution of the SMP exists, is independent of the initial state of the SMP, and is given by*

$$p_j = \frac{\pi_j w_j}{\sum_{i=1}^{N} \pi_i w_i}, \quad 1 \leq j \leq N, \tag{5.16}$$

where $[\pi_1, \pi_2, \ldots, \pi_N]$ *is a positive solution to the balance equations*

$$\pi_j = \sum_{i=1}^{N} \pi_i p_{i,j}, \quad 1 \leq j \leq N,$$

and w_i, $1 \leq i \leq N$, *is the mean sojourn time in state* i.

Proof. Suppose the SMP starts in state j. We first show that $\{M_j(t), t \geq 0\}$ is a cumulative process as defined in Section 5.3. Let $\hat{S}_n$ be the time of the nth entry into state j. ($\hat{S}_0 = 0$.) Let $\hat{C}_n$ be the total time the SMP spends in state j during the nth cycle $(\hat{S}_{n-1}, \hat{S}_n]$. Thus

$$\hat{C}_n = M_j(\hat{S}_n) - M_j(\hat{S}_{n-1}).$$

Since the SMP has the Markov property at every transition epoch, its behavior from time $\hat{S}_n$ onward depends on the history only via its state at that time. However, the SMP is in state j at time $\hat{S}_n$ for all $n \geq 0$ by definition. Hence its behavior over the intervals $(\hat{S}_{n-1}, \hat{S}_n]$ $(n \geq 1)$ is iid. Thus $\{(\hat{T}_n, \hat{C}_n), n \geq 1\}$ is a sequence of iid bivariate random variables, where $\hat{T}_n = \hat{S}_n - \hat{S}_{n-1}$. Then, from the definition of cumulative processes in Section 5.3, it follows that $\{M_j(t), t \geq 0\}$ is a cumulative process.

Using Theorem 5.3, we get

$$\lim_{t \to \infty} \frac{M_j(t)}{t} = \frac{\mathsf{E}(\hat{C}_1)}{\mathsf{E}(\hat{T}_1)}.$$

Now, $\mathsf{E}(\hat{T}_1)$ is just the mean inter-visit time $m_{j,j}$ of state j and is given by (5.14). Hence

$$\mathsf{E}(\hat{T}_1) = \mathsf{E}(\hat{S}_1) = \frac{\sum_{i=1}^{N} \pi_i w_i}{\pi_j}.$$

Furthermore, during the interval $(0, \hat{S}_1]$, the SMP spends exactly one sojourn time in state j. Hence

$$\mathsf{E}(\hat{C}_1) = w_j.$$

Using these two equations, we get

$$\lim_{t \to \infty} \frac{M_j(t)}{t} = \frac{\pi_j w_j}{\sum_{i=1}^{N} \pi_i w_i}.$$

Thus the theorem is proved if the initial state is j. Now consider the case where the initial state is $i \neq j$. Define $\hat{S}_n$, $\hat{T}_n$, and $\hat{C}_n$ as before. Now, $\{(\hat{T}_n, \hat{C}_n), n \geq 2\}$ is a sequence of iid bivariate random variables and is independent of $(\hat{S}_1, \hat{C}_1)$. We can follow the proof of Theorem 5.3 to show that

$$\lim_{t \to \infty} \frac{M_j(t)}{t} = \frac{E(\hat{C}_2)}{E(\hat{T}_2)}.$$

Since the SMP enters state j at time $\hat{S}_1$, it follows that

$$E(\hat{T}_2) = \frac{\sum_{i=1}^{N} \pi_i w_i}{\pi_j}$$

and

$$E(\hat{C}_2) = w_j.$$

Thus the theorem is valid for all initial states $1 \leq i \leq N$ and hence for any initial distribution. ■

Note that we have defined the occupancy distribution as the long-run fraction of the time spent in state j, $1 \leq j \leq N$. But in the case of CTMCs and DTMCs, the occupancy distribution was defined as the long-run *expected* fraction of the time spent in state j, $1 \leq j \leq N$. So, strictly speaking, we should have studied

$$\lim_{t \to \infty} \frac{E(M_j(t))}{t}$$

to be consistent. It can be shown that, under the hypothesis of Theorem 5.6, the limit above also exists and is given by p_j of (5.16); however, it is beyond the scope of this book to do so. We illustrate the theorem by several examples.

Example 5.25. (Two-State Machine). Compute the long-run fraction of the time the machine of Example 5.19 is up.

Using the results of Example 5.19, we see that the state of the machine is described by a two-state SMP $\{X(t), t \geq 0\}$. The embedded DTMC $\{X_n, n \geq 0\}$ has the P matrix given in (5.9) and is irreducible. Hence Theorem 5.6 can be applied. The relevant quantities are given in Examples 5.19 and 5.23. Hence, the long-run fraction of the time the machine is up is given by

$$p_1 = \frac{\pi_1 w_1}{\pi_0 w_0 + \pi_1 w_1}$$
$$= \frac{E(U)}{E(D) + E(U)}.$$

This agrees with our intuition and with the result in Example 5.17 for a more general two-state machine. ■

Example 5.26. (Series System). Consider the series system described in Example 5.21. Compute the long-run fraction of the time the system is up.

The state of the system is described by an SMP in Example 5.21. The embedded DTMC has a transition probability matrix given in (5.11). The DTMC is irreducible, and the balance equations are given by

$$\pi_0 = \sum_{i=1}^{N} \pi_i,$$
$$\pi_j = \frac{v_j}{v}\pi_0, \ 1 \le j \le N,$$

where $v = \sum_{i=1}^{N} v_i$. A solution to the equations above is given by

$$\pi_0 = 1,$$
$$\pi_j = \frac{v_j}{v}, \ 1 \le j \le N.$$

The mean sojourn times are computed in Example 5.21. Using (5.16), the long-run fraction of the time the system is up is given by

$$p_0 = \frac{\pi_0 w_0}{\sum_{i=0}^{N} \pi_i w_i}$$
$$= \frac{1/v}{1/v + \sum_{i=1}^{N}(v_i/v)\tau_i}$$
$$= \frac{1}{1 + \sum_{i=1}^{N} v_i \tau_i}. \ \blacksquare$$

Example 5.27. (Machine Maintenance). Compute the long-run fraction of the time a machine is working if it is maintained by using the policy of Example 5.22.

The state of the machine under the maintenance policy of Example 5.22 is an SMP with the transition matrix of the embedded DTMC as given in (5.12). This DTMC is irreducible, and hence Theorem 5.6 can be used to compute the required quantity as

$$p_1 = \frac{\pi_1 w_1}{\sum_{i=1}^{3} \pi_i w_i}. \tag{5.17}$$

Using the relevant quantities as computed in Examples 5.22 and 5.24, we get

$$p_1 = \frac{a}{a + A(v)b(u) + c(1 - A(v)B(u))}.$$

For the numerical values given in Example 5.24, we get

$$p_1 = \frac{25}{26.5179} = .9428.$$

Thus the machine is up 94% of the time. $\blacksquare$

5.5.3 Long-Run Cost Rates

In this subsection, we shall study cost models for systems modeled by SMPs, as we did for the CTMCs in Section 4.7. In the DTMC and CTMC cases, we studied two types of cost models: the total expected cost over a finite horizon and the long-run expected cost per unit time. In the case of SMPs, we shall only study the latter since the former requires results for the short-term analysis of SMPs, which we have not developed here. We begin with the following cost model.

Let $\{X(t), t \geq 0\}$ be an SMP on state space $\{1, 2, \ldots, N\}$. Suppose that the system incurs costs at a rate of $c(i)$ per unit time while in state i. Now let $C(T)$ be the total cost incurred by the system up to time T. We are interested in the long-run cost rate defined as

$$g = \lim_{T \to \infty} \frac{C(T)}{T},$$

assuming this limit exists with probability 1. The next theorem shows when this limit exists and how to compute it if it does exist.

Theorem 5.7. (Long-Run Cost Rates). *Suppose the hypothesis of Theorem 5.6 holds, and let $[p_1, p_2, \ldots, p_N]$ be the occupancy distribution of the SMP. Then the long-run cost rate exists, is independent of the initial state of the SMP, and is given by*

$$g = \sum_{i=1}^{N} p_i c(i). \tag{5.18}$$

Proof. Let $M_i(T)$ be the total amount of time the SMP spends in state i during $[0, T]$. Since the SMP incurs costs at a rate of $c(i)$ per unit time while in state i, it follows that

$$C(T) = \sum_{i=1}^{N} c(i) M_i(T). \tag{5.19}$$

Hence

$$\lim_{T \to \infty} \frac{C(T)}{T} = \lim_{T \to \infty} \sum_{i=1}^{N} c(i) \frac{M_i(T)}{T}$$

$$= \sum_{i=1}^{N} c(i) p_i,$$

where the last equation follows from Theorem 5.6. This proves the theorem. ∎

Although we have assumed a specific cost structure in the derivation above, Theorem 5.7 can be used for a variety of cost structures. For example, suppose the system incurs a lump sum cost d_i whenever it visits state i. We can convert this into the cost structure of Theorem 5.7 by assuming that the cost rate $c(i)$ per unit time in state i is such that the expected total cost incurred in one sojourn time in state i is d_i. Since the expected sojourn time in state i is w_i, we must have $c(i)w_i = d_i$, which yields

$$c(i) = \frac{d_i}{w_i}.$$

Hence the long-run cost rate under the lump sum cost structure can be derived from Theorem 5.7 to be

$$g = \sum_{i=1}^{N} p_i \frac{d_i}{w_i}.$$

Substituting for p_i from (5.16), we get

$$g = \frac{\sum_{i=1}^{N} \pi_i d_i}{\sum_{i=1}^{N} \pi_i w_i}. \tag{5.20}$$

We illustrate this with several examples.

Example 5.28. (Two-State Machine). Consider the two-state machine of Example 5.19. Suppose the machine produces net revenue of $\$A$ per unit time when the machine is up, while it costs $\$B$ per unit time to repair the machine when it is down. Compute the long-run cost rate of the machine.

Using the results of Example 5.19 we see that the state of the machine is described by a two-state SMP $\{X(t), t \geq 0\}$. We have $c(0) = B$ and $c(1) = -A$. From Example 5.25, we have

$$p_0 = \frac{E(D)}{E(D) + E(U)}$$

and

$$p_1 = \frac{E(U)}{E(D) + E(U)}$$

Substituting in (5.18), we get as the long-run cost rate

$$g = Bp_0 - Ap_1 = \frac{BE(D) - AE(U)}{E(D) + E(U)}.$$

This is consistent with the results of Example 4.30. ∎

Example 5.29. (Series System). Consider the series system of Example 5.21 with three components. Suppose the mean lifetime of the first component is 5 days, that of the second component is 4 days, and that of the third component is 8 days. The mean repair time is 1 day for component 1, .5 days for component 2, and 2 days for component 3. It costs $200 to repair component 1, $150 to repair component 2, and $500 to repair component 3. When the system is working, it produces revenue at a rate of R per unit time. What is the minimum value of R that makes it worthwhile to operate the system?

The state of the system is described by an SMP in Example 5.21 with state space $\{0, 1, 2, 3\}$. Recall that the lifetime of component i is an $\text{Exp}(v_i)$ random variable with mean $1/v_i$. Using the data given above, we get

$$v_1 = .2, v_2 = .25, v_3 = .125.$$

The mean repair times are

$$\tau_1 = 1, \tau_2 = .5, \tau_3 = 2.$$

The cost structure is a mixture of lump sum costs and continuous costs. We first convert it to the standard cost structure. We have

$$c(0) = -R, c(1) = 200/1 = 200, c(2) = 150/.5 = 300, c(3) = 500/2 = 250.$$

Using the results of Example 5.26, we have

$$p_0 = \frac{1}{1 + \sum_{i=1}^{3} v_i \tau_i} = \frac{1}{1.575},$$

$$p_1 = \frac{v_1 \tau_1}{1 + \sum_{i=1}^{3} v_i \tau_i} = \frac{.2}{1.575},$$

$$p_2 = \frac{v_2 \tau_2}{1 + \sum_{i=1}^{3} v_i \tau_i} = \frac{.125}{1.575},$$

$$p_3 = \frac{v_3 \tau_3}{1 + \sum_{i=1}^{3} v_i \tau_i} = \frac{.25}{1.575}.$$

Substituting in (5.18), we get

$$g = \sum_{i=0}^{3} c(i) p_i = \frac{-R + 40 + 37.5 + 62.5}{1.575}.$$

To break even, we must have $g \leq 0$. Hence we must have

$$R \geq 140$$

dollars per day. ∎

5.6 Case Study: Healthy Heart Coronary Care Facility

This case is inspired by a paper by Kao 1974.

Healthy Heart is a medical facility that caters to heart patients. It has two main parts: the outpatient clinic and an inpatient unit. The doctors see the patients in the outpatient clinic. Patients needing various heart-related surgical procedures, such as angiograms, angioplasty, valve replacement, heart bypass heart surgery, replacement, etc., are treated at the inpatient facility. The inpatient facility can be roughly divided into six parts:

1. ER: the emergency room, where patients are brought in after suffering a heart attack;
2. POW: the pre-op ward, where patients are prepared for surgery;
3. SU: the surgical unit, basically the operating rooms where the actual surgeries take place;
4. POU: the post-operative unit, where the patients stay immediately after the surgery until they stabilize;
5. ICU: the intensive care unit, where the patients stay as long as they need continuous attention;
6. ECU: the extended care unit, where the patients stay until they can be discharged.

The patients coming to the inpatient facility can be classified into two categories: scheduled patients and emergency patients. The scheduled patients arrive at the POW, while the emergency patients arrive at the ER. An emergency patient spends some time in the ER and is then transferred to either the POW, the ICU, the ECU. A patient coming to the POW is transferred to the SU after the pre-op preparation is complete. After the surgery is done, the patient is moved to the POU. In rare cases, the patient may have to be rushed back to the SU from the POU, but most often the patient is moved to the ICU after he or she stabilizes. Similarly, the patient can stay in the ICU for a random amount of time and then move back to the SU or ECU. From the ECU, a patient is discharged after the doctors deem it appropriate.

Healthy Heart is interested in studying the movement of patients among the six units of the facility and the amount of time the patients spend in various units of the inpatient facility during one visit (between admission and discharge). We begin by studying the movement of patients through the six units at the inpatient facility. After collecting enough data, we conclude that the patient movement can be adequately described by a semi-Markov process. However, the behavior of the emergency patients is substantially different from that of the scheduled patients.

We model the movement of an emergency patient as an SMP on state space $\{1 = \text{ER}, 2 = \text{POW}, 3 = \text{SU}, 4 = \text{POU}, 5 = \text{ICU}, 6 = \text{ECU}, 7 = \text{Discharged}\}$. An emergency patient enters the clinic in state 1. Let $X_e(t)$ be the state of (i.e., the unit occupied by) an emergency patient at time t. Thus $X_e(0) = 1$ and $\{X_e(t), t \geq 0\}$ is an SMP. Our data suggest that the embedded transition probability matrix is

$$P_e = \begin{bmatrix} 0 & .32 & 0 & 0 & .38 & .05 & .25 \\ 0 & 0 & 1 & 0 & 0 & 0 & 0 \\ 0 & 0 & 0 & 1 & 0 & 0 & 0 \\ 0 & 0 & 0.06 & 0 & .94 & 0 & 0 \\ 0 & 0 & 0.02 & 0 & 0 & .98 & 0 \\ 0 & 0 & 0 & 0 & .12 & 0 & .88 \\ 0 & 0 & 0 & 0 & 0 & 0 & 1 \end{bmatrix}, \tag{5.21}$$

and mean sojourn times (in hours) are

$$w_e = [4.5, \ 2.4, \ 6.8, \ 4.4, \ 36.7, \ 118.0, \ \infty]. \tag{5.22}$$

Note that state 7 (Discharged) is an absorbing state. If a discharged patient later returns to the clinic, we treat her as a new patient.

Similarly, the movement of a scheduled patient is an SMP on state space $\{2 = \text{POW}, 3 = \text{SU}, 4 = \text{POU}, 5 = \text{ICU}, 6 = \text{ECU}, 7 = \text{Discharged}\}$. A scheduled patient enters the clinic in state 2. Let $X_s(t)$ be the state of (i.e., the unit occupied by) a scheduled patient at time t. Thus $X_s(0) = 2$ and $\{X_s(t), t \geq 0\}$ is an SMP with embedded transition probability matrix

$$P_s = \begin{bmatrix} 0 & 1 & 0 & 0 & 0 & 0 \\ 0 & 0 & 1 & 0 & 0 & 0 \\ 0 & 0.02 & 0 & .98 & 0 & 0 \\ 0 & 0.02 & 0 & 0 & .98 & 0 \\ 0 & 0 & 0 & .09 & 0 & .91 \\ 0 & 0 & 0 & 0 & 0 & 1 \end{bmatrix} \tag{5.23}$$

and mean sojourn times (in hours)

$$w_s = [2.4, \ 5.3, \ 3.4, \ 32.8, \ 98.7, \ \infty]. \tag{5.24}$$

The first metric the management wants to know is how much time on average the patient spends in the clinic before getting discharged. Let m_{ij} be the mean first-passage time from state i to state $j \neq i$ as defined in Subsection 5.5.1. Thus we need to compute m_{17} for the emergency patients and m_{27} for the scheduled patients. Using Theorem 5.4, we get the following equations for the emergency patients:

$$m_{17} = 4.5 + .32m_{27} + .38m_{57} + .05m_{67},$$
$$m_{27} = 2.4 + m_{37},$$
$$m_{37} = 6.8 + m_{47},$$
$$m_{47} = 4.4 + .06m_{37} + .94m_{57},$$
$$m_{57} = 36.7 + .02m_{37} + .98m_{67},$$
$$m_{67} = 118.0 + .12m_{57}.$$

This can be solved easily to get

$$m_{17} = 139.8883 \text{ hours} = 5.8287 \text{ days}.$$

Similarly, the equations for the scheduled patients are

$$m_{27} = 2.4 + m_{37},$$
$$m_{37} = 5.3 + m_{47},$$

$$m_{47} = 3.4 + .02m_{37} + .98m_{57},$$
$$m_{57} = 32.8 + .02m_{37} + .98m_{67},$$
$$m_{67} = 98.7 + .09m_{57}.$$

This can be solved easily to get

$$m_{27} = 156.7177 \text{ hours} = 6.5299 \text{ days}.$$

Thus, an emergency patient spends 5.83 days in the clinic on average, while a scheduled patient spends 6.53 days in the clinic on average. This seems counterintuitive but is a result of the fact that not all emergency patients actually undergo surgery and a quarter of them are discharged immediately after their visit to the ER.

Next the management wants to know the fraction of the time these patients spend in various units. To do this for the emergency patients, we assume that there is always exactly one emergency patient in the system. We ensure this by replacing an emergency patient, immediately upon discharge, by a new independent and stochastically identical emergency patient in state 1. Let $Y_e(t)$ be the unit occupied by this circulating emergency patient at time t. We see that $\{Y_e(t), t \geq 0\}$ is an SMP on state space $\{1 = \text{ER}, 2 = \text{POW}, 3 = \text{SU}, 4 = \text{POU}, 5 = \text{ICU}, 6 = \text{ECU}\}$ with embedded transition probability matrix

$$P_{ec} = \begin{bmatrix} .25 & .32 & 0 & 0 & .38 & .05 \\ 0 & 0 & 1 & 0 & 0 & 0 \\ 0 & 0 & 0 & 1 & 0 & 0 \\ 0 & 0 & 0.06 & 0 & .94 & 0 \\ 0 & 0 & 0.02 & 0 & 0 & .98 \\ .88 & 0 & 0 & 0 & .12 & 0 \end{bmatrix} \tag{5.25}$$

and mean sojourn times (in hours)

$$w_{ec} = [4.5, \ 2.4, \ 6.8, \ 4.4, \ 36.7, \ 118.0]. \tag{5.26}$$

Note that this SMP has an irreducible embedded DTMC with transition probability matrix P_{ec}. The limiting distribution of the embedded DTMC is given by

$$\pi_{ec} = [0.2698, \ 0.0863, \ 0.0965, \ 0.0965, \ 0.2209, \ 0.2299].$$

The occupancy distribution of the SMP $\{Y_e(t), t \geq 0\}$ can now be computed from Theorem 5.6 as

$$p_{ec} = [0.0322, \ 0.0055, \ 0.0174, \ 0.0113, \ 0.2148, \ 0.7189]. \tag{5.27}$$

Thus an emergency patient spends 71.89% of his time in the clinic in the extended care center, while only 1.74% of the time is spent in actual surgery. Since a typical emergency patient spends on average 139.8883 hours in the clinic, we see that the average times spent in various units are given by

$$139.8883 * [0.0322, \ 0.0055, \ 0.0174, \ 0.0113, \ 0.2148, \ 0.7189]$$

$$= [4.5000, \ 0.7680, \ 2.4333, \ 1.5745, \ 30.0443, \ 100.5682].$$

Thus an emergency patient spends 4.5 hours in the ER (this is to be expected). Note that a typical emergency patient spends 2.4 hours in surgery, although the sojourn time in surgery is 6.8 hours. This is because only about a third of the emergency patients visit the surgical unit during their stay at the clinic.

A similar analysis can be done for the scheduled patients. We create a circulating scheduled patient by immediately replacing a scheduled patient, upon discharge, by another one in state 2. Let $Y_s(t)$ be the unit occupied by this circulating scheduled patient at time t. We see that $\{Y_s(t), t \geq 0\}$ is an SMP on state space $\{2 = \text{POW}, 3 = \text{SU}, 4 = \text{POU}, 5 = \text{ICU}, 6 = \text{ECU}\}$ with embedded transition probability matrix

$$P_{sc} = \begin{bmatrix} 0 & 1 & 0 & 0 & 0 \\ 0 & 0 & 1 & 0 & 0 \\ 0 & 0.02 & 0 & .98 & 0 \\ 0 & 0.02 & 0 & 0 & .98 \\ .91 & 0 & 0 & .09 & 0 \end{bmatrix} \tag{5.28}$$

and mean sojourn times (in hours)

$$w_{sc} = [2.4, \ 5.3, \ 3.4, \ 32.8, \ 98.7]. \tag{5.29}$$

Note that this SMP has an irreducible embedded DTMC with transition probability matrix P_{sc}. The limiting distribution of the embedded DTMC is given by

$$\pi_{sc} = [0.1884, \ 0.1966, \ 0.1966, \ 0.2113, \ 0.2071].$$

The occupancy distribution of the SMP $\{Y_s(t), t \geq 0\}$ can now be computed from Theorem 5.6 as

$$p_{sc} = [0.0153, \ 0.0353, \ 0.0226, \ 0.2347, \ 0.6921]. \tag{5.30}$$

Thus a scheduled patient spends 69.21% of his time in the clinic in the extended care center, while only 3.53% of the time is spent in actual surgery. Since a typical scheduled patient spends on average 156.7177 hours in the clinic, we see that the average times spent in various units are given by

$$156.7177 * [0, \ 0.0153, \ 0.0353, \ 0.0226, \ 0.2347, \ 0.6921]$$

$$= [0, \ 2.4000, \ 5.5294, \ 3.5472, \ 36.7795, \ 108.4615].$$

Thus a scheduled patient spends 5.53 hours in surgery. This is more than 5.3, the average sojourn time in the SU, because a patient may visit the SU more than once during a stay in the clinic. A typical scheduled patient spends 108.46 hours, or approximately four and a half days, in the ECU during his entire stay in the clinic.

How can the information above help us design the clinic (i.e., decide the capacities of each unit)? Unfortunately, it cannot, since we do not know how to account for the arrival process of the patients. An attempt to account for the arrival process will lead us into a queueing model of the clinic, and we will have to wait until the next chapter to complete this analysis.

We end this section with the Matlab function that was used to compute the results given above.

```
*************************************
function [M L] = smpcase(P,W)
%P = tr pr matrix for a single patient
%W = sojourn times for a single patient
%Output M: M(i) = = expected time to hit the absorbing state from
%state i in P
%Output L: L(i) = limiting probability that the circulating patient is in
%unit i
[m,n] = size(P)
Pc = P
Pc(:,1) = P(:,1) + P(:,n); Pc = Pc(1:n − 1,1:n − 1); %tr pr matrix of the circulating
patient
M = inv(eye(m − 1) − P(1:m − 1,1:m − 1))*W(1:m − 1)'
Pcinf = Pc^1000
pic = Pcinf(1,:)
%embedded limiting distribution of the circulating patient
L = pic.*W(1:m − 1)/sum(pic.*W(1:m − 1))
```

5.7 Problems

CONCEPTUAL PROBLEMS

5.1. Consider the battery replacement problem of Example 5.2. Let $N(t)$ be the number of planned replacements up to time t. Show that $\{N(t), t \geq 0\}$ is a renewal process.

5.2. Consider the battery replacement problem of Example 5.2. Let $N(t)$ be the number of unplanned replacements up to time t. Show that $\{N(t), t \geq 0\}$ is a renewal process.

5.3. Let $\{N(t), t \geq 0\}$ be a renewal process, and let $X(t)$ be the integer part of $N(t)/2$. Is $\{X(t), t \geq 0\}$ a renewal process?

Definition 5.5. (Delayed Renewal Process). A counting process generated by a sequence of nonnegative random variables $\{T_n, n \geq 1\}$ is called a delayed renewal process if $\{T_n, n \geq 2\}$ is a sequence of iid random variables and is independent of T_1.

5.4. Let $\{N(t), t \geq 0\}$ be a renewal process, and let $X(t)$ be the integer part of $(N(t) + 1)/2$. Is $\{X(t), t \geq 0\}$ a renewal process? Is it a delayed renewal process?

5.5. Let $\{X(t), t \geq 0\}$ be a CTMC on state space $\{1, 2, \ldots, N\}$. Let $N(t)$ be the number of times the CTMC enters state i over $(0, t]$. Show that $\{N(t), t \geq 0\}$ is a delayed renewal process if $X(0) \neq i$. (We have shown in Example 5.4 that it is a renewal process if $X(0) = i$.)

5.6. Prove Theorem 5.2 for a delayed renewal process with $\tau = E(T_n), n \geq 2$.

5.7. Tasks arrive at a receiving station one at a time, the inter-arrival times being iid with common mean τ. As soon as K tasks are accumulated, they are instantly dispatched to a workshop. Let $Y(t)$ be the number of tasks received by the receiving station by time t. Show that $\{Y(t), t \geq 0\}$ is a renewal process. Let $Z(t)$ be the number of tasks received by the workshop by time t. Show that $\{Z(t), t \geq 0\}$ is a cumulative process. Compute the long-run rate at which jobs are received by the receiving station and the workshop.

5.8. Suppose the tasks in Conceptual Problem 5.7 arrive according to a PP(λ). Does this make the $\{Z(t), t \geq 0\}$ process defined there a compound Poisson process? Why or why not?

5.9. A machine produces parts in a deterministic fashion at a rate of one per hour. They are stored in a warehouse. Trucks leave from the warehouse according to a Poisson process (i.e., the inter-departure times are iid exponential random variables) with rate λ. The trucks are sufficiently large and the parts are small so that each truck can carry away all the parts produced since the previous truck's departure. Let $Z(t)$ be the number of parts that have left the warehouse by time t. Is $\{Z(t), t \geq 0\}$ a CPP? Why or why not?

5.10. Consider Conceptual Problem 5.7. Let $X(t)$ be the number of tasks in the receiving station at time t. Show that $\{X(t), t \geq 0\}$ is an SMP, and compute its transition probability matrix and the vector of expected sojourn times.

5.11. Consider the series system described in Example 5.21. Compute the long-run fraction of the time that the component i is down.

5.12. Consider the series system described in Example 5.21. Let $Y(t)$ be 1 if the system is up and 0 if it is down at time t. Show that $\{Y(t), t \geq 0\}$ is an SMP. Compute the transition probability matrix of the embedded DTMC and the vector of expected sojourn times.

5.13. Consider the series system described in Example 5.21. Compute the expected time between two consecutive failures of component i, $1 \leq i \leq N$.

5.14. (Parallel System). Consider a parallel system of N components. The system fails as soon as all the components fail. Suppose the lifetimes of the components are iid $\text{Exp}(v)$ random variables. The system is repaired when it fails. The mean repair time of the system is τ. Let $X(t)$ be the number of functioning components at time t. Show that $\{X(t), t \geq 0\}$ is an SMP. Compute the transition matrix P and the sojourn time vector w.

5.15. Consider the parallel system of Conceptual Problem 5.14. Compute the long-run fraction of the time that the system is down.

5.16. Consider the parallel system described in Conceptual Problem 5.14. Compute the expected time between two consecutive failures of the system.

5.17. Consider the parallel system described in Conceptual Problem 5.14. Let $Y(t)$ be 1 if the system is up and 0 if it is down at time t. Show that $\{Y(t), t \geq 0\}$ is an SMP. Compute the transition probability matrix of the embedded DTMC and the vector of expected sojourn times.

5.18. Customers arrive at a public telephone booth according to a $\text{PP}(\lambda)$. If the telephone is available, an arriving customer starts using it. The call durations are iid random variables with common distribution $A(\cdot)$. If the telephone is busy when a customer arrives, he or she simply goes away in search of another public telephone. Let $X(t)$ be 0 if the phone is idle and 1 if it is busy at time t. Show that $\{X(t), t \geq 0\}$ is an SMP. Compute the P matrix and the w vector.

5.19. A critical part of a machine is available from two suppliers. The lifetimes of the parts from the ith ($i = 1, 2$) supplier are iid random variables with common cdf $A_i(\cdot)$. Suppose the following replacement policy is followed. If the component currently in use lasts longer than T amount of time (T is a fixed positive constant), it is replaced upon failure by another component from the same supplier that provided the current component. Otherwise it is replaced upon failure by a component from the other supplier. Replacement is instantaneous. Let $X(t) = i$ if the component in use at time t is from supplier i ($i = 1, 2$). Show that $\{X(t), t \geq 0\}$ is an SMP, and compute its P matrix and w vector.

5.20. A department in a university has several full-time faculty positions. Whenever a faculty member leaves the department, he or she is replaced by a new member at the assistant professor level. Approximately 20% of the assistant professors leave (or are asked to leave) at the end of the fourth year (without being considered for tenure), and 30% of the assistant professors are considered for tenure at the end of the fifth year, 30% at the end of the sixth year, and the remainder at the end of the seventh year. The probability of getting tenure at the end of 5 years is .4, at the end of 6 years it is .5, and at the end of 7 years it is .6. If tenure is granted, the assistant professor becomes an associate professor with tenure, otherwise he or she leaves the department. An associate professor spends a minimum of 3 years and a maximum of 7 years in that position before becoming a full professor. At the end of each of years 3, 4, 5, 6, and 7, there is a 20% probability that the associate professor is promoted to full professor, independent of everything else. If the promotion does not come through in a given year, the associate professor leaves with probability .2 and continues for another year with probability .8. If no promotion comes through even at the end of the seventh year, the associate professor leaves the department. A full professor stays with the department for 6 years on average and then leaves. Let $X(t)$ be the position of a faculty member at time t (1 if assistant professor, 2 if associate professor, and 3 if full professor). Note that if the faculty member leaves at time t, then $X(t) = 1$ since the new faculty member replacing the departing one starts as an assistant professor. Show that $\{X(t), t \geq 0\}$ is an SMP. Compute its P matrix and w vector.

5.21. Redo Conceptual Problem 5.20 with the following modification: a departing faculty member is replaced by an assistant professor with probability .6, an associate professor with probability .3, and a full professor with probability .1.

COMPUTATIONAL PROBLEMS

5.1. Consider the battery replacement problem of Example 5.2. Suppose the battery lifetimes are iid Erlang with parameters $k = 3$ and $\lambda = 1$ (per year). Compute the long-run replacement rate if Mr. Smith replaces the battery upon failure.

5.2. Consider Computational Problem 5.1. Suppose Mr. Smith replaces the battery upon failure or upon reaching the expected lifetime of the battery. Compute the long-run rate of replacement.

5.3. Consider Computational Problem 5.2. Compute the long-run rate of planned replacement.

5.4. Consider Computational Problem 5.2. Compute the long-run rate of unplanned replacement.

5.5. Consider the machine reliability problem of Example 2.2. Suppose a repair has just been completed at time 0 and the machine is up. Compute the number of repair completions per day in the long run.

5.6. Consider the inventory system of Example 2.4. Compute the number of orders placed per week (regardless of the size of the orders) in the long run.

5.7. Consider the five-processor computer system described in Conceptual Problem 4.3 and Computational Problem 4.6. Suppose the computer system is replaced by a new one upon failure. Replacement time is negligible. Successive systems are independent. Compute the number of replacements per unit time in the long run.

5.8. Consider the two-component system described in Conceptual Problem 4.6 and Computational Problem 4.8. Suppose the system is replaced by a new one instantaneously upon failure. Successive systems are independent. Compute the number of replacements per unit time in the long run.

5.9. Consider Computational Problem 5.2. Suppose it costs \$75 to replace a battery. It costs an additional \$75 if the battery fails during operation. Compute the long-run cost rate of the planned replacement policy followed by Mr. Smith. Compare the cost rate of the planned replacement policy with the "replace upon failure" policy.

5.10. The lifetime of a machine is an Erlang random variable with parameters $k = 2$ and $\lambda = .2$ (per day). When the machine fails, it is repaired. The repair times are $\text{Exp}(\mu)$ with mean 1 day. Suppose the repair costs \$10 per hour. The machine produces revenues at a rate of \$200 per day when it is working. Compute the long-run net revenue per day.

5.11. In Computational Problem 5.10, what minimum revenue per working day of the machine is needed to make the operation profitable?

5.12. The Quick-Toast brand of toaster oven sells for \$80 per unit. The lifetime (in years) of the oven has the following pdf:

$$f(t) = \frac{t^2}{21}, \ 1 \le t \le 4.$$

The Eat Well restaurant uses this oven and replaces it by a new Quick-Toast oven upon failure. Compute the yearly cost in the long run.

5.13. The Eat Well restaurant of Computational Problem 5.12 faces the following decision: The toaster oven manufacturer has introduced a maintenance policy on the Quick-Toast toaster oven. For \$10 a year, the manufacturer will replace, free of cost, any toaster oven that fails in the first 3 years of its life. If a toaster oven fails after 3 years, a new one has to be purchased for the full price. Is it worth signing up for the maintenance contract?

5.14. The Quick-Toast toaster oven of Computational Problem 5.12 can be purchased with a prorated warranty that works as follows. If the oven fails after T years, the manufacturer will buy it back for $\$80*(4 - T)/3$. The price of the toaster oven with the warranty is \$90. Suppose the Eat Well restaurant decides to buy the oven with this warranty. What is the yearly cost of this policy?

5.15. Consider the general two-state machine of Example 5.17. The mean uptime of the machine is 5 days, while the mean downtime is 1 day. The machine produces 100 items per day when it is working, and each item sells for \$5. The repair costs \$10 per hour. The owner of this machine is considering replacing the current repair person by another who works twice as fast but costs \$20 per hour of repair time. Is it worth switching to the more efficient repair person?

5.16. The lifetime of an oil pump is an exponential random variable with mean 5 days, and the repair time is an exponential random variable with mean 1 day. When the pump is working, it produces 500 gallons of oil per day. The oil is consumed at a rate of 100 gallons a day. The excess oil is stored in a tank of unlimited capacity. Thus the oil reserves increase by 400 gallons a day when the pump is working. When the pump fails, the oil reserves decrease by 100 gallons a day. When the pump is repaired, it is turned on as soon as the tank is empty. Suppose it costs 5 cents per gallon per day to store the oil. Compute the storage cost per day in the long run.

5.17. Consider Computational Problem 5.16 with the following modification: the oil tank has a finite capacity of 1000 gallons. When the tank becomes full, the output of the oil pump is reduced to 100 gallons per day, so that the tank remains full until the pump fails. The lifetime of the oil pump is unaffected by the output rate of the pump. Now compute the storage cost per day in the long run.

5.18. In Computational Problem 5.16, what fraction of the time is the pump idle (repaired but not turned on) in the long run? (*Hint:* Suppose it costs $1 per day to keep the pump idle. Then the required quantity is the long-run cost per day.)

5.19. In Computational Problem 5.17, what fraction of the time is the pump idle (repaired but not turned on) in the long run? (See the hint for Computational Problem 5.18.)

5.20. In Computational Problem 5.17, what fraction of the time is the tank full in the long run? (See the hint for Computational Problem 5.18.)

5.21. Customers arrive at a bank according to a Poisson process at the rate of 20 per hour. Sixty percent of the customers make a deposit, while 40% of the customers make a withdrawal. The deposits are distributed uniformly over [50, 500] dollars, while the withdrawals are distributed uniformly over [100, 200] dollars. Let $Z(t)$ be the total amount of funds (deposits − withdrawals) received by the bank up to time t. (This can be a negative quantity.) What assumptions are needed to make $\{Z(t), t \geq 0\}$ a CPP? Using those assumptions, compute the expected funds received by the bank in an 8-hour day.

5.22. Customers arrive at a department store according to a Poisson process at the rate of 80 per hour. The average purchase is $35. Let $Z(t)$ be the amount of money spent by all the customers who arrived by time t. What assumptions are needed to make $\{Z(t), t \geq 0\}$ a CPP? Using those assumptions, compute the expected revenue during a 16-hour day.

5.23. Travelers arrive at an airport in batches. The successive batch sizes are iid random variables with the following pmfs: $p(1) = .4, p(2) = .2, p(3) = .2, p(4) = .1, p(5) = .05, p(6) = .03, p(8) = .01, p(10) = .01$. The batches arrive according to a Poisson process at the rate of 35 per hour. Compute the expected number of travelers that arrive during 3 hours.

5.24. Consider Computational Problem 5.16. Let $X(t)$ be the state of the oil pump at time t (1 if it is up, 2 if it is under repair, and 3 if it is repaired but idle). Is $\{X(t), t \geq 0\}$ an SMP?

5.25. Consider the series system described in Example 5.21 with three components. The mean lifetimes of the three components are 2, 3, and 4 days, respectively, while the mean repair times are .1, .2, and .3 days. Compute the long-run fraction of the time that the system is up.

5.26. Consider the series system described in Computational Problem 5.25. Compute the expected time between two consecutive failures of the first component.

5.27. Consider the series system of Computational Problem 5.25. Compute the long-run fraction of the time that the third component is down.

5.28. Consider the series system of Computational Problem 5.25. Suppose the three components produce revenues at the rate of $10, $20, and $30 per day (respectively) when the system is up. The repair costs of the three components are $15, $20, and $25 per day (respectively) when the component is down. Compute the long-run net revenue per day for the system.

5.29. Consider the parallel system described in Conceptual Problem 5.14 with three components. The mean lifetime of each component is 4 days, while the mean repair time is 1 day. Compute the long-run fraction of the time that the system is down.

5.30. Consider the parallel system described in Computational Problem 5.29. Compute the long-run fraction of the time that all three components are working.

5.31. Consider the parallel system described in Computational Problem 5.29. Suppose each of the three components produce revenue at the rate of $30 per day that the system is up. The repair cost is $100 per day. Compute the long-run net revenue per day for the system.

5.32. Functionally equivalent machines are available from two suppliers. The machines from supplier 1 have iid lifetimes that are uniformly distributed over $(6, 8)$ days, and installation times are uniformly distributed over $(1, 2)$ days. The machines from supplier 2 have iid Erl$(2, .5)$ lifetimes and iid Exp(1) installation times. A single machine is maintained as follows. When a machine fails, it is replaced by a machine from supplier 1 with probability .4 and from supplier 2 with probability .6. Let $X(t)$ be 1 if a machine from supplier 1 is under installation at time t, 2 if a machine from supplier 1 is in operation at time t, 3 if a machine from supplier 2 is under installation at time t, and 4 if a machine from supplier 2 is in operation at time t. Show that $\{X(t), t \geq 0\}$ is an SMP, and compute its P matrix and w vector.

5.33. In the system described in Computational Problem 5.32, compute the long-run fraction of the time that a functioning machine is in place.

5.34. Consider the following modification of Computational Problem 5.32: the machines from the two suppliers are used in an alternating fashion; i.e., when the machine from supplier 1 fails, it is replaced by one from supplier 2, and vice versa. Let $X(t)$ be as in Computational Problem 5.32. Show that $\{X(t), t \geq 0\}$ is an SMP, and compute its P matrix and w vector.

5.35. Compute the long-run fraction of the time that a machine from supplier 1 is in use if the replacement policy of Computational Problem 5.34 is followed.

5.36. Consider Computational Problem 5.32. Suppose the machines from supplier 1 cost $2000, while those from supplier 2 cost $1800. These prices include the cost of installation. The revenue from a working machine is $1000 per day. Compute the long-run net revenue per day.

5.37. Do Computational Problem 5.36 if the alternating replacement policy described in Computational Problem 5.34 is followed.

5.38. Consider the machine maintenance policy described in Example 5.22. Suppose the lifetimes (in days) are iid $Erl(3, .1)$ random variables, repair times (in days) are iid $Exp(1)$ random variables, and replacement times are exactly 2 days. The other policy constants are $v = 30$ days, $u = 1$ day. Compute the long-run fraction of the time that the machine is working.

5.39. In Computational Problem 5.38, suppose the repair cost is $100 per day, while the cost of replacement is $100. The new machine costs $2500. Compute the long-run cost rate of following the given maintenance policy.

5.40. Compute the expected number of full professors in the department modeled in Conceptual Problem 5.20. Assume that the department has 12 positions.

5.41. Compute the rate of turnover (expected number of faculty departures per year) in the department modeled in Conceptual Problem 5.20. Assume that the department has 12 positions.

5.42. Compute the expected number of full professors in the department modeled in Conceptual Problem 5.21. Assume that the department has 12 positions.

5.43. Compute the rate of turnover (expected number of faculty departures per year) in the department modeled in Conceptual Problem 5.21. Assume that the department has 12 positions.

Case Study Problems. You may use the Matlab program of Section 5.6 to do the following problems.

5.44. It is instructive to further classify the patients based on age as those below 60, and those above 60. The behavior of emergency patients below 60 is identical to that of the emergency patients of Section 5.6 in all units except the ECU. The patient stays in the ECU for 96 hours on average and gets discharged with probability .94. What is the expected time spent in the clinic by such a patient?

5.45. The behavior of emergency patients above 60 is identical to that of the emergency patients of Section 5.6 in all units except the ECU. An emergency patient above 60 stays in the ECU for 148 hours on average and gets discharged with probability .82. What fraction of the time does such a patient spend in the ECU?

5.46. Suppose the hourly expense of keeping a patient is $200 in the ER, $80 in the POW, $300 in the SU, $150 in the POP, $100 in the ICU, and $10 in the ECU. What is the expected total cost of treating an emergency patient in the inpatient clinic?

5.47. With the cost structure given above, what is the expected total cost of treating a scheduled patient?

Chapter 6
Queueing Models

6.1 Queueing Systems

Queues (or waiting lines) are an unavoidable component of modern life. We are required to stand physically in queues in grocery stores, banks, department stores, amusement parks, movie theaters, etc. Although we don't like standing in a queue, we appreciate the fairness that it imposes. Even when we use phones to conduct business, often we are put on hold and served in a first-come–first-served fashion. Thus we face a queue even if we are in our own home!

Queues are not just for humans, however. Modern communication systems transmit messages (like emails) from one computer to another by queueing them up inside the network in a complicated fashion. Modern manufacturing systems maintain queues (called inventories) of raw materials, partly finished goods, and finished goods throughout the manufacturing process. "Supply chain management" is nothing but the management of these queues!

In this chapter, we shall study some simple models of queues. Typically, a queueing system consists of a stream of customers (humans, finished goods, messages) that arrive at a service facility, get served according to a given service discipline, and then depart. The service facility may have one or more servers and finite or infinite capacity. In practice, we are interested in designing a queueing system, namely its capacity, number of servers, service discipline, etc. These models will help us do this by answering the following questions (and many others):

(1) How many customers are there in the queue on average?
(2) How long does a typical customer spend in the queue?
(3) How many customers are rejected or lost due to capacity limitations?
(4) How busy are the servers?

We start by introducing standard nomenclature for single-station queues; i.e., queueing systems where customers form a single queue. Such a queue is described as follows:

1. Arrival Process
   We assume that customers arrive one at a time and that the successive inter-arrival times are iid. (Thus the arrival process is a renewal process, studied

V.G. Kulkarni, *Introduction to Modeling and Analysis of Stochastic Systems*, Springer Texts in Statistics, DOI 10.1007/978-1-4419-1772-0_6, © Springer Science+Business Media, LLC 2011

in Chapter 5.) It is described by the distribution of the inter-arrival times, represented by special symbols as follows:

- M: exponential;
- G: general;
- D: deterministic;
- E_k: Erlang with k phases, etc.

Note that the Poisson arrival process is represented by M (for memoryless; i.e., exponential inter-arrival times).

2. Service Times
   We assume that the service times of successive customers are iid. They are represented by the same letters as the inter-arrival times.

3. Number of Servers
   Typically denoted by s. All the servers are assumed to be identical, and it is assumed that any customer can be served by any server.

4. Capacity
   Typically denoted by K. It includes the customers in service. If an arriving customer finds K customers in the system, he or she is permanently lost. If capacity is not mentioned, it is assumed to be infinite.

Example 6.1. (Nomenclature). In an $M/M/1$ queue, customers arrive according to a Poisson process, request iid exponential service times, and are served by a single server. The capacity is infinite. In an $M/G/2/10$ queue, the customers arrive according to a Poisson process and demand iid service times with a general distribution. The service facility has two servers and the capacity to hold ten customers. ■

In this chapter, we shall always assume a first-come–first-served (FCFS) service discipline.

6.2 Single-Station Queues: General Results

In this section, we state several general results for single-station queueing systems. Let A_n be the time of the nth arrival. We allow for the possibility that not all arrivals may actually enter the system. Hence we define E_n to be the time of the nth entry. (If all arriving customers enter the system, we have $A_n = E_n$ for all $n = 0, 1, 2, \ldots$.) Let D_n be the time of the nth departure. If the service discipline is FCFS, then the nth entering customer enters at time E_n and leaves at time D_n. Hence the time spent by the nth customer in the system is given by $W_n = D_n - E_n$. (See Figure 6.1.)

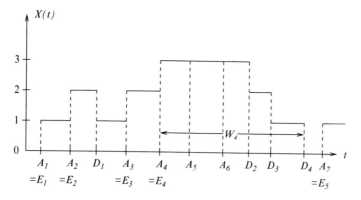

Fig. 6.1 A typical sample path of the queue-length process.

Let $N(t)$ be the number of customers that arrived by time t, let $D(t)$ be the number of customers that departed by time t, and let $X(t)$ be the number of customers in the system at time t. Furthermore, let

$$X_n = X(D_n+),$$
$$X_n^* = X(E_n-),$$
$$\hat{X}_n = X(A_n-).$$

Thus, X_n is the number of customers left behind by the nth departure, X_n^* is the number of customers in the system as seen by the nth *entry* (not including the entering customer) into the system, and $\hat{X}_n$ is the number of customers in the system as seen by the nth *arrival* (not including the arriving customer) to the system. Now (for $j = 0, 1, 2, \ldots$) define the following limits, assuming they exist:

$$p_j = \lim_{t \to \infty} P(X(t) = j), \tag{6.1}$$

$$\pi_j = \lim_{n \to \infty} P(X_n = j), \tag{6.2}$$

$$\pi_j^* = \lim_{n \to \infty} P(X_n^* = j), \tag{6.3}$$

$$\hat{\pi}_j = \lim_{n \to \infty} P(\hat{X}_n = j). \tag{6.4}$$

We can interpret p_j as the long-run fraction of the time that the system has j customers in it, π_j as the long-run fraction of departures that leave behind j customers in the system, π_j^* as the long-run fraction of entering customers that see j customers ahead of them in the system, and $\hat{\pi}_j$ as the long-run fraction of arrivals that see j customers ahead of them in the system. Since all these quantities relate to a common queueing system, we expect them to be related to each other. The next two theorems state these relations.

Theorem 6.1. (When is $\pi_j = \pi_j^*$?). *Suppose the arrivals and departures take place one at a time. If the limits in (6.2) or (6.3) exist,*

$$\pi_j = \pi_j^*, \quad j \geq 0.$$

Idea of Proof. The theorem is a deterministic result with no probabilistic assumptions. Suppose $X(0) = i$. Then, using a careful counting argument (using the fact that the arrivals and departures take place one at a time), we can show that

$$\{X_{n+i} \leq j\} \Leftrightarrow \{X_{n+j+1}^* \leq j\}$$

for all $n \geq 0$ and all $j \geq 0$. (See Kulkarni [2010].) Hence

$$P(X_{n+i} \leq j) = P(X_{n+j+1}^* \leq j).$$

Assuming the limit exists on either the left or the right side, we get

$$\lim_{n \to \infty} P(X_{n+i} \leq j) = \lim_{n \to \infty} P(X_{n+j+1}^* \leq j),$$

which yields

$$\lim_{n \to \infty} P(X_n \leq j) = \lim_{n \to \infty} P(X_n^* \leq j).$$

However, this implies

$$\sum_{i=0}^{j} \pi_i = \sum_{i=0}^{j} \pi_i^*, \quad j \geq 0,$$

and hence

$$\pi_j = \pi_j^*, \quad j \geq 0.$$

Hence the theorem follows. ∎

Theorem 6.2. (Poisson Arrivals See Time Averages: PASTA). *Suppose the arrival process is Poisson and $\{N(t + s), s \geq 0\}$ is independent of $\{X(u), 0 \leq u \leq t\}$. If the limits in (6.1) or (6.4) exist,*

$$p_j = \hat{\pi}_j, \quad j \geq 0.$$

Intuition. The technical condition about independence says that the arrival process from any point t onward is independent of the history of the system up to time t. Now, assuming this condition and using the properties of the Poisson processes given in Section 3.3, we see that the probability that exactly one arrival occurs in the interval $[t, t + h]$ is independent of the history up to time t. If h is sufficiently small, we can say that the arrival occurs at time t. By assumption, $X(t)$ is independent of what happens to the arrival process over $(t, t+h)$. Hence the conditional distribution of $X(t)$, given that there is an arrival in the interval $(t, t + h)$, is the same as the distribution of $X(t)$. Thus the distribution of the state as seen by an arrival at time t

is the same as the distribution of $X(t)$. Thus the long-run fraction of the arrivals that see the system in state j (namely $\hat{\pi}_j$) must be the same as the fraction of the time the system spends in state j (namely p_j). This result is sometimes called PASTA (Poisson Arrivals See Time Averages). ∎

We illustrate the theorems above by means of an example.

Example 6.2. ($M/M/1/1$ Queue). Consider an $M/M/1/1$ queue with arrival rate λ and service rate μ. In such a queue, the arrival process is Poisson with rate λ and service times are iid exponential with parameter μ. However, since the capacity is 1, an arriving customer can enter the system only if the server is idle; otherwise the arrival is lost. The $\{X(t), t \geq 0\}$ process for such a queue is modeled as a CTMC on state space $\{0, 1\}$ in Example 4.7 with $K = 1$. The steady-state distribution can be computed from the results of Example 4.25 as

$$p_0 = \frac{\mu}{\lambda + \mu}, \quad p_1 = \frac{\lambda}{\lambda + \mu}.$$

Since the arrival process is Poisson, it follows from Theorem 6.2 that

$$\hat{\pi}_0 = p_0, \quad \hat{\pi}_1 = p_1.$$

Thus the long-run probability that a customer is turned away is given by $\hat{\pi}_1 = \lambda/(\lambda + \mu)$. Since a departing customer always leaves an empty system, we have $X_n = 0$ for all $n \geq 0$. Hence

$$\pi_0 = 1, \quad \pi_1 = 0.$$

Similarly, since a customer can enter only if the system is empty, we have $X_n^* = 0$ for all $n \geq 0$. Hence,

$$\pi_0^* = 1, \quad \pi_1^* = 0.$$

Thus Theorem 6.1 is verified. Note that the last two observations are valid for any $G/G/1/1$ queue. ∎

Now define

L = the expected number of customers in the system in steady state;
λ = the arrival rate of customers;
W = the expected time spent in the system by a customer in steady state.

The relation between these quantities is the celebrated Little's Law and is stated in the following theorem.

Theorem 6.3. (Little's Law). *If the quantities L, λ, and W defined above exist and are finite, they satisfy*

$$L = \lambda W.$$

Intuition. Suppose each arriving customer pays the system \$1 per unit time that the customer spends in the system. First we compute the long-run rate, $R_{customer}$, at which the money is spent by the customers. Customers arrive at the rate of λ per unit time. Each customer spends on average W time in the system and hence pays \$$W$ to the system. Hence we must have

$$R_{customer} = \lambda W.$$

Next we compute the long-run rate, R_{system}, at which the system earns money. Since each customer pays the system \$1 per unit time and there are L customers in the system on average, we must have

$$R_{system} = L.$$

Now we expect that $R_{customer}$ will equal R_{system} since no money is created or destroyed in the transaction. Hence we get

$$L = \lambda W,$$

which is Little's Law. ∎

Little's Law is also a sample path result; i.e., it does not make any probabilistic assumptions about the system. Care has to be taken in applying it in a consistent manner, as described in the following example.

Example 6.3. (Little's Law and $M/M/1/1$ Queue). Consider the $M/M/1/1$ queue of Example 6.2. The expected number of customers in the system in steady state can be computed as

$$L = 0 \cdot p_0 + 1 \cdot p_1 = \frac{\lambda}{\lambda + \mu}. \tag{6.5}$$

Little's Law can be applied to the stream of arriving customers or entering customers. We discuss these two cases below.

Arriving Customers. With probability p_0, the arriving customer finds the server idle, in which case he or she spends $1/\mu$ amount of time in the system; with probability p_1, the arriving customer finds the server busy, in which case he spends 0 amount of time in the system (since he leaves instantaneously). Hence the expected time spent by a customer in the system is

$$W = \frac{1}{\mu} \cdot p_0 + 0 \cdot p_1 = \frac{1}{\lambda + \mu}. \tag{6.6}$$

The customers arrive according to a PP(λ); hence the arrival rate is λ. Hence, from (6.5) and (6.6), we see that $L = \lambda W$; i.e., Little's Law holds.

Entering Customers. An entering customer always spends $1/\mu$ amount of time in the system on average. Hence

$$W_e = \frac{1}{\mu}. \tag{6.7}$$

In order to verify Little's Law, we need to compute λ_e, the rate at which customers enter the system. Since the successive inter-entry times are iid with mean $\tau = 1/\mu + 1/\lambda = (\lambda + \mu)/\lambda\mu$ (mean service time followed by mean idle time), we see that the rate at which the customers enter the system per unit time is given by (see Theorem 5.2) $\lambda_e = 1/\tau = \lambda\mu/(\lambda + \mu)$. Hence, we have

$$L = \frac{\lambda}{\lambda + \mu} = \frac{\lambda\mu}{\lambda + \mu} \cdot \frac{1}{\mu} = \lambda_e W_e.$$

Thus Little's Law is verified. ∎

Example 6.4. (Variation of Little's Law). We can use the intuition behind Little's Law to create many variations by changing the payment rules for the customers. For example, suppose each customer pays \$1 per unit time that he spends in the queue (but not in service). Then the expected amount paid by a customer is W_q, the expected time spent in the queue (not including service). Also, the rate at which the system collects revenue is L_q, the expected number of customers in the queue (not including those in service). Hence the same intuitive argument that yielded Little's Law yields

$$L_q = \lambda W_q,$$

where λ is the arrival rate of customers. ∎

Example 6.5. (Another Variation of Little's Law). Now consider the following payment scheme for the customers. Each customer pays \$1 per unit time that he spends in service. Then the expected amount paid by a customer is τ, the expected time spent in service. Also, the rate at which the system collects revenue is B, the expected number of customers in service, which is the same as the expected number of busy servers. (In a single-server queue, this is also the fraction of the time that the server is busy.) Hence the same intuitive argument that yielded Little's Law yields

$$B = \lambda\tau,$$

where λ is the arrival rate of customers. However, B cannot exceed s, the number of servers at the station. Hence we must have

$$B = \min(s, \lambda\tau). ∎$$

The example above creates an interesting question: what happens when $\lambda\tau > s$? We discuss this question in Theorem 6.4 below. When there is unlimited waiting room space (i.e., when the capacity is infinite), the state space of the $\{X(t), t \geq 0\}$ process is $\{0, 1, 2, \ldots\}$. So far in this book we have always considered finite state spaces. New issues arise in the study of stochastic processes with infinite state space that we did not encounter in the finite state space stochastic processes. In particular, it is possible that $X(t)$ may tend to infinity as t goes to infinity. In such a case, customers may end up waiting an infinite amount of time in the system in the long run. If this occurs, we call the process unstable. For example, if the arrival rate of customers at a queue is larger than the server's ability to serve, we would expect the queue length to build up without bounds, thus making the queue-length process unstable. We formally define the concept below.

Definition 6.1. (Stability). A single-station queue with unlimited capacity is called *stable* if

$$\sum_{j=0}^{\infty} p_j = 1,$$

where p_j is as defined in (6.1). Otherwise it is called *unstable*.

In an unstable queue, we have

$$\sum_{j=0}^{\infty} p_j < 1.$$

This can be interpreted to mean that, in the long run, the queue length goes to infinity with a positive probability given by $1 - \sum_{j=0}^{\infty} p_j$. Obviously, having infinite queue lengths is bad for the customers. Hence it makes sense to call such a system unstable. The next theorem gives a condition for stability.

Theorem 6.4. (Condition of Stability). *Consider a single-station queue with s servers and infinite capacity. Suppose the customers enter at rate λ and the mean service time is τ. The queue is stable if*

$$\lambda \tau < s. \tag{6.8}$$

Intuition. From Example 6.5, we get

$$B = \text{expected number of busy servers} = \lambda \tau, \tag{6.9}$$

where λ is the arrival rate of entering customers. However, the number of busy servers cannot exceed s, the total number of servers. Hence, for the argument to be valid, we must have

$$\lambda \tau \leq s.$$

What goes wrong when $\lambda \tau > s$? It was implicitly assumed in the derivation of (6.9) that each entering customer eventually enters service. Intuitively, when $\lambda \tau > s$, some of the customers never get to enter service; i.e., they wait in the system indefinitely. This implies that the number in the system must be infinity; i.e., the system must be unstable. It is more difficult to prove that the system can be unstable even if $\lambda \tau = s$. Essentially, this is due to the variance of either the inter-arrival times or the service times. Thus we can safely say that the queue is stable if $\lambda \tau < s$. ∎

6.3 Birth and Death Queues with Finite Capacity

Let $X_K(t)$ be the number of customers at time t in a single-station queue with finite capacity K. In this section, we consider the case where $\{X_K(t), t \geq 0\}$ is a birth and death process on state space $\{0, 1, 2, \ldots, K\}$ with birth parameters $\{\lambda_i, i = 0, 1, 2, \ldots, K - 1\}$ and death parameters $\{\mu_i, i = 1, 2, 3, \ldots, K\}$. We have studied such processes in Example 4.8 and their limiting behavior in Example 4.23. We restate the result about the limiting probabilities as

$$p_i(K) = \lim_{t \to \infty} P(X_K(t) = i), \quad 0 \leq i \leq K.$$

Let $\rho_0 = 1$ and

$$\rho_i = \frac{\lambda_0 \lambda_1 \cdots \lambda_{i-1}}{\mu_1 \mu_2 \cdots \mu_i}, \quad 1 \leq i \leq K. \tag{6.10}$$

Then, from Example 4.23, we get

$$p_i(K) = \frac{\rho_i}{\sum_{j=0}^{K} \rho_j}, \quad 0 \leq i \leq K. \tag{6.11}$$

We consider several special cases below.

6.3.1 M/M/1/K Queue

Consider a service station where customers arrive according to a PP(λ). The service times are iid random variables with Exp(μ) distribution. (Equivalently, we say the service rate is μ.) They are served in an FCFS fashion by a single server. The system has the capacity to hold K customers. Thus an incoming customer who sees K customers in the system ahead of him is permanently lost.

We have already seen this system in Example 4.7 as a model of a bank queue. Let $X_K(t)$ be the number of customers in the system at time t. Then we know (see the discussion following Example 4.8) that $\{X_K(t), t \geq 0\}$ is a birth and death process on state space $\{0, 1, \ldots, K\}$ with birth parameters

$$\lambda_i = \lambda, \quad i = 0, 1, \ldots, K - 1,$$

and death parameters

$$\mu_i = \mu, \quad i = 1, 2, \ldots, K.$$

Substituting in (6.10), we get

$$\rho_i = \left(\frac{\lambda}{\mu}\right)^i, \quad 0 \leq i \leq K. \tag{6.12}$$

Using

$$\rho = \frac{\lambda}{\mu},$$

we get

$$\sum_{i=0}^{K} \rho_i = \begin{cases} \dfrac{1 - \rho^{K+1}}{1 - \rho} & \text{if } \rho \neq 1, \\[2mm] K + 1 & \text{if } \rho = 1. \end{cases}$$

Substituting in (6.11), we get

$$p_i(K) = \begin{cases} \dfrac{1 - \rho}{1 - \rho^{K+1}} \rho^i & \text{if } \rho \neq 1, \\[2mm] \dfrac{1}{K + 1} & \text{if } \rho = 1, \end{cases} \qquad 0 \leq i \leq K. \qquad (6.13)$$

We compute several relevant quantities with the help of this limiting distribution.

Server Idleness. The long-run fraction of the time that the server is idle is given by

$$p_0(K) = \frac{1 - \rho}{1 - \rho^{K+1}}.$$

Blocking Probability. The long-run fraction of the time that the system is full is given by

$$p_K(K) = \frac{1 - \rho}{1 - \rho^{K+1}} \rho^K. \qquad (6.14)$$

Since the arrival process is Poisson, we can use Theorem 6.2 to get the system distribution as seen by an arriving (not necessarily entering) customer as

$$\hat{\pi}_i(K) = p_i(K), \quad i = 0, 1, 2, \ldots, K.$$

Thus an arriving customer sees the system full with probability $p_K(K)$. Hence, $p_K(K)$ also represents the long-run blocking probability and long-run fraction of the arriving customers that are blocked (or lost).

Entering Customers. Now we compute $\pi_i^*(K)$, the probability that an entering customer sees i customers in the system ahead of him. We have $\pi_K^*(K) = 0$ since a customer cannot enter if the system is full. For $0 \leq i \leq K - 1$, we have

$$\pi_i^*(K) = \mathsf{P}(\text{entering customer sees } i \text{ in the system})$$

$$= \mathsf{P}(\text{arriving customer sees } i \text{ in the system} | \text{arriving customer enters})$$

$$= \frac{\mathsf{P}(\text{arriving customer sees } i \text{ in the system and arriving customer enters})}{\mathsf{P}(\text{arriving customer enters})}$$

$$= \frac{P(\text{arriving customer sees } i \text{ in the system})}{P(\text{arriving customer enters})}$$

$$= \frac{p_i(K)}{1 - p_K(K)}, \quad 0 \le i \le K - 1,$$

where $p_i(K)$ are as given in (6.13).

Expected Number in the System. The expected number in the system in steady state is computed as follows. If $\rho = 1$, we have

$$L = \sum_{i=0}^{K} i p_i(K) = \frac{1}{K+1} \sum_{i=0}^{K} i = \frac{K}{2}.$$

If $\rho \ne 1$, more tedious calculations yield

$$L = \sum_{i=0}^{K} i p_i(K)$$

$$= \frac{1 - \rho}{1 - \rho^{K+1}} \sum_{i=0}^{K} i \rho^i$$

$$= \frac{\rho}{1 - \rho} \cdot \frac{1 - (K+1)\rho^K + K\rho^{K+1}}{1 - \rho^{K+1}}. \tag{6.15}$$

Note that L increases from 0 at $\rho = 0$ to K as $\rho \to \infty$. This stands to reason since as $\rho \to \infty$ the arrival rate is increasing relative to the service rate. Hence, eventually, when the arrival rate becomes very large, the system will always be full, thus leading to $L = K$.

Expected Waiting Time: Arrivals. We can compute the expected time spent by an arrival in the system by using Little's Law as stated in Theorem 6.3. Here the arrival rate is λ and the mean number in the system is L as given in (6.15). Hence W, the expected time spent in the system by an arriving customer, is given by

$$W = L/\lambda.$$

Note that an arriving customer spends no time in the system if he finds the system full.

Expected Waiting Time: Entries. Next we compute the expected time spent by an entering customer in the system by using Little's Law as stated in Theorem 6.3. Here the arrival rate of entering customers is

$$\lambda P(\text{arriving customer enters}) = \lambda(1 - p_K(K)).$$

The mean number in the system is L as given in (6.15). Hence W, the expected time spent in the system by an entering customer, is given by

$$W = L/(\lambda(1 - p_K(K))). \tag{6.16}$$

Example 6.6. (ATM Queue). Consider the model of the queue in front of an ATM as described in Example 4.7. The data given in Example 4.20 imply that this is an $M/M/1/5$ queue with an arrival rate of $\lambda = 10$ per hour and a service rate of $\mu = 15$ per hour. Thus

$$\rho = \lambda/\mu = 2/3.$$

1. What is the probability that an incoming customer finds the system full?
   Using (6.14), we get the desired answer

$$p_5(5) = .0481.$$

2. What is the expected amount of time an entering customer spends in the system?
   Using (6.16), we get the desired answer

$$W = .1494 \text{ hr} = 8.97 \text{ min. } \blacksquare$$

6.3.2 M/M/s/K Queue

Let $X_K(t)$ be the number of customers in an $M/M/s/K$ queue at time t. (We shall assume that $K \geq s$ since it does not make sense to have more servers than there is room for customers.) We saw this queue in Example 4.10 as a model of a call center. We saw there that $\{X_K(t), t \geq 0\}$ is a birth and death process on $\{0, 1, \ldots, K\}$ with birth parameters

$$\lambda_i = \lambda, \ \ 0 \leq i \leq K - 1,$$

and death parameters

$$\mu_i = \min(i, s)\mu, \ \ 0 \leq i \leq K.$$

The next theorem gives the limiting distribution of an $M/M/s/K$ queue in terms of the dimensionless quantity defined below,

$$\rho = \frac{\lambda}{s\mu}. \tag{6.17}$$

Theorem 6.5. (Limiting Distribution of an $M/M/s/K$ Queue). *The limiting distribution of an $M/M/s/K$ queue is given by*

$$p_i(K) = p_0(K)\rho_i, \ \ i = 0, 1, 2, \ldots, K,$$

where

$$\rho_i = \begin{cases} \dfrac{1}{i!}\left(\dfrac{\lambda}{\mu}\right)^i, & \text{if } \ 0 \le i \le s-1, \\[3mm] \dfrac{s^s}{s!}\rho^i, & \text{if } \ s \le i \le K, \end{cases}$$

and

$$p_0(K) = \left[\sum_{i=0}^{s-1} \frac{1}{i!}\left(\frac{\lambda}{\mu}\right)^i + \frac{1}{s!}\left(\frac{\lambda}{\mu}\right)^s \cdot \frac{1-\rho^{K-s+1}}{1-\rho} \right]^{-1}.$$

Proof. The expressions for ρ_i follow from (6.10), and that for $p_i(K)$ follows from (6.11). Furthermore, we have

$$\begin{aligned} \sum_{i=0}^{K} \rho_i &= \sum_{i=0}^{s-1} \frac{1}{i!}\left(\frac{\lambda}{\mu}\right)^i + \sum_{i=s}^{K} \frac{s^s}{s!}\rho^i \\ &= \sum_{i=0}^{s-1} \frac{1}{i!}\left(\frac{\lambda}{\mu}\right)^i + \frac{s^s}{s!}\rho^s \sum_{i=s}^{K} \rho^{i-s} \\ &= \sum_{i=0}^{s-1} \frac{1}{i!}\left(\frac{\lambda}{\mu}\right)^i + \frac{s^s}{s!}\rho^s \sum_{i=0}^{K-s} \rho^i \\ &= \sum_{i=0}^{s-1} \frac{1}{i!}\left(\frac{\lambda}{\mu}\right)^i + \frac{s^s}{s!}\rho^s \frac{1-\rho^{K-s+1}}{1-\rho}. \end{aligned}$$

Substituting in (6.11), we get the theorem. ∎

Using the limiting distribution above, we can study the same quantities as in the $M/M/1/K$ case.

Example 6.7. (Call Center). Consider the call center described in Example 4.10. Using the data in Computational Problem 4.9, we get an $M/M/6/10$ queue with an arrival rate of $\lambda = 60$ per hour and a service rate of $\mu = 10$ per hour per server.

1. Compute the limiting distribution of the number of calls in the system in steady state.
   From Theorem 6.5, we get

$$\begin{aligned} p(K) = [&.0020 \ .0119 \ .0357 \ .0715 \ .1072 \ .1286 \ .1286 \ .1286 \\ &.1286 \ .1286 \ .1286]. \end{aligned}$$

2. Compute the expected number of calls on hold.
   This is the same as the number of customers in the system not being served. It is given by

$$\sum_{i=7}^{10}(i-6)p_i(K) = 1.2862.$$

3. What fraction of the calls are lost?
   The desired quantity is given by

$$p_{10}(10) = .1286.$$

Thus 12.86% of the incoming calls are lost. ∎

6.3.3 M/M/K/K Queue

An $M/M/s/K$ queue with $s = K$ produces the $M/M/K/K$ queue as a special case. Such a system is called a *loss system* since any customer that cannot enter service immediately upon arrival is lost. We have seen such a queue as a model of a telephone switch in Example 4.9. By setting $s = K$ in Theorem 6.5, we get the limiting distribution of an $M/M/K/K$ queue as given in the following theorem.

Theorem 6.6. (Limiting Distribution of an $M/M/K/K$ Queue). *Let $X(t)$ be the number of customers in an $M/M/K/K$ queue at time t and*

$$p_i(K) = \lim_{t\to\infty} P(X(t) = i), \quad 0 \le i \le K.$$

Then

$$p_i(K) = \frac{\rho_i}{\sum_{j=0}^{K}\rho_j}, \quad 0 \le i \le K,$$

where

$$\rho_i = \frac{(\lambda/\mu)^i}{i!}, \quad 0 \le i \le K.$$

It is possible to interpret the limiting distribution in the theorem above as the distribution of a truncated (at K) Poisson random variable. To be precise, let Y be a Poisson random variable with parameter λ/μ. Then, for $0 \le i \le K$,

$$P(Y = i | Y \le K) = \frac{P(Y = i, Y \le K)}{P(Y \le K)}$$

$$= \frac{P(Y = i)}{P(Y \le K)}$$

$$= \frac{e^{-\lambda/\mu}(\lambda/\mu)^i/i!}{\sum_{j=0}^{K} e^{-\lambda/\mu}(\lambda/\mu)^j/j!}$$

$$= \frac{(\lambda/\mu)^i/i!}{\sum_{j=0}^{K}(\lambda/\mu)^j/j!} = p_i(K).$$

The quantity $p_K(K)$ is an important quantity, called the blocking probability. It is the fraction of the time that the system is full, or the fraction of the customers that are lost, in the long run.

Example 6.8. (Telephone Switch). Recall the model of a telephone switch developed in Example 4.9. Using the data in Example 4.31, we see that the number of calls handled by the switch is an $M/M/6/6$ queue with an arrival rate of $\lambda = 4$ per minute and a service rate of $\mu = .5$ per minute per server.

1. What is the expected number of calls in the switch in steady state?
   Using Theorem 6.6, we compute the limiting distribution as

$$p(K) = [.0011 \ .0086 \ .0343 \ .0913 \ .1827 \ .2923 \ .3898].$$

   Hence the expected number of calls is given by

$$\sum_{i=0}^{6} i p_i(K) = 4.8820.$$

2. How large should the switch capacity be if we desire to lose no more than 10% of the incoming calls?
   The fraction of calls lost is given by $p_K(K)$. At $K = 6$, this quantity is .3898. By computing $p_K(K)$ for larger values of K, we get

$$p_7(7) = .3082, \quad p_8(8) = .2356, \quad p_9(9) = .1731,$$
$$p_{10}(10) = .1217, \quad p_{11}(11) = .0813.$$

   Hence the telephone switch needs at least 11 lines to lose less than 10% of the traffic. ∎

Remark. One can show that Theorem 6.6 remains valid for the $M/G/K/K$ queue where the service times are iid generally distributed random variables with mean $1/\mu$. The proof of this remarkable result is beyond the scope of this book, and we refer the reader to any advanced textbook on queueing theory.

6.4 Birth and Death Queues with Infinite Capacity

Let $X(t)$ be the number of customers in a single-station queue with infinite capacity. Here we consider the cases where $\{X(t), t \geq 0\}$ is a birth and death process on state space $\{0, 1, 2, \ldots\}$ with birth parameters $\{\lambda_i, i \geq 0\}$ and death parameters $\{\mu_i, i \geq 1\}$. We are interested in the condition under which such a queue is stable. If the queue is stable, we are interested in its limiting distribution

$$p_i = \lim_{t \to \infty} P(X(t) = i), \quad i \geq 0.$$

We have seen a birth and death queue with finite state space $\{0, 1, 2, \ldots, K\}$ in the previous section. We treat the infinite-state birth and death queue as the limiting case of it as $K \to \infty$.

Define $\rho_0 = 1$ and

$$\rho_i = \frac{\lambda_0 \lambda_1 \cdots \lambda_{i-1}}{\mu_1 \mu_2 \cdots \mu_i}, \quad i \geq 1. \tag{6.18}$$

Theorem 6.7. (Infinite Birth and Death Process). *A birth and death queue with infinite state space is stable if and only if*

$$\sum_{j=0}^{\infty} \rho_j < \infty. \tag{6.19}$$

When the queue is stable, its limiting distribution is given by

$$p_i = \frac{\rho_i}{\sum_{j=0}^{\infty} \rho_j}, \quad i \geq 0. \tag{6.20}$$

Proof. Let $p_i(K)$ be as in (6.11). We have

$$
\begin{aligned}
p_i &= \lim_{K \to \infty} p_i(K) \\
&= \begin{cases} \dfrac{\rho_i}{\sum_{j=0}^{\infty} \rho_j} & \text{if } \sum_{j=0}^{\infty} \rho_j < \infty, \\[2mm] 0 & \text{if } \sum_{j=0}^{\infty} \rho_j = \infty. \end{cases}
\end{aligned}
$$

Thus we have

$$
\begin{aligned}
\sum_{i=0}^{\infty} p_i &= \begin{cases} \sum_{i=0}^{\infty} \rho_i / \sum_{j=0}^{\infty} \rho_j & \text{if } \sum_{j=0}^{\infty} \rho_j < \infty, \\[2mm] 0 & \text{if } \sum_{j=0}^{\infty} \rho_j = \infty, \end{cases} \\[2mm]
&= \begin{cases} 1 & \text{if } \sum_{j=0}^{\infty} \rho_j < \infty, \\[2mm] 0 & \text{if } \sum_{j=0}^{\infty} \rho_j = \infty. \end{cases}
\end{aligned}
$$

Hence it follows from Definition 6.1 that the queue is stable if and only if (6.19) holds and that the limiting distribution of a stable birth and death queue is given by (6.20). ∎

We study several special cases of the birth and death queues with infinite state spaces below.

6.4.1 M/M/1 Queue

In an $M/M/1$ queueing system, the customers arrive according to a PP(λ) and the service times are iid Exp(μ) random variables. There is a single server and an unlimited waiting room capacity. The next theorem gives the main result for the $M/M/1$ queue. Define

$$\rho = \frac{\lambda}{\mu}. \tag{6.21}$$

The quantity ρ is called the *traffic intensity* of the $M/M/1$ queue.

Theorem 6.8. (Limiting Distribution of an $M/M/1$ Queue). *An $M/M/1$ queue with traffic intensity ρ is stable if and only if*

$$\rho < 1. \tag{6.22}$$

If the queue is stable, its limiting distribution is given by

$$p_i = (1 - \rho)\rho^i, \quad i \geq 0. \tag{6.23}$$

Proof. Let $X(t)$ be the number of customers in the system at time t. $\{X(t), t \geq 0\}$ is a birth and death process on $\{0, 1, 2, \ldots\}$ with birth parameters

$$\lambda_i = \lambda, \quad i \geq 0,$$

and death parameters

$$\mu_i = \mu, \quad i \geq 1.$$

Substituting in (6.18), we get

$$\rho_i = \left(\frac{\lambda}{\mu}\right)^i = \rho^i, \quad i \geq 0.$$

Now

$$\sum_{i=0}^{\infty} \rho_i = \sum_{i=0}^{\infty} \rho^i = \begin{cases} \frac{1}{1-\rho} & \text{if } \rho < 1, \\ \infty & \text{if } \rho \geq 1. \end{cases}$$

Thus, from Theorem 6.7, we see that the condition of stability is as given in (6.22). Substituting in (6.20), we get the limiting distribution in (6.23). ∎

Note that the condition of stability in (6.22) is consistent with the general condition in (6.8) (using $\tau = 1/\mu$). Using the definition of traffic intensity ρ, the condition of stability can also be written as

$$\lambda < \mu.$$

In this form, it makes intuitive sense: if the arrival rate (λ) is less than the service rate (μ), the system is able to handle the load and the queue does not build up to infinity. Hence it is stable. If not, the system is unable to handle the load and the queue builds to infinity! Hence it is unstable. It is curious that even in the case $\lambda = \mu$ (or, equivalently, $\rho = 1$) the queue is unstable. This is due to the effect of randomness. Even if the average service capacity of the system is perfectly matched with the average load on the system, it has no leftover capacity to absorb the random buildups that will invariably occur in a stochastic service system.

We shall compute several quantities using the limiting distribution given in Theorem 6.8.

Server Idleness. The probability that the server is idle is given by

$$p_0 = 1 - \rho, \tag{6.24}$$

and the probability that the server is busy is given by

$$1 - p_0 = \rho.$$

Hence the expected number of busy servers is ρ. This is consistent with the general result in Example 6.5.

Embedded Distributions. The queue length in an $M/M/1$ queue changes by ± 1 at a time. The arrival process is Poisson, and all arriving customers enter the system. Hence Theorems 6.1 and 6.2 imply that

$$\hat{\pi}_i = \pi_i^* = \pi_i = p_i = (1 - \rho)\rho^i, \quad i \geq 0.$$

Expected Number in the System. The expected number in the system is given by

$$L = \sum_{i=0}^{\infty} i p_i$$

$$= \sum_{i=0}^{\infty} i(1 - \rho)\rho^i$$

$$= \frac{\rho}{1 - \rho}. \tag{6.25}$$

The last equation is a result of tedious calculations.

Expected Waiting Time. Using Little's Law, we get

$$W = \frac{L}{\lambda} = \frac{1}{\mu} \frac{1}{1-\rho} = \frac{1}{\mu - \lambda}. \tag{6.26}$$

Note that, as expected, both L and W tend to infinity as $\rho \to 1$.

Example 6.9. (Taxi Stand). Customers arrive at a taxi stand according to a Poisson process at a rate of 15 per hour and form a queue for the taxis. Taxis arrive at the stand according to a Poisson process at a rate of 18 per hour. Each taxi carries away one passenger. If no customers are waiting when a taxi arrives, it leaves right away without picking up any customers. Let $X(t)$ be the number of customers waiting at the taxi stand at time t. Due to the memoryless property of the exponential distribution, we see that the customer at the head of the line has to wait for an exp(18) amount of time for the next taxi. Thus we can think of $\{X(t), t \geq 0\}$ as an $M/M/1$ queue with an arrival rate of $\lambda = 15$ per hour and a service rate of $\mu = 18$ per hour.

1. Is this queue stable?
   The traffic intensity is

$$\rho = \lambda/\mu = 15/18 = 5/6.$$

Since this is less than 1, the queue is stable, from Theorem 6.8.
2. On average, how long does a customer wait before getting into a taxi?
   This is given by the average waiting time

$$W = 1/(\mu - \lambda) = .3333 \text{ hours.}$$

Thus the average wait for a taxi is 20 minutes. ∎

6.4.2 M/M/s Queue

Next we study an $M/M/s$ queueing system. Customers arrive according to a PP(λ). The service times are iid Exp(μ) random variables. There are s identical servers and there is unlimited waiting room capacity. The customers form a single line and get served by the next available server on an FCFS basis. The next theorem gives the main result for the $M/M/s$ queue.

Theorem 6.9. (Limiting Distribution of an $M/M/s$ Queue). *An $M/M/s$ queue with an arrival rate of λ and s identical servers each with service rate μ is stable if and only if*

$$\rho = \frac{\lambda}{s\mu} < 1. \tag{6.27}$$

If the queue is stable, its limiting distribution is given by

$$p_i = p_0 \rho_i, \quad i \geq 0,$$

where

$$\rho_i = \begin{cases} \dfrac{1}{i!}\left(\dfrac{\lambda}{\mu}\right)^i & \text{if } 0 \leq i \leq s-1, \\[3mm] \dfrac{s^s}{s!}\rho^i & \text{if } i \geq s, \end{cases} \tag{6.28}$$

and

$$p_0 = \left[\sum_{i=0}^{s-1} \frac{1}{i!}\left(\frac{\lambda}{\mu}\right)^i + \frac{s^s}{s!} \cdot \frac{\rho^s}{1-\rho}\right]^{-1}.$$

Proof. Let $X(t)$ be the number of customers in the $M/M/s$ queue at time t. $\{X(t), t \geq 0\}$ is a birth and death process on $\{0, 1, 2, \ldots\}$ with birth parameters

$$\lambda_i = \lambda, \quad i \geq 0,$$

and death parameters

$$\mu_i = \min(s, i)\mu, \quad i \geq 1.$$

Using

$$\rho = \frac{\lambda}{s\mu} \tag{6.29}$$

in (6.18), we get (6.28). Now

$$\sum_{i=0}^{\infty} \rho_i = \sum_{i=0}^{s-1} \rho_i + \sum_{i=s}^{\infty} \rho_i$$

$$= \sum_{i=0}^{s-1} \rho_i + \frac{s^s}{s!}\frac{\rho^s}{1-\rho} \quad \text{if } \rho < 1.$$

If $\rho \geq 1$, the sum above diverges. Thus, from Theorem 6.7, we see that the condition of stability is as given in (6.27). The limiting distribution then follows from (6.20). ∎

Note that the condition of stability in (6.27) can also be written as

$$\lambda < s\mu.$$

In this form, it makes intuitive sense: if the arrival rate λ is less than the maximum service rate $s\mu$ (which occurs when all servers are busy), the system is able to handle the load and the queue does not build up to infinity. Hence it is stable. If not, the system is unable to handle the load and the queue builds to infinity! Hence it is unstable. As in the $M/M/1$ queue, the queue is unstable when $\lambda = s\mu$.

We shall compute several quantities using the limiting distribution given in Theorem 6.9.

Embedded Distributions. The queue length in an $M/M/s$ queue changes by ± 1 at a time. The arrival process is Poisson, and all arriving customers enter the system. Hence Theorems 6.1 and 6.2 imply that

$$\hat{\pi}_i = \pi_i^* = \pi_i = p_i.$$

Probability of Waiting. An incoming customer has to wait for service if all the servers are busy. This probability is given by

$$
\begin{aligned}
\sum_{i=s}^{\infty} p_i &= \sum_{i=s}^{\infty} p_0 \frac{s^s}{s!} \rho^i \\
&= p_0 \frac{s^s}{s!} \rho^s \sum_{i=s}^{\infty} \rho^{i-s} \\
&= p_s \sum_{i=0}^{\infty} \rho^i \\
&= \frac{p_s}{1-\rho}.
\end{aligned}
$$

Expected Number of Busy Servers. If there are i customers in the system, the number of busy servers is given by $\min(i, s)$. Hence B, the expected number of busy servers, is given by

$$B = \sum_{i=0}^{\infty} \min(i, s) p_i.$$

The sum above can be simplified to

$$B = \frac{\lambda}{\mu}. \tag{6.30}$$

This matches the general result in Example 6.5.

Expected Number in the System. The expected number in the system is given by

$$L = \sum_{i=0}^{\infty} i p_i.$$

This can be simplified to

$$L = \frac{\lambda}{\mu} + p_s \frac{\rho}{(1-\rho)^2}. \tag{6.31}$$

Expected Waiting Time. Using Little's Law, we get

$$W = \frac{L}{\lambda} = \frac{1}{\mu} + \frac{1}{s\mu} \frac{p_s}{(1-\rho)^2}. \tag{6.32}$$

Note that, as expected, both L and W tend to infinity as $\rho \to 1$.

Example 6.10. (How Many Service Windows?). The U.S. Postal Service has announced that no customer will have to wait more than 5 minutes in line in a post office. It wants to know how many service windows should be kept open to keep this promise. You are hired as a consultant to help decide this. You realize immediately that this is an impossible promise to keep since the service times and arrival rates of customers are beyond the control of the post office. Hence you decide to concentrate on the average wait in the queue. Thus you want to decide how many service windows to keep open so that the mean queueing time (excluding service time) is less than 5 minutes.

Suppose the arrival rate is λ, service times are iid $\text{Exp}(\mu)$, and s windows are kept open. Then, using (6.32), the queueing time is given by

$$W_q(s) = W - 1/\mu = \frac{1}{s\mu} \frac{p_s}{(1-\rho)^2}.$$

You estimate the average service time to be 3.8 minutes and the arrival rate to be 47.2 per hour. Thus

$$\mu = 60/3.8 = 15.7895, \quad \lambda = 47.2.$$

Using the stability condition of Theorem 6.9, we see that the queue is unstable for $s = 1$ and 2. Using the formula above for $s \geq 3$, we get

$$W_q(3) = 5.8977 \text{ hr} = 353.86 \text{ min}, \ W_q(4) = .0316 \text{ hr} = 1.8986 \text{ min}.$$

Hence you recommend that the post office should keep four windows open. ∎

6.4.3 M/M/∞ Queue

By letting $s \to \infty$, we get the $M/M/\infty$ queue as a limiting case of the $M/M/s$ queue. The limiting distribution of the $M/M/\infty$ queue thus follows from Theorem 6.9 as given below.

Theorem 6.10. (Limiting Distribution of an $M/M/\infty$ Queue). *An $M/M/\infty$ queue with an arrival rate of λ and an unlimited number of identical servers each with service rate μ is always stable, and its limiting distribution is given by*

$$p_i = e^{-\lambda/\mu} \frac{(\lambda/\mu)^i}{i!}, \quad i \geq 0.$$

Thus, in steady state, the number of customers in an $M/M/\infty$ queue is a Poisson random variable with parameter λ/μ. Hence the expected number of customers in steady state is given by

$$L = \frac{\lambda}{\mu}.$$

Then, using Little's Law, we get

$$W = \frac{1}{\mu},$$

which is equal to the mean service time since customers enter service immediately upon arrival.

Example 6.11. (Library). Users arrive at a library according to a Poisson process at a rate of 200 per hour. Each user is in the library for an average of 24 minutes. Assume that the time spent by a user in the library is exponentially distributed and independent of that of the other users. (This implies that the time spent in the check-out line is negligible.) How many users are there in the library on average?

Let $X(t)$ be the number of users in the library at time t. Since each user stays in the library independently of each other user, $\{X(t), t \geq 0\}$ can be thought of as an $M/M/\infty$ queue. The parameters are

$$\lambda = 200 \text{ per hour}, \quad \mu = 60/24 \text{ per hour}.$$

Hence, using Theorem 6.10, the number of users in the library in steady state is a Poisson random variable with parameter $\lambda/\mu = 80$. Hence the expected number of users in the library in steady state is 80. ∎

Remark. One can also think of an $M/M/\infty$ queue as the limit of the $M/M/K/K$ queue of Subsection 6.3.3 as $K \to \infty$. It follows from the remark there that Theorem 6.10 remains valid for an $M/G/\infty$ queue when the service times are iid random variables with mean $1/\mu$ and a general distribution. We shall use this fact in the case study in Section 6.8.

6.5 M/G/1 Queue

Consider a single-station queueing system where customers arrive according to a PP(λ) and require iid service times with mean τ, variance σ^2 and second moment $s^2 = \sigma^2 + \tau^2$. The service times may not be exponentially distributed. The queue is serviced by a single server and has infinite waiting room. Such a queue is called an $M/G/1$ queue.

Let $X(t)$ be the number of customers in the system at time t. $\{X(t), t \geq 0\}$ is a continuous-time stochastic process with state space $\{0, 1, 2, \ldots\}$. Knowing the current state $X(t)$ does not provide enough information about the remaining service time of the customer in service (unless the service times are exponentially distributed), and hence we cannot predict the future based solely on $X(t)$. Hence $\{X(t), t \geq 0\}$ is not a CTMC. Hence we will not be able to study an $M/G/1$ queue in as much detail as the $M/M/1$ queue. Instead we shall satisfy ourselves with results about the expected number and expected waiting time of customers in the $M/G/1$ queue in steady state.

Stability. We begin with the question of stability. Let

$$\rho = \lambda \tau \tag{6.33}$$

be the traffic intensity. Then, from Theorem 6.4, it follows that the $M/G/1$ queue is stable if

$$\rho < 1.$$

Indeed, it is possible to show that the $M/G/1$ queue is unstable if $\rho \geq 1$. Thus $\rho < 1$ is a necessary and sufficient condition of stability. We shall assume that the queue is stable.

Server Idleness. Suppose the queue is stable. It follows from Example 6.5 that, in steady state, the expected number of busy servers is ρ. Since the system is served by a single server, in the long run we must have

$$P(\text{server is busy}) = \rho$$

and

$$P(\text{server is idle}) = 1 - \rho.$$

Remaining Service Time. Consider the following payment scheme. Each customer pays money to the system at rate 0 while waiting for service and at a rate equal to the remaining service time while in service. For example, consider a customer with service time x. Suppose this customer arrives at time s, enters service at time t, and departs at time $t + x$. The customer payment rate $c(\cdot)$ is shown in Figure 6.2 as a function of time.

It is clear from the figure that the total amount paid by a customer with service time x is

$$\int_{u=t}^{t+x} c(u)du = x^2/2.$$

Hence the expected payment made by a customer with random service time X is

$$E(X^2/2) = \frac{s^2}{2}.$$

Since the arrival rate is λ, the rate at which customers pay the system is

$$R_{\text{customer}} = \lambda \frac{s^2}{2}.$$

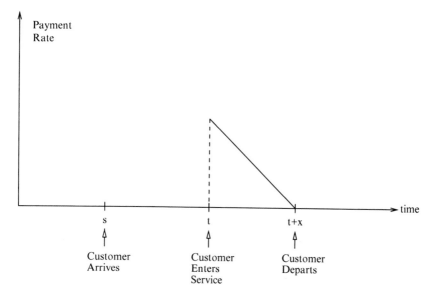

Fig. 6.2 Payment rate for a typical customer.

Now let us compute the rate at which the system collects revenue. Note that if the server is idle, the system collects no revenue. If the server is busy, the system collects revenue at a rate equal to the remaining service time of the customer in service. Let $S(t)$ be the rate at which the system collects revenue at time t. Then the long-run rate at which the server collects revenue is given by

$$R_{\text{system}} = \lim_{t \to \infty} \mathsf{E}(S(t)) = \mathsf{E}(S).$$

Since the queue is assumed to be stable, we expect these two rates to be equal; i.e.,

$$\mathsf{E}(S) = \lambda \frac{s^2}{2}. \tag{6.34}$$

Note that we can also interpret $S(t)$ as follows: $t + S(t)$ is the first time after t that the server can start serving a new customer. This interpretation is important in the next calculation.

Expected Time in the Queue. Assume that the service discipline is first-come–first-served. Let L_q be the expected number of customers in the queue (not including those in service), and let W_q be the expected waiting time in the queue (not including the time spent in service). Then, from Example 6.4, we get

$$L_q = \lambda W_q. \tag{6.35}$$

An arriving customer's queueing time has two components: the time until the current customer in service finishes followed by the service times of all the customers in the queue. Now, by using the interpretation of $S(t)$ given earlier, we see that the expected time until the current customer in finishes service is given by $E(S)$ in steady state. The second component is the sum of the service times of the random number of customers waiting in the queue at the time of arrival of this customer. Using Theorems 6.1 and 6.2, we see that L_q is the number of customers in the queue as seen by an arrival in steady state. Since the service times are iid and independent, and also independent of the number of customers in the queue, we can use Equation (D.9) to see that the expected value of the sum of service times of all the customers in the queue is given by $L_q \tau$. Hence, we have

$$W_q = E(S) + L_q \tau.$$

Substituting from (6.33), (6.34), and (6.35), we get

$$W_q = \frac{\lambda s^2}{2} + \lambda W_q \tau = \frac{\lambda s^2}{2} + \rho W_q,$$

which can be simplified to

$$W_q = \frac{\lambda s^2}{2(1 - \rho)}. \tag{6.36}$$

Expected Number in the Queue. Using (6.35) and (6.36), we get as the expected number in the queue

$$L_q = \frac{\lambda^2 s^2}{2(1 - \rho)}. \tag{6.37}$$

Expected Time in the System. Let W be the expected time spent in the system by a customer in steady state. Since the time in the system is simply the time in the queue plus the time in service, we get

$$W = \tau + W_q. \tag{6.38}$$

Expected Number in the System. Using Little's Law and (6.38), we get as the expected number in the queue

$$L = \lambda(\tau + W_q) = \rho + L_q = \rho + \frac{\lambda^2 s^2}{2(1 - \rho)}. \tag{6.39}$$

Equations (6.36), (6.37), (6.38), and (6.39) are variations of what is known as the *Pollaczec–Khintchine formula*, named after the two probabilists who first studied the $M/G/1$ queue. It is interesting that the first moments of the queue length and the waiting times are affected by the second moments of the service times! This is highly counterintuitive. We can think of the mean service time as a measure of the *efficiency* of the server: the more efficient the server, the smaller the mean service time τ. Similarly, the variance of the service time is an indicator of the *consistency* of the server: the more consistent the server, the smaller the variance σ^2. Thus, among two equally efficient servers, the more consistent server will lead to smaller mean waiting times and smaller mean queue lengths!

Example 6.12. (Telecommunications). Packets arrive at an infinite-capacity buffer according to a Poisson process at a rate of 400 per second. All packets are exactly 512 bytes long. The buffer is emptied at a rate of 2 megabits per second. Compute the expected amount of time a packet waits in the buffer before transmission. (A byte equals eight bits.)

The time to transmit a single packet is

$$(512 * 8)/2000000 = .002048 \text{ seconds.}$$

Since the packet lengths are identical, the variance of the transmission time is zero. Thus the buffer can be modeled as an $M/G/1$ queue with arrival rate

$$\lambda = 400 \text{ per second}$$

and iid service times with mean

$$\tau = .002048 \text{ seconds}$$

and variance 0; i.e., second moment

$$s^2 = (.002048)^2.$$

Hence, the traffic intensity is given by

$$\rho = \lambda \tau = .8192.$$

Since this is less than 1, the queue is stable. Using (6.36), we get as the expected time in the queue

$$W_q = \frac{\lambda s^2}{2(1-\rho)} = \frac{400 * (.002048)^2}{2 * .8192} = .00464 \text{ seconds.}$$

Thus the expected wait before transmission is 4.64 milliseconds. The expected number of packets waiting for transmission can be computed using (6.37) as

$$L_q = \frac{\lambda^2 s^2}{2(1-\rho)} = 1.8559.$$

The expected number of packets in the buffer (including any in transmission) can be computed as

$$L = \rho + L_q = .8192 + 1.8559 = 2.6751. \quad \blacksquare$$

Example 6.13. (*M/M/1* as a Special Case of *M/G/1*). Consider an *M/G/1* queue with iid Exp(μ) service times. In this case, the *M/G/1* queue reduces to an *M/M/1* queue with

$$\tau = 1/\mu, \ \sigma^2 = 1/\mu^2, \ s^2 = 2/\mu^2.$$

Using (6.36), (6.37), (6.38), and (6.39), we get

$$W = \frac{1}{\mu} \frac{1}{1-\rho},$$

$$L = \frac{\rho}{1-\rho}.$$

These results match with the corresponding results for the *M/M/1* queue derived in Section 6.4. ∎

6.6 G/M/1 Queue

Consider a single-station queueing system where customers arrive according to a renewal process with iid inter-arrival times having common cdf $G(\cdot)$ and mean $1/\lambda$. The inter-arrival times may not be exponentially distributed. The queue is serviced by a single server and has an infinite waiting room. The service times are iid Exp(μ) random variables. Such a queue is called a *G/M/1* queue. If the inter-arrival times are exponentially distributed, it reduces to an *M/M/1* queue.

Let $X(t)$ be the number of customers in the system at time t. $\{X(t), t \geq 0\}$ is a continuous-time stochastic process with state space $\{0, 1, 2, \ldots\}$. Knowing the current state $X(t)$ does not provide enough information about the time until the next arrival (unless the inter-arrival times are exponentially distributed), and hence we cannot predict the future based solely on $X(t)$. Hence $\{X(t), t \geq 0\}$ is not a CTMC. Hence we will state some of the main results without proof and derive others from it. First define

$$\rho = \frac{\lambda}{\mu} \tag{6.40}$$

as the traffic intensity of the *G/M/1* queue.

Stability of the *G/M/1* Queue. We begin with the study of stability of the *G/M/1* queue. From Theorem 6.4, it follows that the *G/M/1* queue is stable if

$$\rho < 1.$$

Indeed, it is possible to show that (we refer the reader to advanced books on queueing theory for a proof) the *G/M/1* queue is unstable if $\rho \geq 1$. Thus $\rho < 1$ is a necessary and sufficient condition of stability. We shall assume that the queue is stable in the remaining analysis.

Functional Equation. Here we study a key functional equation that arises in the study of the limiting behavior of a $G/M/1$ queue. Let A represent an inter-arrival time; i.e., A is a random variable with cdf G. Define, for $s \geq 0$,

$$\tilde{G}(s) = \mathsf{E}(e^{-sA}).$$

The key functional equation of a $G/M/1$ queue is

$$u = \tilde{G}(\mu(1 - u)). \tag{6.41}$$

Before we discuss the solutions of this key functional equation, we describe how to compute $\tilde{G}(s)$. If A is a continuous random variable with pdf $g(x)$,

$$\tilde{G}(s) = \int_0^\infty e^{-sx} g(x) dx,$$

and if A is a discrete random variable with pmf $p(x_i) = \mathsf{P}(X = x_i)$, $i = 1, 2, \ldots,$

$$\tilde{G}(s) = \sum_{i=1}^\infty e^{-sx_i} p(x_i).$$

We illustrate this with an example below.

Example 6.14. (Computation of $\tilde{G}$).

1. Exponential Distribution: Suppose the inter-arrival times are $\mathrm{Exp}(\lambda)$. Then the pdf is
$$g(x) = \lambda e^{-\lambda x}, \quad x \geq 0.$$
Hence

$$\tilde{G}(s) = \int_0^\infty e^{-sx} g(x) dx$$

$$= \int_0^\infty e^{-sx} \lambda e^{-\lambda x} dx$$

$$= \frac{\lambda}{s + \lambda}.$$

2. Erlang Distribution: Suppose the inter-arrival times are $\mathrm{Erl}(k, \lambda)$. Then the pdf is
$$g(x) = \lambda e^{-\lambda x} \frac{(\lambda x)^{k-1}}{(k-1)!}, \quad x \geq 0.$$

Hence

$$\tilde{G}(s) = \int_0^\infty e^{-sx} g(x) dx$$

$$= \int_0^\infty e^{-sx} \lambda e^{-\lambda x} \frac{(\lambda x)^{k-1}}{(k-1)!} dx$$

$$= \left(\frac{\lambda}{s+\lambda} \right)^k,$$

where the last integral follows from a standard table of integrals.

3. Hyperexponential Distribution: Suppose the inter-arrival times are Hex (k, λ, p). Then the pdf is

$$g(x) = \sum_{i=1}^k p_i \lambda_i e^{-\lambda_i x}, \quad x \geq 0.$$

Hence

$$\tilde{G}(s) = \int_0^\infty e^{-sx} g(x) dx$$

$$= \int_0^\infty e^{-sx} \sum_{i=1}^k p_i \lambda_i e^{-\lambda_i x} dx$$

$$= \sum_{i=1}^k \frac{p_i \lambda_i}{s + \lambda_i}.$$

4. Degenerate Distribution: Suppose the inter-arrival times are constant, equal to c. Then A takes a value c with probability 1. Hence,

$$\tilde{G}(s) = \sum_{i=1}^\infty e^{-sx_i} p(x_i) = e^{-sc}.$$

5. Geometric Distribution: Suppose the inter-arrival times are iid $G(p)$ random variables. Then the pmf is given by

$$p(i) = P(A = i) = (1 - p)^{i-1} p, \quad i = 1, 2, \ldots.$$

Hence

$$\tilde{G}(s) = \sum_{i=1}^\infty e^{-si} (1 - p)^{i-1} p$$

$$= pe^{-s} \sum_{i=1}^{\infty} (e^{-s}(1-p))^{i-1}$$

$$= pe^{-s} \sum_{i=0}^{\infty} (e^{-s}(1-p))^{i}$$

$$= \frac{pe^{-s}}{1 - e^{-s}(1-p)}. \quad \blacksquare$$

Now we discuss the solutions to the key functional equation. Note that $u = 1$ is always a solution to the key functional equation. The next theorem discusses other solutions to this equation in the interval $(0, 1)$. We refer the reader to advanced texts on queueing theroy for a proof.

Theorem 6.11. (Solution of the Key Functional Equation). *If $\rho \geq 1$, there is no solution to the key functional equation in the interval $(0, 1)$. If $\rho < 1$, there is a unique solution $\alpha \in (0, 1)$ to the key functional equation.*

Example 6.15. (Exponential Distribution). Suppose the inter-arrival times are iid $\text{Exp}(\mu)$. Then, using the results of Example 6.14, the key functional equation becomes

$$u = \frac{\lambda}{\mu(1-u) + \lambda}.$$

This can be rearranged to get

$$(1-u)(u\mu - \lambda) = 0.$$

Thus there are two solutions to the key functional equation

$$u = \rho, \quad u = 1.$$

This verifies Theorem 6.11. $\blacksquare$

Unlike in the example above, it is not always possible to solve the key functional equation of the $G/M/1$ queue analytically. In such cases, the solution can be obtained numerically by using the following recursive computation:

$$u_0 = 0, \quad u_{n+1} = \tilde{G}(\mu(1 - u_n)), \quad n \geq 0.$$

Then, if $\rho < 1$,

$$\lim_{n \to \infty} u_n = \alpha,$$

where α is the unique solution in $(0, 1)$ to the key functional equation.

Example 6.16. (Degenerate Distribution). Suppose the inter-arrival times are constant, equal to $1/\lambda$. Using the results from Example 6.14, the key functional equation becomes

$$u = e^{-\mu(1-u)(1/\lambda)} = e^{-(1-u)/\rho}.$$

This is a nonlinear equation, and no closed-form solution is available. It is, however, easy to obtain a numerical solution using the recursive method described above. For example, suppose $\lambda = 5$ and $\mu = 8$. Then $\rho = \frac{5}{8} = .625$ and the recursion above yields $\alpha = .3580$. ∎

The solution to the key functional equation plays an important part in the computation of the limiting distributions, as is shown next.

Embedded Distributions. We next study the distribution of the number of customers in a $G/M/1$ queue as seen by an arrival in steady state. Since the $\{X(t), t \geq 0\}$ process jumps by ± 1, it follows from Theorem 6.1 that the arrival time distribution is the same as the departure time distribution,

$$\pi_j = \pi_j^*, \quad j \geq 0.$$

Using the solution to the key functional equation, we give the main results in the next theorem.

Theorem 6.12. (Arrival Time Distribution). *In a stable $G/M/1$ queue, the limiting distribution of the number of customers as seen by an arrival is given by*

$$\pi_j^* = (1 - \alpha)\alpha^j, \quad j \geq 0,$$

where α is the unique solution in $(0, 1)$ to the key functional equation (6.41).

Idea of Proof. We refer the reader to an advanced text for the complete proof. The main idea is as follows. Define X_n^* as the number of customers in the $G/M/1$ queue as seen by the nth arrival (not including the arrival itself). Then we first show that $\{X_n^*, n \geq 0\}$ is an irreducible and aperiodic DTMC on state space $\{0, 1, 2, \ldots\}$ and compute the transition probabilities

$$p_{i,j} = P(X_{n+1}^* = j \mid X_n^* = i), \quad i, j \geq 0,$$

in terms of μ and $G(\cdot)$. Then we show that $\pi_j^* = (1 - \alpha)\alpha^j$ satisfies the balance equations

$$\pi_j^* = \sum_{i=0}^{\infty} \pi_i^* p_{i,j}, \quad j \geq 0.$$

This, along with aperiodicity of the DTMC, shows that

$$\pi_j^* = \lim_{n \to \infty} P(X_n^* = j), \quad j \geq 0.$$

This proves the theorem. ∎

Limiting Distribution. Next we study the limiting distribution

$$p_j = \lim_{t \to \infty} P(X(t) = j), \quad j \geq 0.$$

Since the arrival process in a $G/M/1$ queue is not Poisson, PASTA is not applicable and $p_j \neq \pi_j^*$. The next theorem gives the main result.

Theorem 6.13. (Limiting Distribution). *In a stable $G/M/1$ queue, the limiting distribution of the number of customers in the system is given by*

$$p_0 = 1 - \rho, \; p_j = \rho \pi^*_{j-1}, \; j \geq 1,$$

where π^*_j are as in Theorem 6.12.

Idea of Proof. It follows from Example 6.5 that, in steady state, the expected number of busy servers is ρ. Since the system is served by a single server, in the long run we must have

$$P(\text{server is busy}) = \rho,$$

and hence

$$P(\text{server is idle}) = p_0 = 1 - \rho.$$

Now, fix $j \geq 1$. Consider the following cost structure. Whenever the number of customers in the system jumps from $j - 1$ to j, the system earns \$1, and whenever it jumps from j to $j - 1$, it loses \$1. Then R_u, the long-run rate at which it earns dollars, is given by the product of λ, the long-run arrival rate of customers, and π^*_{j-1}, the long-run fraction of customers that see $j - 1$ customers ahead of them in the system. Thus

$$R_u = \lambda \pi^*_{j-1}.$$

On the other hand, R_d, the long-run rate at which it loses dollars, is given by the product of μ, the long-run departure rate of customers (when the system is in state j), and p_j, the long-run fraction of the time that the system is in state j. Thus

$$R_d = \mu p_j.$$

Since the system is stable, we expect these two rates to be the same. Hence we have

$$\lambda \pi^*_{j-1} = \mu p_j,$$

which yields the result in Theorem 6.13. ∎

Expected Number in the System. The expected number in the system is given by

$$
\begin{aligned}
L &= \sum_{i=0}^{\infty} i p_i \\
&= \sum_{i=1}^{\infty} i \pi^*_{i-1} \\
&= \sum_{i=1}^{\infty} i \rho (1 - \alpha) \alpha^{i-1} \\
&= \frac{\rho}{1 - \alpha}.
\end{aligned}
\tag{6.42}
$$

Expected Waiting Time. Using Little's Law, we get

$$W = \frac{L}{\lambda} = \frac{1}{\mu}\frac{1}{1-\alpha}. \tag{6.43}$$

Example 6.17. (Manufacturing). A machine manufactures two items per hour in a deterministic fashion; i.e., the manufacturing time is exactly .5 hours per item. The manufactured items are stored in a warehouse with infinite capacity. Demand for the items arises according to a Poisson process with a rate of 2.4 per hour. Any demand that cannot be satisfied instantly is lost.

Let $X(t)$ be the number of items in the warehouse at time t. Then $X(t)$ increases by one every 30 minutes, while it decreases by one whenever a demand occurs. If $X(t)$ is zero when a demand occurs, it stays zero. Thus $\{X(t), t \geq 0\}$ can be thought of as the queue-length process of a $G/M/1$ queue with deterministic inter-arrival times with an arrival rate (production rate) of $\lambda = 2$ per hour and exponential service times with a rate of $\mu = 2.4$ per hour. This makes

$$\rho = \lambda/\mu = 2/2.4 = .8333.$$

Thus the queue is stable. The key functional equation for the $G/M/1$ queue in this case is (see Example 6.16)

$$u = e^{-2.4*.5*(1-u)} = e^{-1.2(1-u)}.$$

Solving the equation above numerically, we get

$$\alpha = .6863.$$

Using this parameter, we can answer several questions about the system.

1. What is the average number of items in the warehouse in steady state?
   The answer is given by

$$L = \rho/(1-\alpha) = 5.4545.$$

2. On average, how long does an item stay in the warehouse?
   The answer is given by

$$W = 1/\mu(1-\alpha) = 2.7273 \text{ hours}.$$

3. What is the probability that the warehouse has more than four items in it?
   The answer is given by

$$\sum_{j=5}^{\infty} p_j = \rho(1-\alpha)\sum_{j=5}^{\infty} \alpha^{j-1} = \rho\alpha^4 = .4019. \blacksquare$$

6.7 Networks of Queues

So far in this chapter we have studied single-station queueing systems. In practice, service systems can consist of multiple service stations. Customers can move from service station to service station many times before leaving the system. Such multiple service-station systems are called *networks of queues*. For example, patients wait at several different queues inside a hospital during a single visit to the hospital. An electronic message may wait at many intermediate nodes on the Internet before it is delivered to its destination. Material in a manufacturing facility may wait at many different work stations until the final product emerges at the end. Paperwork in many organizations has to wait at different offices undergoing scrutiny and collecting signatures until it is finally done. Students registering for a new semester may wait in multiple queues during the registration process.

6.7.1 Jackson Networks

In this subsection, we study a special class of queueing networks, called the *Jackson networks*. A network of queues is called a Jackson network if it satisfies all of the following assumptions:

(1) The network has N single-station queues.
(2) The ith station has s_i servers.
(3) There is an unlimited waiting room at each station.
(4) Customers arrive at station i from outside the network according to $PP(\lambda_i)$. All arrival processes are independent of each other.
(5) Service times of customers at station i are iid $Exp(\mu_i)$ random variables.
(6) Customers finishing service at station i join the queue at station j with probability $p_{i,j}$ or leave the network altogether with probability r_i, independently of each other.

The *routing probabilities* $p_{i,j}$ can be put in matrix form as follows:

$$P = \begin{bmatrix} p_{1,1} & p_{1,2} & p_{1,3} & \cdots & p_{1,N} \\ p_{2,1} & p_{2,2} & p_{2,3} & \cdots & p_{2,N} \\ p_{3,1} & p_{3,2} & p_{3,3} & \cdots & p_{3,N} \\ \vdots & \vdots & \vdots & \ddots & \vdots \\ p_{N,1} & p_{N,2} & p_{N,3} & \cdots & p_{N,N} \end{bmatrix}. \tag{6.44}$$

The matrix P is called the *routing matrix*. Since a customer leaving station i either joins some other station in the network or leaves, we must have

$$\sum_{j=1}^{N} p_{i,j} + r_i = 1, \quad 1 \le i \le N.$$

We illustrate this with two examples.

Fig. 6.3 A tandem queueing network.

Example 6.18. (Tandem Queue). Consider a queueing network with four service stations as shown in Figure 6.3. Such linear networks are called tandem queueing networks.

Customers arrive at station 1 according to a Poisson process at a rate of 10 per hour and get served at stations 1, 2, 3, and 4 in that order. They depart after getting served at station 4. The mean service time is 10 minutes at station 1, 15 minutes at station 2, 5 minutes at station 3, and 20 minutes at station 4. The first station has two servers, the second has three servers, the third has a single server, and the last has four servers.

In order to model this as a Jackson network, we shall assume that the service times are independent exponential random variables with the means given above and that there is an infinite waiting room at each service station. Thus, if the mean service time is 20 minutes, that translates into a service rate of three per hour. Then the parameters of this Jackson network are as follows:

$$N = 4,$$

$$s_1 = 2, s_2 = 3, s_3 = 1, s_4 = 4,$$

$$\lambda_1 = 10, \lambda_2 = 0, \lambda_3 = 0, \lambda_4 = 0,$$

$$\mu_1 = 6, \mu_2 = 4, \mu_3 = 12, \mu_4 = 3,$$

$$P = \begin{bmatrix} 0 & 1 & 0 & 0 \\ 0 & 0 & 1 & 0 \\ 0 & 0 & 0 & 1 \\ 0 & 0 & 0 & 0 \end{bmatrix},$$

$$r_1 = 0, r_2 = 0, r_3 = 0, r_4 = 1. \quad \blacksquare$$

Example 6.19. (Amusement Park). An amusement park has six rides: a roller coaster, a merry-go-round, a water-tube ride, a fantasy ride, a ghostmountain ride, and a journey to the moon ride. (See Figure 6.4.)

Visitors arrive at the park according to a Poisson process at a rate of 600 per hour and immediately fan out to the six rides. A newly arriving visitor is equally likely to go to any of the six rides first. From then on, each customer visits the rides in a random fashion and then departs. We have the following data: the roller coaster ride lasts 2 minutes (including the loading and unloading times). Two roller coaster cars, each carrying 12 riders, are on the track at any one time. The merry-go-round can handle 35 riders at one time, and the ride lasts 3 minutes. The water-tube ride lasts 1.5 minutes, and there can be ten tubes on the waterway at any one time, each carrying two riders. The fantasy ride is a 5-minute ride that can carry 60 persons at

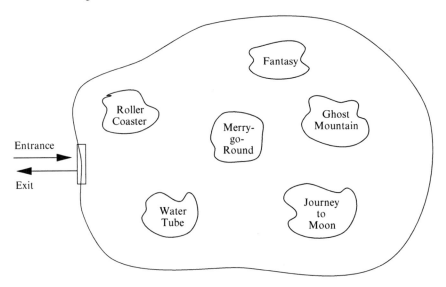

Fig. 6.4 The amusement park.

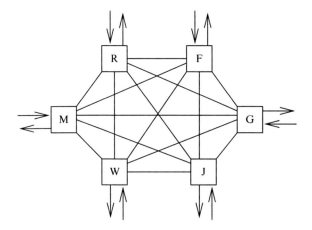

Fig. 6.5 Jackson network model of the amusement park.

any one time. The ghost-mountain ride can handle 16 persons at a time and lasts for 90 seconds. Finally, the journey to the moon ride takes 100 seconds, and 20 riders can be on it at any one time.

We shall model this as a Jackson network with six service stations: 1. roller coaster (R); 2. merry-go-round (M); 3. water-tube (W); 4. fantasy (F); 5. ghost-mountain (G); and 6. journey to the moon (J). (See Figure 6.5.)

Clearly, we need several assumptions. We shall assume that the ride times are exponentially distributed, although this is typically untrue. Thus, if a ride takes 5 minutes, this translates to a service rate of 12 per hour. If a ride can handle s riders simultaneously, we shall assume the corresponding service station has s servers. Although the service times of all riders sharing a ride must be the same, we shall assume the service times are iid. Routing of riders is especially problematic. We are given that new customers go to any of the six rides with equal probability. Hence the rate of new arrivals at each ride is 100 per hour. We assume that, after each ride, the riders choose to join one of the remaining five rides or leave in a completely random fashion. Note that under this model some riders may not visit all the rides, while others may visit some rides more than once! Also, we assume that riders behave independently of each other, while in practice they move in small groups. With these assumptions, we get a Jackson network with the following parameters:

$$N = 6,$$

$$s_1 = 24, s_2 = 35, s_3 = 20, s_4 = 60, s_5 = 16, s_6 = 20,$$

$$\lambda_1 = \lambda_2 = \lambda_3 = \lambda_4 = \lambda_5 = \lambda_6 = 100,$$

$$\mu_1 = 30, \mu_2 = 20, \mu_3 = 40, \mu_4 = 12, \mu_5 = 40, \mu_6 = 36,$$

$$P = \begin{bmatrix} 0 & 1/6 & 1/6 & 1/6 & 1/6 & 1/6 \\ 1/6 & 0 & 1/6 & 1/6 & 1/6 & 1/6 \\ 1/6 & 1/6 & 0 & 1/6 & 1/6 & 1/6 \\ 1/6 & 1/6 & 1/6 & 0 & 1/6 & 1/6 \\ 1/6 & 1/6 & 1/6 & 1/6 & 0 & 1/6 \\ 1/6 & 1/6 & 1/6 & 1/6 & 1/6 & 0 \end{bmatrix},$$

$$r_1 = r_2 = r_3 = r_4 = r_5 = r_6 = 1/6. \quad \blacksquare$$

6.7.2 Stability

Since the stations in a Jackson network have infinite capacities, we are faced with the possibility of instability as discussed in the context of infinite capacity queues earlier. Hence, in this subsection we establish conditions under which a Jackson network is stable.

Consider the jth station in a Jackson network. Customers arrive at this station from outside at rate λ_j. Customers also arrive at this station from other stations in the network. These are called internal arrivals. Let b_j be the rate of internal arrivals at station j. Then the total arrival rate at station j, denoted by a_j, is given by

$$a_j = \lambda_j + b_j, \quad 1 \le j \le N.$$

Now, if the service stations are all stable, the departure rate of customers from station i will be the same as the total arrival rate at station i, namely a_i. A fraction $p_{i,j}$ of these departing customers go to station j. Hence the arrival rate of internal customers from station i at station j is $a_i p_{i,j}$. Hence the internal arrival rate at station j from all the stations in the network is given by

$$b_j = \sum_{i=1}^{N} a_i p_{i,j}, \quad 1 \le j \le N.$$

Substituting in the previous equation, we get

$$a_j = \lambda_j + \sum_{i=1}^{N} a_i p_{i,j}, \quad 1 \le j \le N. \tag{6.45}$$

These simultaneous equations for a_j are called the *traffic equations*. Using

$$a = [a_1 \; a_2 \; \cdots \; a_N]$$

and

$$\lambda = [\lambda_1 \; \lambda_2 \; \cdots \; \lambda_N],$$

the traffic equations can be written in matrix form as

$$a = \lambda + aP$$

or

$$a(I - P) = \lambda.$$

Suppose the matrix $I - P$ is invertible. Then the equation above has a unique solution given by

$$a = \lambda(I - P)^{-1}. \tag{6.46}$$

The next theorem states the stability condition for Jackson networks in terms of the solution above.

Theorem 6.14. (Stability of Jackson Networks). *A Jackson network with external arrival rate vector λ and routing matrix P is stable if $I - P$ is invertible and*

$$a_i < s_i \mu_i$$

for all $i = 1, 2, \ldots, N$, where $a = [a_1, a_2, \ldots, a_N]$ is as given in (6.46).

Intuition. If $I - P$ is invertible, the traffic equations have a unique solution given in (6.46). Now, the ith service station is a single-station queue with s_i servers, arrival rate a_i, and mean service time $1/\mu_i$. Hence, from Theorem 6.4, it follows that the queue is stable if

$$a_i / \mu_i < s_i.$$

This equation has to hold for all $i = 1, 2, \ldots, N$ for the traffic equations to be valid. This yields the theorem above. ∎

We illustrate this with two examples.

Example 6.20. (Tandem Queue). Consider the tandem queueing network of Example 6.18. The traffic equations (6.45) for this example are

$$a_1 = 10,$$
$$a_2 = a_1,$$
$$a_3 = a_2,$$
$$a_4 = a_3.$$

These can be solved to get

$$a_1 = a_2 = a_3 = a_4 = 10.$$

Using the parameters derived in Example 6.18, we see that

$$s_1\mu_1 = 12 > a_1,$$
$$s_2\mu_2 = 12 > a_2,$$
$$s_3\mu_3 = 12 > a_3,$$
$$s_4\mu_4 = 12 > a_4.$$

Thus, from Theorem 6.14, the network is stable. ■

Example 6.21. (Amusement Park). Consider the Jackson network model of the amusement park of Example 6.19. Using the arrival rates given there, we get

$$\lambda = [100 \ 100 \ 100 \ 100 \ 100 \ 100].$$

Using the routing matrix given there we see that the solution to the traffic equations is given by

$$a = [600 \ 600 \ 600 \ 600 \ 600 \ 600].$$

The total arrival rate at each ride is the same due to the symmetry of the problem. The fact that the total arrival rate at each ride is the same as the total external arrival rate at the park can be thought of as an indication that each customer visits each ride once on average! Using the other parameters given there, we see that

$$s_1\mu_1 = 720 > a_1,$$
$$s_2\mu_2 = 700 > a_2,$$
$$s_3\mu_3 = 800 > a_3,$$
$$s_4\mu_4 = 720 > a_4,$$
$$s_5\mu_5 = 640 > a_5,$$
$$s_6\mu_6 = 720 > a_6.$$

Thus, from Theorem 6.14, we see that this network is stable. ■

6.7.3 Limiting Behavior

Having established the conditions of stability for Jackson networks, we proceed with the study of the limiting behavior of the Jackson networks in this subsection.

First we introduce the relevant notation. Let $X_i(t)$ be the number of customers in the ith service station in a Jackson network at time t. Then the state of the network at time t is given by $X(t) = [X_1(t), X_2(t), \ldots, X_N(t)]$. Now suppose the Jackson network is stable, with a as the unique solution to the traffic equations. Let

$$p(n_1, n_2, \ldots, n_N) = \lim_{t \to \infty} P(X_1(t) = n_1, X_2(t) = n_2, \ldots, X_N(t) = n_N)$$

be the limiting distribution of the state of the network. Let $p_i(n)$ be the limiting probability that there are n customers in an $M/M/s_i$ queue with arrival rate a_i and service rate μ_i. From Theorem 6.9, we get

$$p_i(n) = p_i(0)\rho_i(n), \quad i \geq 0,$$

where

$$\rho_i(n) = \begin{cases} \dfrac{1}{n!}\left(\dfrac{a_i}{\mu_i}\right)^n & \text{if } 0 \leq n \leq s_i - 1, \\[3mm] \dfrac{s_i^{s_i}}{s_i!}\left(\dfrac{a_i}{s_i \mu_i}\right)^n & \text{if } n \geq s_i, \end{cases}$$

and

$$p_i(0) = \left[\sum_{n=0}^{s_i-1} \frac{1}{n!}\left(\frac{a_i}{\mu_i}\right)^n + \frac{(a_i/\mu_i)^{s_i}}{s_i!} \cdot \frac{1}{1 - a_i/(s_i \mu_i)}\right]^{-1}.$$

The main result is given in the next theorem.

Theorem 6.15. (Limiting Behavior of Jackson Networks). *The limiting distribution of a stable Jackson network is given by*

$$p(n_1, n_2, \ldots, n_N) = p_1(n_1) p_2(n_2) \cdots p_N(n_N)$$

for $n_i = 0, 1, 2, \ldots$ and $i = 1, 2, \ldots, N$

Idea of Proof. The idea is to show that $\{X(t), t \geq 0\}$ is an N-dimensional CTMC and that the limiting distribution displayed above satisfies the balance equations. The reader is referred to an advanced book on queueing theory for the details. ■

The theorem above has many interesting implications. We mention three important ones:

(1) The limiting marginal distribution of $X_i(t)$ is given by $p_i(\cdot)$; i.e.,

$$\lim_{t \to \infty} P(X_i(t) = n) = p_i(n), \quad n \geq 0.$$

Thus the limiting distribution of the number of customers in the ith station is the same as that in an $M/M/s_i$ queue with arrival rate a_i and service rate μ_i.

(2) Since the limiting joint distribution of $[X_1(t), X_2(t), \ldots, X_N(t)]$ is a product of the limiting marginal distributions of $X_i(t)$, it follows that, in the limit, the queue lengths at various stations in a Jackson network are independent of each other.

(3) In steady state, we can treat a Jackson network as N independent multiserver birth and death queues. They are coupled via the traffic equations (6.45).

We illustrate this with several examples.

Example 6.22. (Jackson Network of Single-Server Queues). Consider a Jackson network with $s_i = 1$ for all $i = 1, 2, \ldots, N$. Thus all service stations are single-server queues. Let a be the solution to the traffic equations. Assume that the network is stable; i.e.,

$$a_i < \mu_i, i = 1, 2, \ldots, N.$$

We shall answer several questions regarding such a network:

1. What is the expected number of customers in the network in steady state?
   Let L_i be the expected number of customers in station i in steady state. Since station i behaves like an $M/M/1$ queue in steady state, L_i can be computed from (6.25) as

$$L_i = \frac{\rho_i}{1 - \rho_i},$$

where

$$\rho_i = a_i / \mu_i.$$

Then the expected number of customers in the network is given by

$$L = \sum_{i=1}^{N} L_i = \sum_{i=1}^{N} \frac{\rho_i}{1 - \rho_i}.$$

2. What is the probability that the network is empty?
   Using Theorem 6.15, the required probability is given by

$$p(0, 0, \ldots, 0) = p_1(0) p_2(0) \cdots p_N(0) = \prod_{i=1}^{N} (1 - \rho_i),$$

where we have used

$$p_i(0) = 1 - \rho_i$$

from (6.24). ∎

Example 6.23. (Tandem Queue). Consider the tandem queueing network of Example 6.18. We solved the corresponding traffic equations in Example 6.20 and saw that the network is stable. Using the results of Theorem 6.15, we see that, in steady state, station i behaves as an $M/M/s_i$ queue with arrival rate a_i and service rate μ_i, where s_i and μ_i are as given in Example 6.18. Using this and the results of the $M/M/s$ queue, we compute the expected number in the network here. We get

$$L_1 = 5.4545, \ L_2 = 6.0112, \ L_3 = 5, \ L_4 = 6.6219,$$
$$L = L_1 + L_2 + L_3 + L_4 = 23.0877. \quad \blacksquare$$

Example 6.24. (Amusement Park). Consider the Jackson network model of the amusement park in Example 6.19. We solved the corresponding traffic equations in Example 6.21 and saw that the network is stable. Using the results of Theorem 6.15, we see that, in steady state, station i behaves as an $M/M/s_i$ queue with arrival rate a_i and service rate μ_i, where s_i and μ_i are as given in Example 6.19.

1. Compute the expected number of visitors in the park in steady state.
   Using (6.31), we compute L_i, the expected number of customers in the ith ride. We get
   $$L_1 = 21.4904, \ L_2 = 31.7075, \ L_3 = 15.4813,$$
   $$L_4 = 50.5865, \ L_5 = 25.9511, \ L_6 = 18.3573.$$

Hence the expected total number of customers in the park is given by

$$L = L_1 + L_2 + L_3 + L_4 + L_5 + L_6 = 163.5741.$$

2. Which ride has the longest queue?
   Here we need to compute L_i^q, the expected number of customers in the queue (not including those in service). This is given by

$$L_i^q = L_i - a_i/\mu_i.$$

Numerical calculations yield

$$L_1^q = 1.4904, \ L_2^q = 1.7075, \ L_3^q = 0.4813,$$
$$L_4^q = 0.5865, \ L_5^q = 10.9511, \ L_6^q = 1.6906.$$

Thus queues are longest in service station 5, the ghost-mountain! Thus, if the management wants to invest money to expand the capacities of the rides, it should start with the ghost-mountain ride! $\quad \blacksquare$

6.8 Case Study: Healthy Heart Coronary Care Facility

We shall continue the study of the Healthy Heart Coronary Care Facility of Section 5.6. Its inpatient facility consists of six units: ER, POW, SU, POU, ICU, and ECU. It tends to patients of two types: emergency and scheduled. The emergency

Fig. 6.6 Queueing network
for the emergency patients.

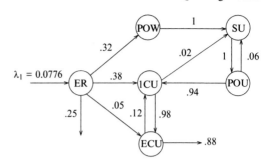

patients enter the facility in the ER, move among the six units according to an SMP, and exit from the ER or the ECU. Similarly, the scheduled patients arrive at the POW and move through the five units (all except the ER) according to an SMP and exit from the ECU.

The management of Healthy Heart wants to know how much capacity it should build in each unit to provide proper service to the patients. In Chapter 5, we did not have enough tools to answer this question properly. However, now we do.

First we need more data about patient demand. The management has historical data that suggest that the inpatient facility treats on average 9.8 patients per day. We have already been told that 19% of these are emergency patients and the remainder are scheduled patients. We begin with the assumption that the capacity of each unit is infinite, so that all the patients can move independently of each other. We shall see later how this will allow us to decide upon the actual capacities.

We shall study the two types of patients separately. For each type, we shall model the inpatient clinic as a Jackson network with one node for each unit, each node having an infinite number of servers.

Figure 6.6 shows the schematic diagram of the routes followed by the emergency patients. The network has six nodes: {1 = ER, 2 = POW, 3 = SU, 4 = POU, 5 = ICU, 6 = ECU}. We assume that the patients arrive at the ER according to a Poisson process with rate 9.8*.19 = 1.862 per day, or .0776 per hour. Alternatively, we can assume that the overall patient arrival process is Poisson with rate 9.8 per day and each patient is an emergency patient with probability .19, independent of everything else. Thus the arrival process of the emergency patients is a thinning of the overall arrival process, and hence Theorem 3.9 implies that it is a Poisson process with rate 1.862 per day. The semi-Markov nature of the movement of a patient implies that the routing of the patients among the six units is Markovian. We assume that the patients move independently of each other. Although we have not assumed the sojourn times in each state (which are equivalent to the service times in each node) are exponential, it is known that the product form solution of the Jackson network holds regardless if the nodes have infinite servers. Using the data from Section 5.6, we see that the Jackson network model of emergency patient movements has the following parameters:

$$N = 6,$$

$$s_i = \infty, \quad 1 \le i \le 6,$$

$$\lambda_1 = 0.0776, \quad \lambda_i = 0, \quad 2 \le i \le 6,$$

$$\mu_1 = 1/4.5, \; \mu_2 = 1/2.4, \; \mu_3 = 1/6.8, \; \mu_4 = 1/4.4, \; \mu_5 = 1/36.7, \; \mu_6 = 1/118,$$

$$P = \begin{bmatrix} 0 & .32 & 0 & 0 & .38 & .05 \\ 0 & 0 & 1 & 0 & 0 & 0 \\ 0 & 0 & 0 & 1 & 0 & 0 \\ 0 & 0 & 0.06 & 0 & .94 & 0 \\ 0 & 0 & 0.02 & 0 & 0 & .98 \\ 0 & 0 & 0 & 0 & .12 & 0 \end{bmatrix},$$

$$r_1 = .25, \; r_2 = 0, \; r_3 = 0, \; r_4 = 0, \; r_5 = 0, \; r_6 = .88.$$

Since there are infinite servers at each node, the network is always stable. Next we solve the traffic equations (6.45) to obtain the total arrival rate vector $a = [a_1, a_2, \cdots, a_6]$ in terms of the external arrival vector $\lambda = [\lambda_1, \cdots, \lambda_6]$ as

$$a = \lambda(I - P)^{-1} = [0.0776, \, 0.0248, \, 0.0278, \, 0.0278, \, 0.0635, \, 0.0661].$$

From Theorem 6.15, we see that, in steady state, each of the six nodes behaves as an $M/M/\infty$ (or $M/G/\infty$) queue. From Theorem 6.10, we see that in steady state the number of patients in node i is a Poisson random variable with parameter a_i/μ_i. Let α_i be the expected number of emergency patients in unit i. Then the vector $\alpha = [\alpha_1, \cdots, \alpha_6]$ is given by

$$\alpha = [0.3491, \, 0.0596, \, 0.1888, \, 0.1222, \, 2.3309, \, 7.8024].$$

We can do a similar analysis for the scheduled patients. The network now has five nodes $\{2 = \text{POW}, 3 = \text{SU}, 4 = \text{POU}, 5 = \text{ICU}, 6 = \text{ECU}\}$. Scheduled patients arrive at node 2 (POW) according to a Poisson process with rate $9.8*.81 = 7.9380$ per day, or 0.3308 per hour. Figure 6.7 shows the schematic diagram of the movements of the scheduled patients. The parameters of the five-node Jackson network are

$$N = 5,$$

$$s_i = \infty, \quad 2 \le i \le 6,$$

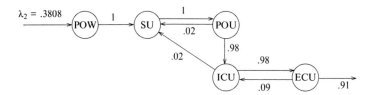

Fig. 6.7 Queueing network for the scheduled patients.

$$\lambda_2 = 0.3308, \quad \lambda_i = 0, \quad 3 \leq i \leq 6,$$

$$\mu_2 = 1/2.4, \quad \mu_3 = 1/5.3, \quad \mu_4 = 1/3.4, \quad \mu_5 = 1/33.8, \quad \mu_6 = 1/98.7,$$

$$P = \begin{bmatrix} 0 & 1 & 0 & 0 & 0 \\ 0 & 0 & 1 & 0 & 0 \\ 0 & 0.02 & 0 & .98 & 0 \\ 0 & 0.02 & 0 & 0 & .98 \\ 0 & 0 & 0 & .09 & 0 \end{bmatrix},$$

$$r_2 = 0, \quad r_3 = 0, \quad r_4 = 0, \quad r_5 = 0, \quad r_6 = .91.$$

We compute the total arrival rate vector $a = [a_2, \cdots, a_6]$ as

$$a = [0.3307, \ 0.3451, \ 0.3451, \ 0.3709, \ 0.3635].$$

Thus the number of scheduled patients in unit i in steady state is a Poisson random variable with parameter $\beta_i = a_i / \mu_i$. The vector $\beta = [\beta_2, \cdots, \beta_6]$ is given by

$$\beta = [0.7938, \ 1.8289, \ 1.1732, \ 12.1648, \ 35.8737].$$

Since the capacity in each node is infinite, the emergency and scheduled patients move independently of each other. Hence, in steady state, the number of emergency patients in unit i is independent of that of the scheduled patients at that node. Thus the total number of patients in node i ($1 \leq i \leq 6$) is a Poisson random variable with parameter $\theta_i = \alpha_i + \beta_i$ (with $\beta_1 = 0$). The vector $\theta = [\theta_1, \cdots, \theta_6]$ is given by

$$\theta = [0.3491, \ 0.8534, \ 2.0177, \ 1.2954, \ 14.4957, \ 43.6761]. \tag{6.47}$$

The knowledge of the expected number (and the distribution) of patients in each unit can help us size each unit. We describe two criteria that can be used to size the units: guarantee a certain service level or minimize the cost per unit time.

Service-Level Criterion. We quantify the service level as the probability that a patient is blocked from entering a needed unit due to capacity restrictions. When this happens, the clinic has to make temporary arrangements for the patient, which results in lower quality of care. For example, suppose that the management wants to ensure that the capacity C_i of unit i is such that the probability that a customer is turned away from this unit is .05. Let $N_i \sim P(\theta_i)$ represent that number of patients in unit i in steady state. Thus we want to choose the smallest C_i such that

$$P(N_i > C_i) \leq .05. \tag{6.48}$$

Performing the numerical calculations, we get

$$C = [1, \ 3, \ 5, \ 3, \ 21, \ 55]. \tag{6.49}$$

Thus the inpatient facility should have one bed in the ER, three in the POW, five operating rooms in the SU, three beds in the POU, 21 beds in the ICU, and 55 beds in the ECU.

Cost Criterion. Next we consider a cost criterion. Let d_i be the hourly cost of maintaining one unit of capacity in unit i. Thus it costs $d_i C_i$ dollars per hour to maintain unit i with capacity C_i. If we have to house more than C_i patients in unit i, then each overflow patient above C_i costs $f_i > d_i$ dollars per hour due to the special temporary arrangements we need to make. The management has provided us the following cost vectors for the six units:

$$d = [200, \ 80, \ 300, \ 150, \ 100, \ 10],$$

$$f = [1000, \ 400, \ 1500, \ 750, \ 500, \ 50].$$

Thus the cost of the temporary arrangements for the overflow patients is five times the cost for regular patients.

Let us compute $G_i(C_i)$, the expected cost of operating unit i with capacity C_i. The operating cost is $d_i C_i$ per hour. Recall that $N_i \sim P(\theta_i)$ is the number of patients in unit i in steady state. If $N_i > C_i$, $N_i - C_i$ patients have to be accommodated with temporary arrangements, which costs $f_i(N_i - C_i)$, and if $N_i \le C_i$, the cost of overflow patients is zero. Hence we can write the cost of overflow patients as $f_i \max\{N_i - C_i, 0\}$. Now,

$$\max(N_i - C_i, 0) = N_i - \min(N_i, C_i) \tag{6.50}$$

and

$$E(\min(N_i, C_i)) = \sum_{n=0}^{C_i - 1} P(N_i > n). \tag{6.51}$$

We leave the proof of the equalities above to the reader as Conceptual Problem 6.25. Hence the expected total cost per hour is given by

$$G_i(C_i) = E(d_i C_i + f_i \max(N_i - C_i, 0))$$

$$= E(d_i C_i + f_i N_i - f_i \min(N_i, C_i))$$

$$= d_i C_i + f_i \theta_i - f_i \sum_{n=0}^{C_i - 1} P(N_i > n).$$

In order to find the optimal value of C_i, we need to study $G_i(C_i)$ as a function of C_i. A bit of algebra shows that

$$G_i(C_i + 1) - G_i(C_i) = d_i - f_i P(N_i > C_i).$$

Thus

$$P(N_i > C_i) > \frac{d_i}{f_i} \Rightarrow G_i(C_i + 1) < G_i(C_i)$$

and

$$P(N_i > C_i) \le \frac{d_i}{f_i} \Rightarrow G_i(C_i + 1) \ge G_i(C_i).$$

Since $P(N_i > C_i)$ decreases to 0 as $C_i \to \infty$ and $0 < d_i/f_i < 1$, we can define C_i^* to be the smallest value of C_i for which

$$P(N_i > C_i) \le \frac{d_i}{f_i}.$$

Then

$$G_i(C_i + 1) < G_i(C_i) \text{ for } C_i < C_i^*$$

and

$$G_i(C_i + 1) \ge G_i(C_i) \text{ for } C_i \ge C_i^*.$$

Hence $G_i(C_i)$ is minimized at $C_i = C_i^*$. Hence it is the optimal capacity of unit i.
    With the data given to us, we thus look for the smallest C_i such that

$$P(N_i > C_i) \le .2. \tag{6.52}$$

Doing numerical calculations, we get

$$C = [1, 2, 3, 2, 18, 49]. \tag{6.53}$$

It is to be expected that these recommended capacities are smaller than those in Equation (6.49) based on the service-level criterion.
    As with other case studies in this book, we end this case study with the cautionary note that the results of the analysis are valid as long as the assumptions are valid. If the assumptions are not quite valid, then one needs to treat the results as an approximation.

6.9 Problems

CONCEPTUAL PROBLEMS

6.1. Consider a single-server queueing system that is initially empty. A customer arrives at time 0, and successive inter-arrival times are exactly x minutes. The successive service times are exactly y minutes. Verify Theorem 6.1. Do the limiting probabilities p_j exist? Does PASTA (Theorem 6.2) hold? (*Hint:* Consider the cases $y \le x$ and $y > x$ separately.)

6.2. Verify Little's Law for the system in Conceptual Problem 6.1.

6.3. Show that the system in Conceptual Problem 6.1 is stable if $y \leq x$ and unstable otherwise.

6.4. Consider a stable single-station queue. Suppose it costs $\$c$ to hold a customer in the system for one unit of time. Show that the long-run holding cost rate is given by cL, where L is the mean number of customers in steady state.

6.5. Show that the blocking probability in an $M/M/1/K$ queue is a monotonically decreasing function of K and that

$$\lim_{K \to \infty} p_K(K) = \begin{cases} 0 & \text{if } \rho \leq 1, \\ \dfrac{\rho - 1}{\rho} & \text{if } \rho > 1. \end{cases}$$

What is the intuitive interpretation of this result?

6.6. Show that L_q, the expected number in the queue (not including those in service) in an $M/M/1/K$ queue, is given by

$$L_q = L - \frac{1 - \rho^K}{1 - \rho^{K+1}} \rho,$$

where L is as given in (6.15).

6.7. Show that the blocking probability $p_K(K)$ in an $M/M/K/K$ queue satisfies the recursive relationship $p_0(0) = 1$ and

$$p_K(K) = \frac{(\lambda/\mu) p_{K-1}(K-1)}{K + (\lambda/\mu) p_{K-1}(K-1)}.$$

6.8. A machine produces items one at a time, the production times being iid exponential with mean $1/\lambda$. The items produced are stored in a warehouse of capacity K. When the warehouse is full, the machine is turned off, and it is turned on again when the warehouse has space for at least one item. Demand for the items arises according to a PP(μ). Any demand that cannot be satisfied is lost. Show that the number of items in the warehouse is an $M/M/1/K$ queueing system.

6.9. A bank uses s tellers to handle its walk-in business. All customers form a single line in front of the tellers and get served in a first-come–first-served fashion by the next available teller. Suppose the customer service times are iid Exp(μ) random variables and the arrival process is a PP(λ). The bank lobby can hold at most K customers, including those in service. Model the number of customers in the lobby as an $M/M/s/K$ queue.

6.10. A machine shop has K machines and a single repair person. The lifetimes of the machines are iid Exp(λ) random variables. When the machines fail, they are repaired by the single repair person in a first-failed–first-repaired fashion. The repair times are iid Exp(μ) random variables. Let $X(t)$ be the number of machines under repair at time t. Model $\{X(t), t \geq 0\}$ as a finite birth and death queue.

6.11. Do Conceptual Problem 6.10 if there are $s \leq K$ identical repair persons. Each machine needs exactly one repair person when it is under repair.

6.12. Telecommunication networks often use the following traffic control mechanism, called the leaky bucket. Tokens are generated according to a PP(λ) and stored in a token pool of infinite capacity. Packets arrive from outside according to a PP(μ). If an incoming packet finds a token in the token pool, it takes one token and enters the network. If the token pool is empty when a packet arrives, the packet is dropped. Let $X(t)$ be the number of tokens in the token pool at time t. Show that $\{X(t), t \geq 0\}$ is the queue-length process of an $M/M/1$ queue.

6.13. In Conceptual Problem 6.12, we described a leaky bucket with an infinite-capacity token pool but no buffer for data packets. Another alternative is to use an infinite-capacity buffer for the packets but a zero-capacity token pool. This creates a traffic-smoothing control mechanism: Packets arrive according to a PP(λ), and tokens are generated according to a PP(μ). Packets wait in the buffer for the tokens. As soon as a token is generated, a waiting packet grabs it and enters the network. If there are no packets waiting when a token is generated, the token is lost. Let $X(t)$ be the number of packets in the buffer at time t. Show that $\{X(t), t \geq 0\}$ is the queue-length process of an $M/M/1$ queue.

6.14. In a stable $M/M/1$ queue with arrival rate λ and service rate μ, show that

$$L_q = \frac{\rho^2}{1 - \rho}$$

and

$$W_q = \frac{1}{\mu} \frac{\rho}{1 - \rho}.$$

6.15. Consider an $M/M/s$ queue with arrival rate λ and service rate μ. Suppose it costs \$$r$ per hour to hire a server, an additional \$$b$ per hour if the server is busy, and \$$h$ per hour to keep a customer in the system. Show that the long-run cost per hour is given by

$$rs + b(\lambda/\mu) + hL,$$

where L is the expected number in the queue.

6.16. In a stable $M/M/s$ queue with arrival rate λ and service rate μ, show that

$$L_q = p_s \frac{\rho}{(1 - \rho)^2}$$

and

$$W_q = \frac{p_s}{\mu} \frac{1}{(1 - \rho)^2}.$$

6.17. Show that, among all service time distributions with a given mean τ, the constant service time minimizes the expected number of customers in an $M/G/1$ queue.

6.18. Consider the traffic-smoothing mechanism described in Conceptual Problem 6.13. Suppose the tokens are generated in a deterministic fashion with a constant inter-token time. Let $X(t)$ be as defined there. Is $\{X(t), t \geq 0\}$ the queue-length process in an $M/G/1$ queueing system with deterministic service times?

6.19. Consider the traffic control mechanism described in Conceptual Problem 6.12. Suppose the tokens are generated in a deterministic fashion with a constant inter-token time. Let $X(t)$ be as defined there. Is $\{X(t), t \geq 0\}$ the queue-length process in a $G/M/1$ queueing system with deterministic inter-arrival times?

6.20. Consider a queueing system with two identical servers. Customers arrive according to a Poisson process with rate λ and are assigned to the two servers in an alternating fashion (i.e., numbering the arrivals consecutively, all even-numbered customers go to server 1, and odd-numbered customers go to server 2). Let $X_i(t)$ be the number of customers in front of server i, $i = 1, 2$. (No queue jumping is allowed). Show that the queue in front of each server is a $G/M/1$ queue.

6.21. Show that the key functional equation for a $G/M/1$ queue with $U(a, b)$ inter-arrival times and $\text{Exp}(\mu)$ service times is

$$u = \frac{e^{-a\mu(1-u)} - e^{-b\mu(1-u)}}{\mu(1-u)(b-a)}.$$

6.22. Customers arrive at a department store according to a $PP(\lambda)$. They spend iid $\text{Exp}(\mu)$ amount of time picking their purchases and join one of the K checkout queues at random. (This is typically untrue: customers will join nearer and shorter queues more often, but we will ignore this.) The service times at the checkout queues are iid $\text{Exp}(\theta)$. A small fraction p of the customers go to the customer service station for additional service after leaving the checkout queue, while others leave the store. The service times at the customer service counter are iid $\text{Exp}(\alpha)$. Customers leave the store from the customer service counter. Model this as a Jackson network.

6.23. Derive the condition of stability for the network in Conceptual Problem 6.22.

6.24. Assuming stability, compute the expected number of customers in the department store of Conceptual Problem 6.22.

6.25. Prove Equations (6.50) and (6.51).

COMPUTATIONAL PROBLEMS

6.1. Customers arrive at a single-station queue at a rate of five per hour. Each customer needs 78 minutes of service on average. What is the minimum number of servers needed to keep the system stable?

6.2. Consider the system in Computational Problem 6.1. What is the expected number of busy servers if the system employs s servers, $1 \le s \le 10$.

6.3. Consider the system in Computational Problem 6.1. How many servers are needed if the labor laws stipulate that a server cannot be kept busy more than 80% of the time?

6.4. Consider a single-station queueing system with arrival rate λ. What is the effect on the expected number in the system if the arrival rate is doubled but the service rate is increased so that the expected time spent in the system is left unchanged?

6.5. Consider a single-station queueing system with arrival rate λ and mean service time τ. Discuss the effect on L, W, L_q, and W_q if the arrival rate is doubled and the service times are simultaneously cut in half.

6.6. Customers arrive at a barber shop according to a Poisson process at a rate of eight per hour. Each customer requires 15 minutes on average. The barber shop has four chairs and a single barber. A customer does not wait if all chairs are occupied. Assuming an exponential distribution for service times, compute the expected time an entering customer spends in the barber shop.

6.7. Consider the barber shop of Computational Problem 6.6. Suppose the barber charges $12 for service. Compute the long-run rate of revenue for the barber. (*Hint:* What fraction of the arriving customers enter?)

6.8. Consider the barber shop of Computational Problem 6.6. Suppose the barber hires an assistant, so that now there are two barbers. What is the new rate of revenue?

6.9. Consider the barber shop of Computational Problem 6.6. Suppose the barber installs one more chair for customers to wait in. How much does the revenue increase due to the extra chair?

6.10. Consider the production model of Conceptual Problem 6.8. Suppose the mean manufacturing time is 1 hour and the demand rate is 20 per day. Suppose the warehouse capacity is 10. Compute the fraction of the time the machine is turned off in the long run.

6.11. Consider the production model of Computational Problem 6.10. Compute the fraction of the demand lost in the long run.

6.12. Consider the production model of Computational Problem 6.10. When an item enters the warehouse, its value is $100. However, it loses value at a rate of $1 per hour as it waits in the warehouse. Thus, if an item was in the warehouse for 10 hours when it is sold, it fetches only $90. Compute the long-run revenue per hour.

6.13. Consider the call center of Example 6.7. How many additional holding lines should be installed if the airline desires to lose no more than 5% of the incoming calls? What is the effect of this decision on the queueing time of customers?

6.14. Consider the queueing model of the bank in Conceptual Problem 6.9. Suppose the arrival rate is 18 per hour, the mean service time is 10 minutes, and the lobby can handle 15 customers at one time. How many tellers should be used if the aim is to keep the mean queueing time to less than 5 minutes?

6.15. Consider the queueing model of the bank in Computational Problem 6.14. How many tellers should the bank use if the aim is to lose no more than 5% of the customers?

6.16. Cars arrive at a parking garage according to a Poisson process at a rate of 60 per hour and park there if there is space, otherwise they go away. The garage has the capacity to hold 75 cars. A car stays in the garage on average for 45 minutes. Assume that the times spent by the cars in the garage are iid exponential random variables. What fraction of the incoming cars are turned away in the steady state?

6.17. Consider the garage in Computational Problem 6.16. Suppose each car pays at a rate of $3 per hour based on the exact amount of time spent in the garage. Compute the long-run revenue rate for the garage.

6.18. Consider the taxi stand of Example 6.9. What is the expected number of customers waiting at the stand in steady state? What is the probability that there are at least three customers waiting?

6.19. Consider the leaky bucket traffic control mechanism of Conceptual Problem 6.12. Suppose 150,000 tokens are generated per second, while the packet arrival rate is 200,000 per second. Compute the long-run probability that a packet is dropped.

6.20. Consider the leaky bucket traffic-smoothing mechanism of Conceptual Problem 6.13. Suppose 200,000 tokens are generated per second, while the packet arrival rate is 150,000 per second. Compute the long-run expected time that a packet waits in the buffer.

6.21. Consider the manufacturing operation described in Conceptual Problem 6.8 but with an infinite-capacity warehouse. Suppose the demand rate is 12 per hour. What is the largest mean production time so that at most 10% of the demands are lost? What is the expected number of items in the warehouse under this production rate?

6.22. Consider a queueing system where customers arrive according to a Poisson process at a rate of 25 per hour and demand iid exponential service times. We have a choice of using either two servers, each of whom can process a customer in 4 minutes on average, or a single server who can process a customer in 2 minutes on average. What is the optimal decision if the aim is to minimize the expected wait in the system? In the queue?

6.23. Consider the post office in Example 6.10. How many server windows should be kept open to keep the probability of waiting for service to begin to at most .10?

6.24. Consider a service system where customers arrive according to a Poisson process at a rate of 20 per hour and require exponentially distributed service times with mean 10 minutes. Suppose it costs $15 per hour to hire a server and an additional $5 per hour when the server is busy. It costs $1 per minute to keep a customer in the system. How many servers should be employed to minimize the long-run expected cost rate? (*Hint:* See Conceptual Problem 6.15.)

6.25. Consider the service system of Computational Problem 6.24 with six servers. How much should each customer be charged so that it is profitable to operate the system?

6.26. Customers arrive at the drivers license office according to a Poisson process at a rate of 15 per hour. Each customer takes on average 12 minutes of service. The office is staffed by four employees. Assuming that the service times are iid exponential random variables, what is the expected time that a typical customer spends in the office?

6.27. A computer laboratory has 20 personal computers. Students arrive at this laboratory according to a Poisson process at a rate of 36 per hour. Each student needs on average half an hour at a PC. When all the PCs are occupied, students wait for the next available one. Compute the expected time a student has to wait before getting access to a PC. Assume that the times spent at a PC by students are iid exponential random variables.

6.28. Users arrive at a nature park in cars according to a Poisson process at a rate of 40 cars per hour. They stay in the park for a random amount of time that is exponentially distributed with mean 3 hours and leave. Assuming the parking lot is sufficiently big so that nobody is turned away, compute the expected number of cars in the lot in the long run.

6.29. Students arrive at a bookstore according to a Poisson process at a rate of 80 per hour. Each student spends on average 15 minutes in the store independently of each other. (The time at the checkout counter is negligible since most of them come to browse!) What is the probability that there are more than 25 students in the store?

6.30. Components arrive at a machine shop according to a Poisson process at a rate of 20 per day. It takes exactly 1 hour to process a component. The components are shipped out after processing. Compute the expected number of components in the machine shop in the long run.

6.31. Consider the machine shop of Computational Problem 6.30. Suppose the processing is not always deterministic and 10% of the components require an additional 10 minutes of processing. Discuss the effect of this on the expected number of components in the shop.

6.32. Customers arrive at a single-server service station according to a Poisson process at a rate of three per hour. The service time of each customer consists of two stages. Each stage lasts for an exponential amount of time with mean 5 minutes, the stages being independent of each other and other customers. Compute the mean number of customers in the system in steady state.

6.33. Consider Computational Problem 6.32. Now suppose there are two servers and the two stages of service can be done in parallel, one by each server. However, each server is half as efficient, so that it takes on average 10 minutes to complete one stage. Until both stages are finished, service cannot be started for the next customer. Now compute the expected number of customers in the system in steady state.

6.34. Compute the expected queueing time in an $M/G/1$ queue with an arrival rate of 10 per hour and the following service time distributions, all with mean 5 minutes:

(1) exponential;
(2) uniform over $(0, a)$;
(3) deterministic.

Which distribution produces the smallest W_q and which the largest?

6.35. Customers arrive at a post office with a single server. Fifty percent of the customers buy stamps, and take 2 minutes to do so. Twenty percent come to pick up mail and need 3 minutes to do so. The rest take 5 minutes. Assuming all these times are deterministic and that the arrivals form a Poisson process with a rate of 18 per hour, compute the expected number of customers in the post office in steady state.

6.36. Consider the queueing system described in Conceptual Problem 6.20. Suppose the arrival rate is 10 per hour and the mean service time is 10 minutes. Compute the long-run expected number of customers in front of each server. Will it reduce congestion if the customers form a single line served by the two servers?

6.37. A machine produces items one at a time, the production times being exactly 1 hour. The items produced are stored in a warehouse of infinite capacity. Demand for the items arises according to a Poisson process at a rate of 30 per day. Any demand that cannot be satisfied is lost. Model the number of items in the warehouse as a $G/M/1$ queues, and compute the expected number of items in the warehouse in steady state.

6.38. Compute the fraction of demands lost in the production system of Computational Problem 6.37. Also compute the expected amount of time an item stays in the warehouse.

6.39. Compute the expected queueing time in a $G/M/1$ queue with a service rate of 10 per hour and the following inter-arrival time distributions, all with mean 8 minutes:

(1) exponential;
(2) deterministic.

Which distribution produces the smallest W_q and which the largest?

6.40. Customers arrive in a deterministic fashion at a single-server queue. The arrival rate is 12 per hour. The service times are iid exponential with mean 4 minutes. It costs the system $2 per hour to keep a customer in the system and $40 per hour to keep the server busy. How much should each customer be charged for the system to break even?

6.41. Consider a tandem queue with two stations. Customers arrive at the first station according to a Poisson process at a rate of 24 per hour. The service times at the first station are iid exponential random variables with mean 4 minutes. After service completion at the first station, the customers move to station 2, where the service times are iid exponential with mean 3 minutes. The network manager has six servers at her disposal. How many servers should be stationed at each station so that the expected number of customers in the network is minimized in the long run?

6.42. Redo Computational Problem 6.41 if the mean service times are 6 minutes at station 1 and 2 minutes at station 2.

6.43. In the amusement park of Example 6.21, what is the maximum arrival rate the park is capable of handling?

6.44. Consider Jackson network 1 as shown in Figure 6.8. The external arrival rate is 25 per hour. The probability of returning to the first station from the last is .3. There are two servers at each station. The mean service time at station 1 is 3 minutes, at station 2 it is 2 minutes, and at station 3 it is 2.5 minutes.

1. Is the network stable?
2. What is the expected number of customers in the system?
3. What is the expected wait at station 3?

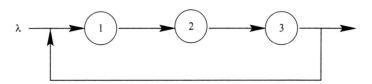

Fig. 6.8 Jackson network 1.

6.45. Consider Jackson network 2 as shown in Figure 6.9. The external arrival rate is 20 per hour. The probability of returning to the first station from the second is .3 and from the third station it is .2. There are four servers at station 1 and two each at stations 2 and 3. Forty percent of the customers from station 1 go to station 2, while the rest go to station 3. The mean service time at station 1 is 5 minutes, at station 2 it is 3 minutes, and at station 3 it is 4 minutes.

1. Is the network stable?
2. What is the expected number of customers in the system?
3. What is the expected wait at station 3?

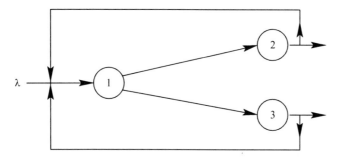

Fig. 6.9 Jackson network 2.

Case Study Problems. Use the data of Section 6.8 to do the following problems.

6.46. What should the cost of treating overflow patients in various units be so that the optimal capacities under the cost-minimization criterion match those under the service-level criterion given by Equation (6.49)?

6.47. What is the distribution of the total number of patients in the inpatient facility in steady state? For what fraction of the time is the ER idle?

6.48. What is the expected hourly cost of operating the inpatient clinic under the optimal capacities given in Equation (6.53)? What if the management uses the capacities from Equation (6.49) instead?

6.49. The facility charges the patients a fixed fee of r per hour spent in the clinic. What is the minimum value of r needed to break even? Assume that the system parameters do not change with r.

Chapter 7
Brownian Motion

So far in this book, we have studied stochastic processes in discrete time and continuous time but always restricted the state space to be discrete and in most cases a finite set. This restriction was necessitated by technical difficulties that arise when dealing with continuous state space. In this chapter, we shall study one very special stochastic process in continuous time and continuous state space. It is called Brownian motion in honor of the biologist Brown, who observed (using a microscope) that small particles suspended in a liquid perform a very frenzied-looking motion. The process is also called the Wiener process in honor of the probabilist who provided the rigorous mathematical framework for its description. We shall see that the normal random variable plays an important role in the analysis of Brownian motion, analogous to the role played by exponential random variables in Poisson processes. Hence we start with the study of normal random variables first.

7.1 Univariate Normal Random Variables

Consider a random variable taking values in $R = (-\infty, \infty)$ with the parameters $\mu \in R$ and $\sigma^2 \in (0, \infty)$. Suppose the pdf is given by

$$f(x) = \frac{1}{\sqrt{2\pi\sigma^2}} \exp\left\{-\frac{1}{2}\left(\frac{x-\mu}{\sigma}\right)^2\right\}, \quad x \in R. \tag{7.1}$$

Such a random variable is called a normal random variable and is denoted by $N(\mu, \sigma^2)$. The density above is called the normal density. There is no closed-form expression for the cdf of this random variable. The normal distribution is also called the *Gaussian distribution*. It occurs frequently in statistics as the limiting case of other distributions. An $N(0,1)$ random variable is called a standard normal random variable, and its density is denoted by $\phi(x)$. From Equation (7.1), we get

$$\phi(x) = \frac{1}{\sqrt{2\pi}} \exp\left\{-\frac{1}{2}x^2\right\}, \quad x \in R. \tag{7.2}$$

V.G. Kulkarni, *Introduction to Modeling and Analysis of Stochastic Systems*, Springer Texts in Statistics, DOI 10.1007/978-1-4419-1772-0_7, © Springer Science+Business Media, LLC 2011

The cdf of a standard normal random variable is denoted by $\Phi(\cdot)$:

$$\Phi(x) = \int_{-\infty}^{x} \frac{1}{\sqrt{2\pi}} \exp\left\{-\frac{1}{2}u^2\right\} du. \tag{7.3}$$

The following property of the Φ function is very useful and can be proved by using the symmetry of the density function of the standard normal around $x = 0$,

$$\Phi(-x) = 1 - \Phi(x). \tag{7.4}$$

The next theorem shows that a standard normal random variable can be obtained as a linear transformation of an $N(\mu, \sigma^2)$ random variable.

Theorem 7.1. (Linear Transformation). *Suppose X is an $N(\mu, \sigma^2)$ random variable, and define*

$$Y = \frac{X - \mu}{\sigma}.$$

Then Y is a standard normal random variable.

Proof. Using Equation (B.8) and f from Equation (7.1), we get

$$f_Y(y) = \sigma f(\sigma y + \mu) = \phi(y).$$

Hence $Y \sim N(0,1)$. ∎

As a corollary to the theorem above, we get the following. Let Y be standard normal, and define

$$X = \mu + \sigma Y. \tag{7.5}$$

Then X is an $N(\mu, \sigma^2)$ random variable. Hence, using Equations (B.7) and (7.3), we see that the cdf of X is given by

$$F_X(x) = \Phi\left(\frac{x - \mu}{\sigma}\right). \tag{7.6}$$

Thus we can compute the cdf of any $N(\mu, \sigma^2)$ random variable if we know the cdf of an $N(0,1)$ random variable. For this reason, it is sufficient to tabulate Φ. Such tables are generally available. We illustrate this with an example.

Example 7.1. (Cdf of $N(\mu, \sigma^2)$). Suppose X is an $N(2, 16)$ random variable. Compute $P(-4 \leq X \leq 8)$.

We know that $Y = (X - 2)/4$ is an $N(0, 1)$ random variable. Hence

$$P(-4 \leq X \leq 8) = P((-4 - 2)/4 \leq (X - 2)/4 \leq (8 - 2)/4)$$
$$= P(-1.5 \leq Y \leq 1.5) = \Phi(1.5) - \Phi(-1.5) = 2\Phi(1.5) - 1 = .8664. \quad \blacksquare$$

The next theorem gives the moments of a standard normal distribution.

Theorem 7.2. (Moments of Normal Distribution). *Let $Y \sim N(0,1)$. Then*

$$E(Y) = 0 \tag{7.7}$$

and

$$\text{Var}(Y) = 1. \tag{7.8}$$

Proof. Note that the standard normal density of Equation (7.2) is an even function of y; that is, $\phi(y) = \phi(-y)$. Hence

$$E(Y) = \int_{-\infty}^{\infty} y\phi(y)dy = 0.$$

Next

$$E(Y^2) = \int_{-\infty}^{\infty} y^2\phi(y)dy$$

$$= \frac{1}{\sqrt{2\pi}} \int_{-\infty}^{\infty} y^2 \exp(-y^2/2)dy$$

$$= \frac{2}{\sqrt{2\pi}} \int_{0}^{\infty} y^2 \exp(-y^2/2)dy \quad \text{(using symmetry)}$$

$$= \frac{2}{\sqrt{\pi}} \int_{0}^{\infty} \sqrt{u}e^{-u}du \quad \text{(substituting } y^2/2 = u).$$

Now, using the standard table of integrals, we recognize the last integral as a special case of the Gamma function defined as

$$\Gamma(z) = \int_{0}^{\infty} t^{z-1}e^{-t}dt.$$

Using the properties

$$\Gamma(1/2) = \sqrt{\pi}, \quad \Gamma(z+1) = z\Gamma(z),$$

we get

$$\int_{0}^{\infty} \sqrt{u}e^{-u}du = \sqrt{\pi}/2.$$

Hence the result follows. ∎

Next let X be as defined in Equation (7.5). Then $X \sim N(\mu, \sigma^2)$. Using Equations (B.15) and (B.19), we see that

$$E(X) = \mu, \quad \text{Var}(X) = \sigma^2. \tag{7.9}$$

We leave it to the reader to show that

$$E(e^{sX}) = \exp\left\{s\mu + \frac{1}{2}s^2\sigma^2\right\}. \tag{7.10}$$

The expectation above is called the moment-generating function of X.

7.2 Multivariate Normal Random Variables

Next we define the multivariate normal random variable. We say $X = (X_1, X_2, \cdots, X_n)$ is a multivariate normal random variable with parameters $\mu = [\mu_i]_{i=1:n}$ and $\Sigma = [\sigma_{ij}]_{i,j=1:n}$ if its joint density is given by

$$f_X(x) = \frac{1}{\sqrt{(2\pi)^n \det(\Sigma)}} \exp\left(-\frac{1}{2}(x-\mu)\Sigma^{-1}(x-\mu)^\top\right), \quad x \in R^n. \tag{7.11}$$

Here the superscript $\top$ denotes transpose. Note that we consider X, μ, and x to be row vectors in the formula above. Sometimes it is useful to think of them as column vectors. In that case, the formula has to be adjusted by interchanging $(x-\mu)$ and $(x-\mu)^\top$.

Example 7.2. (A bivariate normal). A bivariate normal random variable $X = (X_1, X_2)$ with parameters

$$\mu = [\mu_1, \mu_2]$$

and

$$\Sigma = \begin{bmatrix} \sigma_1^2 & \rho\sigma_1\sigma_2 \\ \rho\sigma_1\sigma_2 & \sigma_2^2 \end{bmatrix}$$

has its density given by

$$f(x_1, x_2) = \frac{1}{2\pi\sigma_1\sigma_2\sqrt{1-\rho^2}} \times$$
$$\exp\left\{-\frac{1}{2(1-\rho^2)}\left(\frac{(x_1-\mu_1)^2}{\sigma_1^2} + \frac{(x_2-\mu_2)^2}{\sigma_2^2} - \frac{2\rho(x_1-\mu_1)(x_2-\mu_2)}{\sigma_1\sigma_2}\right)\right\}$$

for $(x_1, x_2) \in R^2$. If $\rho = 0$, we get

$$f(x_1, x_2) = \frac{1}{2\pi\sigma_1\sigma_2} \exp\left\{-\frac{1}{2}\left(\frac{(x_1-\mu_1)^2}{\sigma_1^2} + \frac{(x_2-\mu_2)^2}{\sigma_2^2}\right)\right\}$$
$$= \phi\left(\frac{x_1-\mu_1}{\sigma_1}\right)\phi\left(\frac{x_2-\mu_2}{\sigma_2}\right).$$

Thus X_1 and X_2 are independent normal random variables if $\rho = 0$. ∎

Example 7.3. (Iid Normals). Let X_i, $1 \leq i \leq n$, be iid $N(\theta, \sigma^2)$ random variables. Then $(X_1, X_2, \cdots, X_n)$ is a multivariate random variable with parameters μ and Σ, where

$$\mu_i = \theta, \quad 1 \leq i \leq n,$$

$$\sigma_{ij} = \begin{cases} \sigma^2 & \text{if } i = j, \\ 0 & \text{if } i \neq j. \end{cases}$$

In general, independent normal random variables are multivariate normal random variables with diagonal matrix Σ. ∎

The next theorem gives an important property of the marginal distributions of multivariate normal random variables.

Theorem 7.3. (Marginal Distributions). *Let $X = (X_1, X_2, \cdots, X_n)$ be a multivariate normal random variable with parameters μ and Σ. Then X_i is an $N(\mu_i, \sigma_{ii})$ random variable for $1 \leq i \leq n$, and (X_i, X_j) $(1 \leq i < j \leq n)$ is a bivariate normal random variable with parameters*

$$\mu = [\mu_i, \ \mu_j]$$

and

$$\Sigma = \begin{bmatrix} \sigma_{ii} & \sigma_{ij} \\ \sigma_{ji} & \sigma_{jj} \end{bmatrix}.$$

Proof. The proof follows from tedious integrations using Equation (C.8) to compute the marginal density of X_i. The marginal joint density of (X_i, X_j) $(1 \leq i < j \leq n)$ is similarly given by

$$f_{X_i, X_j}(x_i, x_j) = \int_{x_1} \cdots \int_{x_{i-1}} \int_{x_{i+1}} \cdots \int_{x_{j-1}} \int_{x_{j+1}} \cdots \int_{x_n} f(x_1, \cdots, x_{i-1}, x_i,$$
$$x_{i+1}, \cdots, x_{j-1}, x_j, x_{j+1}, \cdots, x_n) \times dx_1 \ldots dx_{i-1} dx_{i+1} \cdots dx_{j-1} dx_{j+1} \cdots dx_n.$$

∎

Next we consider the conditional distributions.

Theorem 7.4. (Conditional Distributions). *Let (X_1, X_2) be a bivariate normal random variable as in Example 7.2. Given $X_2 = x_2$, X_1 is $N(\mu_1 + \frac{\sigma_{11}}{\sigma_{22}}(x_2 - \mu_2), (1 - \rho^2)\sigma_{11})$.*

Proof. Note that the marginal density $f_{X_2}(\cdot)$ of X_2 is $N(\mu_2, \sigma_{22})$. The result follows by substituting the bivariate density $f(x_1, x_2)$ from Example 7.2 in

$$f_{X_1 | X_2}(x_1 | x_2) = \frac{f(x_1, x_2)}{f_{X_2}(x_2)}. \quad ∎$$

The next theorem gives the first and second moments of a multivariate normal.

Theorem 7.5. (Moments). *Let $X = (X_1, X_2, \cdots, X_n)$ be a multivariate normal random variable with parameters μ and Σ. Then*

$$\mathsf{E}(X_i) = \mu_i, \quad \mathrm{Cov}(X_i, X_j) = \sigma_{ij}, \quad i, j = 1, 2, \cdots, n. \tag{7.12}$$

Proof. From Theorem 7.3, we know that $X_i \sim N(\mu_i, \sigma_{ii})$. Hence, from Equation (7.9), we get

$$\mathsf{E}(X_i) = \mu_i, \quad \mathrm{Var}(X_i) = \mathrm{Cov}(X_i, X_i) = \sigma_{ii}, \quad i = 1, 2, \cdots, n.$$

Theorem 7.3 also says that (X_i, X_j) is a bivariate normal random variable with parameters given there. Hence, by tedious integration, we get

$$\mathrm{Cov}(X_i, X_j) = \mathsf{E}(X_i X_j) - \mathsf{E}(X_i)\mathsf{E}(X_j) = \sigma_{ij}, \quad 1 \leq i < j \leq n.$$

This proves the theorem. ∎

The theorem above is why the matrix Σ is called the variance-covariance matrix of the multivariate normal random variable. As a corollary, we see that Σ must be a symmetric matrix with positive elements along the diagonal.

Example 7.4. (Covariance). Let $X = (X_1, X_2)$ be a bivariate normal random variable as in Example 7.2. From the theorem above, we see that the correlation coefficient of X_1 and X_2 is given by

$$\frac{\mathrm{Cov}(X_1, X_2)}{\sqrt{\mathrm{Var}(X_1)\mathrm{Var}(X_2)}} = \frac{\rho\sigma_1\sigma_2}{\sqrt{\sigma_1^2\sigma_2^2}} = \rho.$$

Thus, if $\rho = 0$, the random variables X_1 and X_2 are uncorrelated. We saw in Example 7.2 that $\rho = 0$ also makes them independent. Thus, uncorrelated bivariate normal random variables are independent. In general, uncorrelated random variables are not necessarily independent. ∎

Next, we consider linear combinations of multivariate normal random variables. The main result is given in the next theorem.

Theorem 7.6. (Linear Transformation). *Let $X = (X_1, X_2, \cdots, X_n)^\top$ be a multivariate normal random variable with parameters μ and Σ. Let $A = [a_{ij}]_{i=1:m, j=1:n}$ be an $m \times n$ matrix of real numbers and c be a column vector in R^m. Then $Y = (Y_1, Y_2, \cdots, Y_n)^\top = AX + c$ is a multivariate normal random variable with parameters*

$$\mu_Y = A\mu + c, \quad \Sigma_Y = A\Sigma A^\top.$$

Proof. See Appendix B.6 in Bickel and Docksum [1976]. ∎

From this we get the following characterization of the variance-covariance matrix of a multivariate normal random variable.

Theorem 7.7. (Variance-Covariance Matrix). *Let Σ be the variance-covariance matrix of a multivariate normal random variable. Then it is symmetric and positive semi-definite; that is,*

$$\sigma_{ij} = \sigma_{ji}$$

and

$$\sum_{i=1}^{n} \sum_{j=1}^{n} a_i \sigma_{ij} a_j \geq 0 \ \text{for all} \ a = [a_1, a_2, \cdots, a_n] \in R^n.$$

Proof. Symmetry follows from Theorem 7.5. See Conceptual Problems 7.8 and 7.9 for the proof of positive semi-definiteness. ∎

With these properties, we are now ready to define Brownian motion in the next section.

7.3 Standard Brownian Motion

We begin by recalling the concept of stationary and independent increments that we discussed in Chapter 3, where we saw that a Poisson process has stationary and independent increments. We start with a formal definition.

Definition 7.1. A stochastic process $\{X(t), t \geq 0\}$ with state space R is said to have stationary increments if the distribution of the increment $X(s + t) - X(s)$ over the interval $(s, s + t]$ depends only on t, the length of the interval. It is said to have independent increments if the increments over nonoverlapping intervals are independent.

With the definition above, we are ready to define Brownian motion.

Definition 7.2. (Standard Brownian Motion). A stochastic process $\{B(t), t \geq 0\}$ with state space R is said to be a standard Brownian motion (SBM) if

1. $\{B(t), t \geq 0\}$ has stationary and independent increments and
2. $B(t) \sim N(0, t)$ for $t \geq 0$.

The next theorem lists some important consequences of the definition above.

Theorem 7.8. (Properties of SBM). *Let $\{B(t), t \geq 0\}$ be an SBM. Then:*

1. *$B(0) = 0$ with probability 1.*
2. *$B(s + t) - B(s) \sim N(0, t)$, $s, t \geq 0$.*
3. *Let $0 < t_1 < t_2 < \cdots < t_n$. Then $(B(t_1), B(t_2), \cdots, B(t_n))$ is a multivariate normal random variable with parameters μ and Σ given by*

$$\mu_i = 0, \quad \sigma_{ij} = \min(t_i, t_j), \quad 1 \leq i, j \leq n.$$

4. *The conditional density of $B(s)$ given $B(s + t) = y$ is $N(\frac{s}{s+t} y, \frac{st}{s+t})$.*

Proof. 1. From the definition, we see that $B(0)$ is an $N(0,0)$ random variable that takes the value 0 with probability 1. Hence $B(0) = 0$ with probability 1.

2. From the stationarity of increments of an SBM, $B(s + t) - B(s)$ has the same distribution as $B(t) - B(0)$ (obtained by setting $s = 0$). However, $B(0) = 0$. Hence $B(t) - B(0) = B(t) \sim N(0, t)$ from the definition of an SBM. The result thus follows.

3. We shall prove the statement for the case $n = 2$. The general proof is similar. The independence of increments implies that $X_1 = B(t_1)$ and $X_2 = B(t_2) - B(t_1)$ are independent. Also, $X_1 \sim N(0, t_1)$ from the definition and $X_2 \sim N(0, t_2 - t_1)$ from the property 2 proved above. Thus (X_1, X_2) is bivariate normal with parameters

$$\mu = \begin{bmatrix} 0 \\ 0 \end{bmatrix}, \quad \Sigma = \begin{bmatrix} t_1 & 0 \\ 0 & t_2 - t_1 \end{bmatrix}.$$

Now we can write

$$\begin{bmatrix} B(t_1) \\ B(t_2) \end{bmatrix} = \begin{bmatrix} 1 & 0 \\ 1 & 1 \end{bmatrix} \begin{bmatrix} X_1 \\ X_2 \end{bmatrix}.$$

Hence, from Theorem 7.6, we see that $(B(t_1), B(t_2))$ is a bivariate normal random variable with mean vector

$$\begin{bmatrix} 1 & 0 \\ 1 & 1 \end{bmatrix} \begin{bmatrix} 0 \\ 0 \end{bmatrix} = \begin{bmatrix} 0 \\ 0 \end{bmatrix}$$

and variance-covariance matrix

$$\begin{bmatrix} 1 & 0 \\ 1 & 1 \end{bmatrix} \begin{bmatrix} t_1 & 0 \\ 0 & t_2 - t_1 \end{bmatrix} \begin{bmatrix} 1 & 1 \\ 0 & 1 \end{bmatrix} = \begin{bmatrix} t_1 & t_1 \\ t_1 & t_2 \end{bmatrix},$$

as desired. This proves the result for $n = 2$. The general case follows similarly.

4. From property 3 above, we see that $(B(s), B(s+t))$ is a bivariate normal random variable with zero mean and variance-covariance matrix

$$\begin{bmatrix} s & s \\ s & s + t \end{bmatrix}.$$

This implies that the correlation coefficient of $B(s)$ and $B(s + t)$ is

$$\rho = \frac{s}{\sqrt{s(s + t)}} = \sqrt{\frac{s}{s + t}}.$$

The result then follows by using Theorem 7.4. ∎

The next theorem gives a very interesting property of the sample paths of an SBM.

Theorem 7.9. (Sample Paths of SBM). *The sample paths of $\{B(t), t \geq 0\}$ are continuous everywhere, but differentiable nowhere, with probability 1.*

Idea of Proof. The proof of this theorem is beyond the scope of this book. Here we give the intuition behind the result. From Property 2 of the previous theorem, $X(t + h) - X(t)$ is an $N(0, h)$ random variable and hence goes to zero as h goes to zero. Thus we can intuitively say that $X(t + h) \to X(t)$ as $h \to 0$, implying continuity at t. We also see that $(X(t + h) - X(t))/h$ is an $N(0, 1/h)$ random variable and thus has no limit as $h \to 0$. Intuitively, this implies that the derivative of $X(t)$ does not exist at t. ∎

It is very hard to imagine (and harder to draw) a sample path that is continuous everywhere but differentiable nowhere. We typically draw a sample path as a very kinky graph, as shown in Figure 7.1. We end this section with several examples.

Example 7.5. Compute a function $a(t) > 0$ such that an SBM lies between $-a(t)$ and $a(t)$ at time $t > 0$ with probability .5.

We are asked to compute a positive number $a(t)$ such that $P(-a(t) < B(t) < a(t)) = .5$. Since $B(t)$ is an $N(0, t)$ random variable, we see that $B(t)/\sqrt{t}$ is an $N(0, 1)$ random variable. Hence

$$
\begin{aligned}
P(-a(t) < B(t) < a(t)) &= P(-a(t)/\sqrt{t} < B(t)/\sqrt{t} < a(t)/\sqrt{t}) \\
&= \Phi(a(t)/\sqrt{t}) - \Phi(-a(t)/\sqrt{t}) \\
&= \Phi(a(t)/\sqrt{t}) - (1 - \Phi(a(t)/\sqrt{t})) \\
&= 2\Phi(a(t)/\sqrt{t}) - 1 = .5.
\end{aligned}
$$

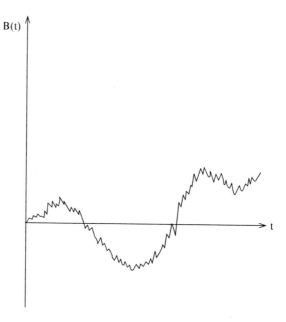

Fig. 7.1 Sample path of a standard Brownian motion.

Hence we get

$$a(t)/\sqrt{t} = \Phi^{-1}(.75),$$

which yields, by using standard tables,

$$a(t) = \Phi^{-1}(.75)\sqrt{t} = .674\sqrt{t}, \quad t > 0.$$

Note that the result above does not mean that the entire sample path of an SBM lies in $(-.674\sqrt{t}, .674\sqrt{t})$ for all $t > 0$ with probability .5. ■

Example 7.6. Compute the probability that an SBM is below zero at time 1 and above zero at time 2.

We are asked to compute $P(B(1) < 0, B(2) > 0)$. First note that, given $B(1) = x$, $B(2) - B(1) = B(2) - x$ is independent of $B(1)$ and is an $N(0,1)$ random variable. Using this observation, we have

$$P(B(1) < 0, B(2) > 0) = \int_{-\infty}^{\infty} P(B(1) < 0, B(2) > 0 | B(1) = x)\phi(x)dx$$

$$= \int_{-\infty}^{0} P(B(2) - x > -x | B(1) = x)\phi(x)dx$$

$$= \int_{-\infty}^{0} (1 - \Phi(-x))\phi(x)dx$$

$$= \int_{-\infty}^{0} \Phi(x)\phi(x)dx \quad \text{(using Equation (7.4))}$$

$$= \int_{0}^{.5} u\,du \quad \text{(substitute } \Phi(x) = u)$$

$$= 1/8. \quad ■$$

Example 7.7. (Standard Brownian Bridge). An SBM $\{B(t), 0 \le t \le 1\}$ is called a standard Brownian bridge if $B(1) = 0$. Compute the distribution of $B(s)$ $(0 \le s \le 1)$ in a Brownian bridge. At what value of $s \in [0, 1]$ is its variance the largest?

We are asked to compute the conditional distribution of $B(s)$ given $B(1) = 0$. This can be computed from property 4 of Theorem 7.8 by using $s + t = 1$ and $y = 0$. Thus $B(s) \sim N(0, s(1 - s))$. The variance $s(1 - s)$ is maximized at $s = .5$. ■

Example 7.8. Let $\{B(t), 0 \le t \le 1\}$ be an SBM. Compute $E(B(t)|B(t) > 0)$. Since $B(t)/\sqrt{t} \sim N(0, 1)$, we have $P(B(t) > 0) = 1/2$. Hence,

$$E(B(t)|B(t) > 0) = \sqrt{t}E(B(t)/\sqrt{t}|B(t)/\sqrt{t} > 0)$$

$$= \sqrt{t}\frac{\int_{0}^{\infty} x\phi(x)dx}{P(B(t)/\sqrt{t} > 0)}$$

$$= \frac{2\sqrt{t}}{\sqrt{2\pi}} \int_{0}^{\infty} x \exp\left(-\frac{1}{2}x^2\right) dx$$

$$= \frac{2\sqrt{t}}{\sqrt{2\pi}} \int_0^\infty \exp(-u)du \quad (\text{substitute } x^2/2 = u)$$

$$= \sqrt{\frac{2t}{\pi}}, \quad t > 0.$$

By symmetry, we get

$$E(B(t)|B(t) < 0) = -\sqrt{\frac{2t}{\pi}}, \quad t > 0. \quad \blacksquare$$

Example 7.9. (Reflected SBM). Let $\{B(t), 0 \le t \le 1\}$ be an SBM. Compute the distribution of $|B(t)|$. We have, for $x > 0$,

$$P(|B(t)| \le x) = P(-x \le B(t) \le x)$$
$$= P(-x/\sqrt{t} \le B(t)/\sqrt{t} \le x/\sqrt{t})$$
$$= \int_{-x/\sqrt{t}}^{x/\sqrt{t}} \phi(x)dx$$
$$= \Phi(x/\sqrt{t}) - \Phi(-x/\sqrt{t})$$
$$= 2\Phi(x/\sqrt{t}) - 1.$$

Hence the density of $|B(t)|$ can be computed by taking the derivative of the expression above with respect to x. This yields the density

$$\sqrt{\frac{2}{\pi t}} \exp\left(-\frac{x^2}{2t}\right), \quad x \ge 0.$$

The state space of the process $\{|B(t)|, t \ge 0\}$ is $[0, \infty)$. It is called a reflected SBM since whenever its sample path hits zero, it gets reflected back into the positive half of the real line. A typical sample path of a reflected SBM is shown in Figure 7.2. $\quad \blacksquare$

Example 7.10. (Geometric SBM). Let $\{B(t), 0 \le t \le 1\}$ be an SBM. Compute $E(e^{rB(t)})$ for a given $t > 0$.

Since $B(t) \sim N(0, t)$, we can use Equation (7.10) to obtain

$$E\left(e^{rB(t)}\right) = \exp\left(\frac{1}{2}r^2 t\right).$$

The process $\{e^{rB(t)}, t \ge 0\}$ has state space $(0, \infty)$ and is called a geometric SBM. It is used often as a model of stock prices. $\quad \blacksquare$

Example 7.11. (Inventory Model). Let $B(t)$ be the inventory level at time t. Positive values imply inventory on hand, and negative values imply backlogs. Thus $B(t) = -3$ is interpreted to mean that three items are on back order, while $B(t) = 5$

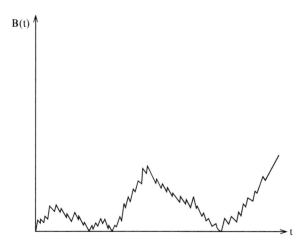

Fig. 7.2 Sample path of a Reflected SBM.

is interpreted to mean that five items are on the shelf. Suppose $\{B(t), t \geq 0\}$ is an SBM. Suppose the inventory is 7.5 at time 10. What is the expected amount of inventory at time 4?

Use property 4 of Theorem 7.8 with $s = 4$, $t = 6$, $y = 7.5$. Thus, given $B(10) = 7.5$, $B(4)$ is a normal random variable with mean $(4/10)*7.5 = 3$ and variance $(4*6)/10 = 2.4$. Hence the expected amount of inventory at time 7.5 is 3.

What is the probability that there was a stockout at time 4? This probability is given by

$$P(B(4) < 0 | B(10) = 7.5) = P(N(3, 2.4) < 0) = P(N(0, 1) < -3/\sqrt{2.4})$$
$$= \Phi(-1.9365) = .0264. \quad \blacksquare$$

7.4 Brownian Motion

In this section, we study a useful generalization of the SBM. The generalized stochastic process is called the Brownian motion (BM). We begin with a formal definition.

Definition 7.3. Let $\{B(t), t \geq 0\}$ be an SBM. A stochastic process $\{X(t), t \geq 0\}$ defined by

$$X(t) = x_0 + \mu t + \sigma B(t), \quad t \geq 0, \tag{7.13}$$

is called a Brownian motion with drift parameter $\mu \in R$, variance parameter $\sigma > 0$, and starting point $x_0 \in R$.

We denote a BM with drift μ and variance parameter σ by BM(μ, σ). Unless otherwise mentioned, we assume that $x_0 = 0$; that is, the BM starts at the origin. The next theorem lists some important consequences of the definition above.

Theorem 7.10. (Properties of BM). *Let $\{X(t), t \geq 0\}$ be a BM(μ, σ) with initial position $X(0) = x_0$. Then:*

1. *$\{X(t), t \geq 0\}$ has stationary and independent increments.*
2. *$X(s + t) - X(s) \sim N(\mu t, \sigma^2 t)$, $s, t \geq 0$, and in particular $X(t) \sim N(x_0 + \mu t, \sigma^2 t)$.*
3. *Let $0 < t_1 < t_2 < \cdots < t_n$. Then $(X(t_1), X(t_2), \cdots, X(t_n))$ is a multivariate normal random variable with parameters θ and Σ given by*

$$\theta_i = x_0 + \mu t_i, \quad \sigma_{ij} = \sigma^2 \min(t_i, t_j), \quad 1 \leq i, j \leq n. \tag{7.14}$$

4. *The conditional density of $X(s)$ given $X(s + t) = y$ is $N(\frac{s}{s+t}(y - x_0), \sigma^2 \frac{st}{s+t})$.*
5. *The sample paths of $\{X(t), t \geq 0\}$ are continuous everywhere, but differentiable nowhere, with probability 1.*

Proof. 1. We have

$$X(t + s) - X(s) = \mu t + \sigma(B(t + s) - B(s)). \tag{7.15}$$

Since the distribution of $B(t + s) - B(s)$ is independent of s, it is clear that the distribution of $X(t + s) - X(s)$ is also independent of s. This proves stationarity of increment. Independence of the increments in the BM follows similarly from the independence of increments of the SBM.

2. We know from Theorem 7.8 that $B(s+t) - B(s) \sim N(0, t)$. Equation (7.15) then implies that $X(s + t) - X(s) \sim N(\mu t, \sigma^2 t)$. As a special case, consider $s = 0$. Then $X(t) - X(0) = X(t) - x_0$ is an $N(\mu t, \sigma^2 t)$ random variable. Hence the result follows.

3. Let $X = (X(t_1), X(t_2), \cdots, X(t_n))$ and $B = (B(t_1), B(t_2), \cdots, B(t_n))$. From the definition of the BM, we have

$$X = \theta + \sigma B.$$

Since B is multivariate normal with parameters given in Theorem 7.8, it follows from Theorem 7.6 that X is multivariate normal with parameters given in Equation (7.14).

4. The conditional density of $X(s)$ given $X(s + t) = y$ is the same as the conditional density $x_0 + \mu s + \sigma B(s)$ given $x_0 + \mu(s + t) + \sigma B(s + t) = y$, which is the same as the conditional density of $x_0 + \mu s + \sigma B(s)$ given $B(s + t) = (y - x_0 - \mu(s + t))/\sigma$. Now, from property 4 of Theorem 7.8, given $B(s + t) = (y - x_0 - \mu(s + t))/\sigma$, $B(s)$ is an $N(\frac{s}{s+t}(y - x_0)/\sigma - s\mu/\sigma, \frac{st}{s+t})$ random variable. Hence, under the same conditioning, $X(s) = x_0 + \mu s + \sigma B(s)$ is an $N(\frac{s}{s+t}(y - x_0), \sigma^2 \frac{st}{s+t})$ random variable. Note that this is independent of μ!

5. This follows from the same property of the paths of the SBM as stated in Theorem 7.9. ■

We illustrate the theorem above with the help of several examples below.

Example 7.12. Let $\{X(t), t \geq 0\}$ be a BM$(-3,2)$ with $X(0) = 10$. What is the probability that the Brownian motion is below zero at time 3?

We know from property 2 of Theorem 7.10 that $X(3)$ is a normal random variable with mean $10 - 3*3 = 1$ and variance $2^2 * 3 = 12$. Hence the required probability is given by

$$P(X(3) < 0) = P((X(3) - 1)/\sqrt{12} < -1/\sqrt{12}) = \Phi(-.2887) = .3864. \; \blacksquare$$

Example 7.13. (Brownian Bridge). A BM $\{X(t), 0 \leq t \leq 1\}$ is called a Brownian bridge if $X(0) = X(1) = 0$. Compute the distribution of $X(s)$ $(0 \leq s \leq 1)$ in a Brownian bridge. At what value of $s \in [0, 1]$ is its variance the largest?

We are asked to compute the conditional distribution of $X(s)$ given $X(1) = 0$. This can be computed from property 4 of Theorem 7.10 by using $s + t = 1$ and $y = 0$. Thus $X(s) \sim N(0, \sigma^2 s(1 - s))$. The variance $\sigma^2 s(1 - s)$ is maximized at $s = .5$. $\blacksquare$

Example 7.14. (Geometric BM). Let $\{X(t), 0 \leq t \leq 1\}$ be a BM with $X(0) = 0$. Compute $\mathsf{E}(e^{rX(t)})$ for a given $t > 0$.

Since $X(t) \sim N(\mu t, \sigma^2 t)$, we can use Equation (7.10) to obtain

$$\mathsf{E}(e^{rX(t)}) = \exp\left(r\mu t + \frac{1}{2}r^2\sigma^2 t\right).$$

As a consequence of the formula above, we see that

$$\mathsf{E}(e^{(-2\mu/\sigma^2)X(t)}) = \exp(0) = 1, \quad t \geq 0. \tag{7.16}$$

The process $\{e^{rX(t)}, t \geq 0\}$ has state space $(0, \infty)$ and is called a geometric BM. It is used often as a model of stock prices. $\blacksquare$

Example 7.15. (Inventory Model). Consider the inventory model of Example 7.11. Suppose we can add to the inventory continuously at a constant rate of 2 per unit time. Let $X(t)$ be the inventory level at time t including this production rate. Suppose the initial inventory is x. What value of x should we choose to make the probability of stockout at any given time $t \in [0, 1]$ be bounded above by .05?

The inventory $X(t)$ is given by

$$X(t) = x + 2t + B(t), \quad t \geq 0.$$

Thus $\{X(t), t \geq 0\}$ is a BM$(2,1)$ with $X(0) = x$. Thus $X(t)$ is an N$(x + 2t, t)$ random variable. The probability of stockout at time t is

$$P(X(t) < 0) = P((X(t) - x - 2t)/\sqrt{t} < -(x + 2t)/\sqrt{t}) = \Phi(-(x + 2t)/\sqrt{t}).$$

Using $\Phi(-1.64) = .05$, we get

$$-(x + 2t)/\sqrt{t} \le -1.64,$$

or

$$x \ge -2t + 1.64\sqrt{t}, \quad 0 \le t \le 1.$$

Now, the right-hand side of the inequality above is maximized at $t = .168$, the maximum value being .3362. Thus, if we choose $x = .3362$, the inequality above is satisfied for all $t \in [0, 1]$. Thus this is the required initial level. ∎

7.5   First-Passage Times in an SBM

We begin with the study of first-passage times in a standard Brownian motion. Let $\{B(t), t \ge 0\}$ be an SBM and $a \in R$ be a given number. Define the first-passage time to a as

$$T(a) = \min\{t \ge 0 : B(t) = a\}. \tag{7.17}$$

This is a well-defined random variable since the sample paths of an SBM are continuous and eventually visit every $a \in R$ with probability 1. This is analogous to the first-passage times we studied in the DTMCs and the CTMCs. The next theorem gives the main result.

Theorem 7.11. (Distribution of the First-Passage Time). *Let $T(a)$ be as in Equation (7.17). Then*

$$P(T(a) \le t) = 2P(B(t) \ge |a|) = 2(1 - \Phi(|a|/\sqrt{t})). \tag{7.18}$$

Proof. Assume $a > 0$. Then, using the Law of Total Probability, we have

$$P(T(a) \le t) = P(T(a) \le t, B(t) \ge a) + P(T(a) \le t, B(t) < a). \tag{7.19}$$

However, $B(t) \ge a$ implies $T(a) \le t$. Hence

$$P(T(a) \le t, B(t) \ge a) = P(B(t) \ge a). \tag{7.20}$$

Next, suppose $T(a) = s \le t$. Then we know that $B(s) = a$. Hence, using independence of increments, we see that $B(t) = B(t) - B(s) + a \sim N(a, (t - s))$. (This implicitly uses the fact that $T(a) = s$ depends only on $\{B(u), 0 \le u \le s\}$.) Thus, using the symmetry of the $N(a, (t - s))$ density around a, we get

$$P(B(t) < a | T(a) = s) = P(B(t) > a | T(a) = s), \quad 0 \le s \le t.$$

Hence,

$$P(B(t) < a, T(a) \leq t) = P(B(t) > a, T(a) \leq t) = P(B(t) \geq a, T(a) \leq t). \tag{7.21}$$

The use of symmetry above is called the reflection principle. Using Equations (7.20) and (7.21) in Equation (7.19), we get

$$\begin{aligned} P(T(a) \leq t) &= P(T(a) \leq t, B(t) \geq a) + P(T(a) \leq t, B(t) < a) \\ &= P(B(t) \geq a) + P(T(a) \leq t, B(t) \geq a) \\ &= P(B(t) \geq a) + P(B(t) \geq a), \end{aligned}$$

thus proving the theorem. The result for $a < 0$ follows by a similar argument. Equation (7.18) combines both these cases. ∎

Note that $T(0) = 0$ with probability 1. When $a \neq 0$, we can differentiate Equation (7.18) to obtain the following density of $T(a)$:

$$f_{T(a)}(t) = \frac{|a|}{\sqrt{2\pi t^3}} \exp\left(\frac{-a^2}{2t}\right), \quad t \geq 0, \ a \neq 0. \tag{7.22}$$

Direct integration shows that

$$E(T(a)) = \int_0^\infty t f_{T(a)}(t) dt = \infty, \quad a \neq 0. \tag{7.23}$$

Thus, although an SBM reaches any level a with probability 1, it takes on average an infinite amount of time to do so.

Using Theorem 7.11, we can easily compute the cdf of the maximum and the minimum of an SBM over $[0, t]$. First define

$$U(t) = \max\{B(s) : 0 \leq s \leq t\} \tag{7.24}$$

and

$$L(t) = \min\{B(s) : 0 \leq s \leq t\}. \tag{7.25}$$

The main result is given in the next theorem.

Theorem 7.12. (Maximum and Minimum of an SBM). *The density of $U(t)$ is given by*

$$f_{U(t)}(a) = \sqrt{\frac{2}{\pi t}} \exp\left(\frac{-a^2}{2t}\right), \quad t \geq 0, \ a \geq 0. \tag{7.26}$$

The density of $L(t)$ is given by

$$f_{L(t)}(a) = \sqrt{\frac{2}{\pi t}} \exp\left(\frac{-a^2}{2t}\right), \quad t \geq 0, \ a \leq 0. \tag{7.27}$$

Proof. For $a \geq 0$, we have

$$\{U(t) \geq a\} \Leftrightarrow \{T(a) \leq t\}.$$

Hence

$$P(U(t) \geq a) = P(T(a) \leq t). \tag{7.28}$$

Using Equation (7.18), we get

$$P(U(t) \geq a) = 2(1 - \Phi(a/\sqrt{t})).$$

Taking the derivative with respect to a, we get Equation (7.26). Similarly, for $a \leq 0$, we have

$$\{L(t) \leq a\} \Leftrightarrow \{T(a) \leq t\}.$$

Hence

$$P(L(t) \leq a) = P(T(a) \leq t).$$

Using Equation (7.18), we get

$$P(L(t) \leq a) = 2(1 - \Phi(-a/\sqrt{t})).$$

Taking the derivative with respect to a, we get Equation (7.27). ∎

The density in Equation (7.26) can be seen to be the density of $|X|$, where X is an $N(0, t)$ random variable, and the density in Equation (7.27) is the density of $-|X|$.

Example 7.16. Compute the probability that an SBM reaches level 3 by time 4.
We are asked to compute $P(T(3) \leq 4)$. Using Equation (7.18), we get

$$P(T(3) \leq 4) = 2(1 - \Phi(3/\sqrt{4})) = 2(1 - .933) = .152. \quad ∎$$

Example 7.17. Let $U(t)$ be as defined in Equation (7.24). Compute $E(U(t))$.
We know that $U(t)$ has the same density as $|X|$, where X is an $N(0, t)$ random variable. Hence, using the result of Conceptual Problem 7.6, we get

$$E(U(t)) = E(|X|) = \sqrt{2t/\pi}. \quad ∎$$

Example 7.18. (Inventory Model). Suppose the inventory level at time t is given by $x + B(t)$ for $t \geq 0$. Here x is the initial inventory level and is under our control. What is the smallest level we need to start with so that the probability of a stockout over $[0, 10]$ is less than .10?
We are interested in the event

$$x + B(t) \geq 0 \quad \text{for} \quad 0 \leq t \leq 10.$$

This is equivalent to

$$T(-x) = \min\{t \geq 0 : B(t) = -x\} > 10.$$

Using Theorem 7.11, we see that we need to ensure that

$$P(T(-x) > 10) = 2\Phi(x/\sqrt{10}) - 1 \geq .90.$$

Thus we need

$$\Phi(x/\sqrt{10}) \geq .55.$$

Using standard normal tables, we see that this can be achieved if we choose

$$x \geq .1257\sqrt{10} = .3975.$$

Hence the initial inventory level should be set to .3975 to ensure that the probability of a stockout over [0,10] is no more than .10. ∎

Example 7.19. (Stock Prices). Suppose the price (in dollars) of a stock at time t (in days) is given by

$$V(t) = e^{\sigma B(t)}, \quad t \geq 0.$$

Here $\sigma > 0$ is called the volatility of the stock. Suppose an investor owns 500 shares of the stock at time 0. He plans to sell the shares as soon as its price reaches $3. What is the probability that he has to wait more than 30 days to sell the stock, assuming that the volatility parameter is 2?

Let

$$T = \min\{t \geq 0 : V(t) = 3\} = \min\{t \geq 0 : B(t) = \ln(3)/2\}.$$

Thus $T = T(.5493)$ of Equation (7.17). Using Theorem 7.11, we get

$$P(T > 30) = P(T(.5493) > 30) = 2\Phi(.5493/\sqrt{30}) - 1 = .4172.$$

Hence the investor has to wait more than 30 days with probability .4172. ∎

Example 7.20. (Stock Prices). Consider the stock price process of Example 7.19. Suppose a dealer offers you an option to buy a financial instrument that pays a dollar if the stock value during the next 30 days goes above 3 at any time and zero dollars otherwise. The dealer wants you to pay 50 cents to buy this option. Should you buy it?

You should buy the option if the expected payoff from the option is more than 50 cents. Let

$$U = \max\{V(t) : 0 \leq t \leq 30\}.$$

The payoff from the option is $1.00 if $U \geq 3$ and 0 otherwise. Hence the expected payoff is

$$1.00P(U \geq 3) + 0.00P(U < 3) = P(U \geq 3).$$

Using the results of Example 7.19, we get

$$P(U \geq 3) = P(T(.5493) \leq 30) = 1 - .4172 = .5828.$$

Thus the expected payoff is 58.28 cents, which is greater than 50 cents. Hence you should buy the option. ∎

7.6 First-Passage Times in a BM

In this section, we study the first-passage times in a BM(μ, σ). This is a much harder problem due to the presence of the drift μ since it destroys the symmetry property that we used so crucially in the analysis of first-passage times in the SBM in Section 7.5. We need tools from advanced calculus or from Martingale theory to derive the results. These are beyond the scope of this book, and we refer the reader to an advanced book (for example, Kulkarni [2010]). Hence we satisfy ourselves with the statement of the results without proof.

Let

$$X(t) = \mu t + \sigma B(t)$$

be a BM(μ, σ) with initial position $X(0) = 0$. Let $a < 0$ and $b > 0$ be two given numbers, and define

$$T(a, b) = \min\{t \geq 0 : X(t) \in \{a, b\}\}. \tag{7.29}$$

Thus $T(a, b)$ is the first time the BM visits a or b. The next theorem gives the main results.

Theorem 7.13. (First-Passage Times in a BM). *Let*

$$\theta = -\frac{2\mu}{\sigma^2}. \tag{7.30}$$

If $\mu \neq 0$, we have

$$P(X(T(a, b)) = b) = \frac{\exp(\theta a) - 1}{\exp(\theta a) - \exp(\theta b)}, \tag{7.31}$$

$$E(T(a, b)) = \frac{b(\exp(\theta a) - 1) - a(\exp(\theta b) - 1)}{\mu(\exp(\theta a) - \exp(\theta b))}. \tag{7.32}$$

If $\mu = 0$, we have

$$P(X(T(a, b)) = b) = \frac{|a|}{|a| + b}, \tag{7.33}$$

$$E(T(a, b)) = \frac{|a|b}{\sigma^2}. \tag{7.34}$$

Equation (7.33) can be deduced from Equation (7.31) (and Equation (7.34) from Equation (7.32)) by letting $\theta \to 0$ and using L'Hopital's Rule to compute the limits. Using Theorem 7.13, we can easily compute the pdf of the maximum and the minimum of a BM over $[0, \infty)$. First define

$$U = \max\{X(s) : 0 \leq s < \infty\} \tag{7.35}$$

and

$$L = \min\{X(s) : 0 \leq s < \infty\}. \tag{7.36}$$

The main result is given in the next theorem. We leave the proof as an exercise for the reader.

Theorem 7.14. (Maximum and Minimum of a BM). *Let $\{X(t), t \geq 0\}$ be a BM(μ, σ), and U and L be as defined in Equations (7.35) and (7.36), respectively. Furthermore, let θ be as in Equation (7.30).*

1. $\mu < 0$. The density of U is given by

$$f_U(b) = \theta e^{-\theta b}, \quad b \geq 0, \tag{7.37}$$

while $L = -\infty$ with probability 1.
2. $\mu > 0$. The density of L is given by

$$f_L(a) = -\theta e^{-\theta a}, \quad a \leq 0, \tag{7.38}$$

while $U = \infty$ with probability 1.
3. $\mu = 0$. $U = \infty$ and $L = -\infty$ with probability 1.

We illustrate the use of the theorem above by several examples.

Example 7.21. Let $\{Y(t), t \geq 0\}$ be a BM(1,2) starting at $Y(0) = 5$. What is the probability that it hits level 8 before hitting zero?

Let $X(t) = Y(t) - 5$. Then we see that $\{X(t), t \geq 0\}$ is a BM(1,2) starting at $X(0) = 0$. Furthermore, the event "$\{Y(t), t \geq 0\}$ hits level 8 before hitting zero" is identical to the event "$\{X(t), t \geq 0\}$ hits level $8 - 5 = 3$ before hitting $0 - 5 = -5$." Thus we are asked to compute $P(X(T) = 3)$, where $T = T(-5, 3)$ is as defined in Equation (7.29). Using

$$\theta = -2\mu/\sigma^2 = -2/4 = -.5$$

in Equation (7.31), we get

$$P(X(T) = 3) = \frac{\exp(5/2) - 1}{\exp(5/2) - \exp(-3/2)} = .9350.$$

What is the expected time when the $\{Y(t), t \geq 0\}$ process hits 8 or 0?

This is given by $E(T(-5, 3))$. Using Equation (7.32), we get

$$E(T) = \frac{3(\exp(5/2) - 1) + 5(\exp(-3/2) - 1)}{(\exp(5/2) - \exp(-3/2))} = 2.8394. \ \blacksquare$$

Example 7.22. (Stock Price). Consider the stock price process of Example 7.19. The initial value of the stock is $1. The investor plans to sell the stock when it reaches $3 (for a profit of $2 per share) or when it reaches $.50 (for a loss of $.50 per share).

(a) What is the probability that the investor ends up selling at a loss?
Let

$$T = \min\{t \geq 0 : V(t) \in \{1/2, 3\}\}.$$

We see that

$$T = \min\{t \geq 0 : B(t) \in \{\ln(1/2)/2, \ln(3)/2\}\}$$
$$= \min\{t \geq 0 : B(t) \in \{-.3466, .5493\}\}.$$

Thus $T = T(-.3466, .5493)$, where $T(a, b)$ is as defined in Equation (7.29). The probability that the investor sells at a loss is given by $P(B(T) = -.3466)$. Using Equation (7.33), we get

$$P(B(T) = -.3466) = 1 - P(B(T) = .5493)$$
$$= 1 - .3466/(.3466 + .5493) = .6131.$$

Hence the probability of selling at a profit is $1 - .6131 = .3869$.
(b) What is the expected net profit from following this policy?
It is given by

$$2 * .3869 - .5 * .6131 = .4672.$$

Thus the investor makes 46.72 cents per share in expected net profit.
(c) What is the expected time when the investor sells the shares?
It is given by

$$E(T) = .3466 * .5493/(1) = .1904.$$

It seems that this is a surefire policy for making a profit of 46.72 cents per share in .19 time units. There is only one problem with this analysis: we have ignored the interest rate on risk-free deposits, effectively assuming that it is zero. The analysis incorporating the interest rate is beyond the scope of this book, and the reader is encouraged to see more advanced books on this topic, such as Kennedy (2010). $\blacksquare$

Example 7.23. (Buffering in Streaming Media). When you listen to a streaming audio file on the Internet, you notice that the media player buffers the data before starting to play it. If the player runs out of data to play, it produces an interruption. On the other hand, if the data come in too rapidly, the buffer can overflow and lead to lost segments of the audio file. Suppose the data are played at a rate of 4 kilobytes per second. Let $Y(t)$ be the amount of data in the buffer at time t. Suppose the size of the buffer is $B = 20$ kilobytes and the audio file starts playing as soon as there are 4 kilobytes of data in the buffer. Assume that the buffer content can be modeled as a BM$(-4, \sqrt{10})$ once the file starts playing. What is the expected time when the buffer becomes full or empty after the file starts playing?

We have $Y(0) = 4$, and we want to compute $\mathsf{E}(T)$, where

$$T = \min\{t \geq 0 : Y(t) \in \{0, 20\}\}.$$

We can see that $T = T(-4, 16)$, where $T(a, b)$ is defined in Equation (7.29). Using

$$\theta = -2\mu/\sigma^2 = 8/10 = .8$$

in Equation (7.32), we get

$$\mathsf{E}(T) = \frac{16(\exp(-3.2) - 1) + 4(\exp(12.8) - 1)}{(-4)(\exp(-3.2) - \exp(12.8))} = 1.$$

Thus, on average, the file will play fine for 1 second before encountering distortion from overflow or underflow. ∎

Example 7.24. (Optimal Buffering). Consider the buffering policy in Example 7.23. Suppose the player starts playing the file when it has x kilobytes of data in the buffer. What is the optimal value of x if the aim is to maximize the expected time until the first distortion?

Following the analysis of Example 7.23, we see that we are asked to find the $x \in [0, 20]$ that maximizes $\mathsf{E}(T(-x, 20 - x))$. Using $\theta = .8$ in Equation (7.32), we get

$$g(x) = \mathsf{E}(T(-x, 20 - x)) = \frac{(20 - x)(\exp(-.8x) - 1) + x(\exp(16 - .8x) - 1)}{4(\exp(16 - .8x) - \exp(-.8x))}.$$

Numerically computing the g function above, we see that it is maximized at $x = 9$. Thus it is optimal to start playing the file when 9 kilobytes have been buffered. This produces uninterrupted play for 24.99 seconds on average. ∎

Example 7.25. (One-Sided First-Passage Time). Let $\{X(t), t \geq 0\}$ be a BM(μ, σ) with $\mu < 0$ and $X(0) = 0$. Let $a < 0$ be given, and define

$$T(a) = \min\{t \geq 0 : X(t) = a\}.$$

Compute $\mathsf{E}(T(a))$.

We use the fact that

$$\mathsf{E}(T(a)) = \lim_{b \to \infty} \mathsf{E}(T(a, b)),$$

where $T(a, b)$ is as defined in Equation (7.29). Taking limits in Equation (7.32) and noting that $\theta > 0$, we get

$$
\begin{aligned}
\mathsf{E}(T(a)) &= \lim_{b \to \infty} \mathsf{E}(T(a, b)) \\
&= \lim_{b \to \infty} \frac{b(\exp(\theta a) - 1) - a(\exp(\theta b) - 1)}{\mu(\exp(\theta a) - \exp(\theta b))} \\
&= \lim_{b \to \infty} \frac{b(\exp(\theta(a - b)) - \exp(-\theta b)) - a(1 - \exp(-\theta b))}{\mu(\exp(\theta(a - b)) - 1)} \\
&= a/\mu.
\end{aligned}
$$

This is independent of the variance parameter! The result above is also interesting since it can be interpreted as follows. We know that

$$
\mathsf{E}(X(t)) = \mu t.
$$

Hence the expected value is a at time $t = a/\mu$. This is also the expected time when the BM reaches a for the first time! Note that this argument is not strictly valid and only happens to produce the correct result as a coincidence. For example, it cannot be applied to $\mathsf{E}(T(a, b))$! ■

7.7 Cost Models

In this section, we shall study a simple cost model associated with a Brownian motion. Let $\{X(t), t \geq 0\}$ be a $\text{BM}(\mu, \sigma)$ starting from $X(0) = x$. Suppose $X(t)$ represents the state of a system at time t; for example, the stock price or the inventory level at time t. Suppose the system incurs costs (or earns rewards) at rate $C(s, X(s))$ at time s. Thus the total cost incurred over the time interval $[0, t]$ is given by

$$
G(t) = \int_0^t C(s, X(s)) ds. \tag{7.39}
$$

Since $X(s)$ is a continuous function of s with probability 1, $G(t)$ is a well-defined integral if we assume that $C(s, u)$ is a continuous function of s and u. Let

$$
g(t) = \mathsf{E}(G(t)) \tag{7.40}
$$

be the expected total cost over $[0, t]$. The next theorem gives the expression for $g(t)$.

Theorem 7.15. (Expected Total Cost). *Let* $\{X(t), t \geq 0\}$ *be a* BM(μ, σ) *starting from* $X(0) = x$, *and let* $g(t)$ *be as defined in Equation (7.40). We have*

$$g(t) = \int_0^t c(s)ds, \tag{7.41}$$

where

$$c(s) = \int_{y=-\infty}^\infty C(s, x + \mu s + \sigma sy)\phi(y)dy. \tag{7.42}$$

Proof. We know that

$$X(t) = x + \mu t + \sigma t Y(t),$$

where $Y(t) \sim N(0, 1)$. Now, the costs are incurred at rate $C(s, X(s)) = C(s, x + \mu s + \sigma sY(s))$ at time s. Hence the expected cost rate at time s is given by

$$c(s) = \mathsf{E}(C(s, x + \mu s + \sigma sY(s))) = \int_{y=-\infty}^\infty C(s, x + \mu s + \sigma sy)\phi(y)dy.$$

Hence the total expected cost over $[0, t]$ is given by

$$g(t) = \int_0^t c(s)ds,$$

which yields Equation (7.41). ∎

We illustrate the theorem above with several examples.

Example 7.26. (Inventory Cost Model). Consider the inventory model of Example 7.11. Suppose it costs \$$h$ to carry one unit of inventory for one unit of time and \$$p$ to carry one unit of backlog for one unit of time. What is the expected cost of carrying the inventory and backlog over the interval $[0, t]$?

The cost rate at time s is given by $hB(s)$ if $B(s) > 0$ and $-pB(s)$ if $B(s) < 0$. Hence the cost rate at time s is given by

$$C(s, B(s)) = h \max(B(s), 0) - p \min(B(s), 0).$$

Hence the expected cost rate at time s is given by

$$c(s) = h\mathsf{E}(B(s)|B(s) > 0)\mathsf{P}(B(s) > 0) - p\mathsf{E}(B(s)|B(s) < 0)\mathsf{P}(B(s) < 0).$$

Using the results of Example 7.8, we can simplify this to

$$c(s) = \frac{1}{2}(h + p)\sqrt{\frac{2s}{\pi}}.$$

The expected cost over $[0, t]$ is given by

$$g(t) = \int_0^t c(s)ds = \frac{h+p}{\sqrt{2\pi}} \int_0^t \sqrt{s}\,ds = (h+p)t\sqrt{\frac{2t}{9\pi}}. \quad \blacksquare$$

Example 7.27. (Stock Price). Consider the stock price process of Example 7.19. Suppose the stock yields dividends continuously at rate $rV(t)$ per share if the stock price is $V(t)$ at time t. Compute the expected total dividend received from one share over $[0, t]$.

The total dividend is given by

$$G(t) = r \int_0^t V(s)ds = r \int_0^t \exp(\sigma B(s))ds.$$

This fits into the cost model of Theorem 7.15 with cost function

$$C(s, B(s)) = r\exp(\sigma B(s)).$$

Hence the expected cost rate is given by

$$c(s) = r\mathsf{E}(e^{\sigma B(s)}) = r\exp(\sigma^2 s/2),$$

where we have used Equation (7.10) to evaluate the expected value. Hence the total expected dividend over $[0, t]$ is given by

$$g(t) = r \int_0^t \exp(\sigma^2 s/2)ds = \frac{2r}{\sigma^2}(\exp(\sigma^2 t/2) - 1).$$

In practice, the payout of dividends will affect the price of the stock. We have ignored this effect entirely in this example. $\quad \blacksquare$

Example 7.28. (Inventory Model). Consider the inventory model of Example 7.18 starting with initial inventory level .3975. What is the expected fraction of the interval $[0, 10]$ for which system faces stockout?

Consider the cost rate function

$$C(s, B(s)) = \begin{cases} 0 & \text{if } .3975 + B(s) \geq 0, \\ 1 & \text{if } .3975 + B(s) < 0. \end{cases}$$

Then the total cost incurred over $[0, 10]$ equals the length of the duration when the inventory is negative; that is, when the system is facing stockouts. The expected cost rate at time s is

$$c(s) = \mathsf{P}(.3975 + B(s) < 0) = \Phi(-.3975/\sqrt{s}).$$

Hence the expected fraction of the time the system faces stockouts is given by

$$g(t) = \frac{1}{10} \int_0^{10} \Phi(-.3975/\sqrt{s})\,ds.$$

This last integral has to be evaluated numerically. Using MATLAB's error function, we can do this to obtain as the expected fraction .40734. Compare this with the result in Example 7.18. The level .3975 was chosen to ensure that the probability of a stockout over $[0, 10]$ is less than .10. However, the calculation above shows that we will face stockouts roughly 40% of the time over $[0, 10]$. There is no inconsistency between these two results since once the system faces a stockout, the length of the stockout period can be large. It may make more sense to set the initial level to ensure the fraction of the stockout period is less than .10. This will involve intensive numerical calculations. ∎

7.8 Black–Scholes Formula

We have seen models of stock prices in Examples 7.19, 7.20, 7.22, and 7.27. In Example 7.20, we analyzed a possible financial deal offered by a dealer that depended on the maximum price reached during a given interval. Such deals are called options, and finding the right price at which to sell them is an important part of financial engineering. Here we study two simple options, called European put options and European call options. The method used to evaluate their fair price is the celebrated Black–Scholes formula. We study it in this section.

Let $V(t)$ be the price of a stock at time t. Assume that $V(t)$ is given by

$$V(t) = V(0)\exp(\theta t + \sigma B(t)), \quad t \geq 0. \tag{7.43}$$

Using Equation (7.10), we see that

$$\mathsf{E}(V(t)) = V(0)\exp\left\{\left(\theta + \frac{1}{2}\sigma^2\right)t\right\}.$$

Thus one dollar invested in this stock at time 0 is worth $\exp\{(\theta + \frac{1}{2}\sigma^2)t\}$ dollars on average at time t. An investor typically has the option of investing in a risk-free instrument, such as a savings account or a treasury bill. Suppose it gives a continuous rate of return of r. That is, one dollar invested in this risk-free instrument will become e^{rt} dollars at time t with certainty. It stands to reason that the value of the option must be made by assuming that these two rates of return are the same; otherwise else positive profits could be made with certainty by judiciously investing the proceeds of selling the option in the stock and cash. Such opportunities for risk-free positive profit, called arbitrage opportunities, do not exist in a competitive market. Hence, from now on, we assume that

$$\theta + \frac{1}{2}\sigma^2 = r.$$

Thus, option pricing is done by assuming that the model of the stock price is

$$V(t) = V(0) \exp\left\{ \left(r - \frac{1}{2}\sigma^2 \right) t + \sigma B(t) \right\}, \quad t \geq 0. \tag{7.44}$$

It is completely described by the risk-free return rate r and the volatility parameter σ.

Now suppose a dealer offers an investor the following deal, called the European call option, on this stock. It gives the investor the right (but not the obligation) to buy one share of this stock at time T in the future (called the maturity time or expiry date) at price K (called the strike price). Now, if the price $V(T)$ is greater than K, the investor will buy the stock at price K and immediately sell it at price $V(T)$ and make a net profit of $V(T) - K$. However, if $V(T) \leq K$, the investor will do nothing. Thus, the investor makes a profit of $\max(V(T) - K, 0)$ at time T. Clearly, the investor never loses on this deal. Hence the dealer asks that the investor pay $\$C$ to enter into this deal. What is the appropriate value of C? This is the famous option-pricing problem.

Clearly, the value of C should be such that C invested in a risk-free instrument should produce the expected return at time T as the expected profit from the call option. This is sometimes called the "no arbitrage" condition. Thus,

$$Ce^{rT} = \mathsf{E}(\max(V(T) - K, 0)). \tag{7.45}$$

The next theorem gives the explicit formula for C in terms of r and σ.

Theorem 7.16. (Black–Scholes Formula: European Call Option). *The value of the European call option with maturity date T and strike price K is given by*

$$C = V(0)\Phi(d_1) - Ke^{-rT}\Phi(d_2), \tag{7.46}$$

where

$$d_1 = \frac{\ln(V(0)/K) + (r + \sigma^2/2)T}{\sigma\sqrt{T}}, \tag{7.47}$$

$$d_2 = \frac{\ln(V(0)/K) + (r - \sigma^2/2)T}{\sigma\sqrt{T}}. \tag{7.48}$$

Proof. Define

$$V(T)^+ = \begin{cases} V(T) & \text{if } V(T) > K, \\ 0 & \text{if } V(T) \leq K, \end{cases}$$

and

$$K^+ = \begin{cases} K & \text{if } V(T) > K, \\ 0 & \text{if } V(T) \leq K. \end{cases}$$

Then

$$\max(V(T) - K, 0) = V(T)^+ - K^+.$$

Using this notation, Equation (7.45) can be written as

$$C = e^{-rT}[\mathsf{E}(V(T)^+) - \mathsf{E}(K^+)].\tag{7.49}$$

Next we compute the two expectations above. Using Equation (7.44), we get

$\mathsf{E}(V(T)^+)$

$$= V(0)\mathsf{E}\left(\exp\left(\left(r - \frac{1}{2}\sigma^2\right)T + \sigma B(T)\right)^+\right)$$

$$= V(0)\mathsf{E}\left(\exp\left(\left(r - \frac{1}{2}\sigma^2\right)T + \sigma B(T)\right) \cdot 1_{\{\ln V(0)+(r-\frac{1}{2}\sigma^2)T+\sigma B(T)>\ln K\}}\right)$$

$$= V(0)\exp\left(\left(r - \frac{1}{2}\sigma^2\right)T\right)\mathsf{E}\left(\exp(\sigma B(T))1_{\{B(T)/\sqrt{T}>c_1\}}\right),$$

where

$$c_1 = \frac{\ln(K/V(0)) - (r - \frac{1}{2}\sigma^2)T}{\sigma\sqrt{T}}.$$

Now $B(T)/\sqrt{T} = Z \sim N(0,1)$, and we can show by direct integration that

$$\mathsf{E}(\exp(aZ)1_{\{Z>b\}}) = e^{a^2/2}\Phi(a - b)\tag{7.50}$$

for positive a. Substituting in the previous equation, we get

$$\mathsf{E}(V(T)^+) = V(0)\exp\left(\left(r - \frac{1}{2}\sigma^2\right)T\right)\exp\left(\frac{1}{2}\sigma^2 T\right)\Phi(d_1) = V(0)e^{rT}\Phi(d_1),$$

where $d_1 = \sigma\sqrt{T} - c_1$ is as given in Equation (7.47). Next, we have

$$\mathsf{E}(K^+) = \mathsf{E}(K1_{V(T)>K}) = K\mathsf{P}(V(T) > K)$$

$$= K\mathsf{P}\left(\ln V(0) + \left(r - \frac{1}{2}\sigma^2\right)T + \sigma B(T) > \ln K\right)$$

$$= K\mathsf{P}(B(T)/\sqrt{T} > c_1) = K\mathsf{P}(B(T)/\sqrt{T} \le d_2),$$

where $d_2 = -c_1$ is as given in Equation (7.48). Substituting in Equation (7.49), we get Equation (7.46). ∎

The European put option works in a similar way. It gives the buyer the right (but not the obligation) to sell one share of the stock at time T at price K. The profit from this option at time T is $\max(K - V(T), 0)$. We leave it to the reader to prove the following theorem.

Theorem 7.17. (Black–Scholes Formula: European Put Option). *The value of the European put option with maturity date T and strike price K is given by*

$$C = Ke^{-rT}\Phi(-d_2) - V(0)\Phi(-d_1), \tag{7.51}$$

where d_1 and d_2 are as in Theorem 7.16.

It is relatively straightforward to implement these formulas in MATLAB or Excel. In fact, it is easy to find Black–Scholes calculators on the Internet that provide easy-to-use applets to compute option prices.

Example 7.29. (Call and Put Options). Suppose a stock is currently trading at $20.00 and its volatility parameter is .5 per year. Compute the Black–Scholes price of a European call option at strike price $15 and maturity date 3 months. Assume the risk-free return rate is $r = .05$ per year.

We are given $r = .05$, $T = .25$ years, $\sigma = .5$, $K = 15$ dollars, and $V(0) = 20$ dollars. Substituting in Equation (7.46), we get $C = \$5.4261$.

Compute the price of a European put option if the strike price is $25. All other parameters are the same as before.

Now we use $K = 25$ and Equation (7.51) to get $C = \$5.3057$. ∎

Example 7.30. (Implied Volatility). One can use the Black–Scholes formula in reverse: one can use the quoted prices of options in the market to compute the volatility parameter so that the Black–Scholes formula produces the same prices. This is called the implied volatility. Consider the same stock as in Example 7.29, but assume that the volatility parameter is unknown. Instead, we see that the call option with 3 month maturity and strike price $15 is quoted in the market at $6. What is the implied volatility?

We are given $r = .05$, $T = .25$ years, $K = 15$ dollars, $V(0) = 20$ dollars, and $C = 6$ dollars. Substituting in Equation (7.46) and solving for σ numerically, we get $\sigma = .769$ per year. This is the implied volatility. ∎

There are many topics of interest in Brownian motion that we have not touched upon here. This is because the mathematics needed to tackle them is beyond what is assumed of the readers of this book. We encourage the reader to read an advanced textbook for more information. One such reference is Kulkarni (2010).

7.9 Problems

CONCEPTUAL PROBLEMS

7.1. Find the mode of an $N(\mu, \sigma^2)$ distribution.

7.2. Find the median of an $N(\mu, \sigma^2)$ distribution.

7.3. Let Y be an $N(0,1)$ random variable. Show that

$$E(Y^n) = \begin{cases} 0 & \text{if } n \text{ is odd,} \\ n!/(2^n(n/2)!) & \text{if } n \text{ is even.} \end{cases}$$

Hint: Follow the proof of Theorem 7.2.

7.4. Let X be an $N(\mu, \sigma^2)$ random variable. Compute its nth central moment $E((X - \mu)^n)$. *Hint:* Use Equation (7.5) and Conceptual Problem 7.3.

7.5. Let X be an $N(0, \sigma^2)$ random variable. Show that the pdf of $|X|$ is given by

$$f(x) = \sqrt{\frac{2}{\pi\sigma^2}} \exp\left(\frac{-x^2}{2\sigma^2}\right), \quad x \geq 0.$$

7.6. Let X be an $N(0, \sigma^2)$ random variable. Show that

$$E(|X|) = \sigma\sqrt{2/\pi}.$$

7.7. Derive Equation (7.10).

7.8. Let $X = (X_1, X_2, \cdots, X_n)^\top$ be a multivariate normal random variable with parameters μ and Σ. Let $a = (a_1, a_2, \cdots, a_n) \in R^n$, and define $Y = aX$. Show that Y is an $N(\mu_Y, \sigma_Y^2)$ random variable, where

$$\mu_Y = a\mu, \quad \sigma_Y^2 = a\Sigma a^\top.$$

7.9. Using Conceptual Problem 7.8 or another method, show that the variance-covariance matrix of a multivariate normal random variable is positive semi-definite.

7.10. Derive the density of $T(a)$ given in Equation (7.22).

7.11. Derive Equation (7.23).

7.12. Let $X \sim N(0, t)$. Using Theorem 7.12 and Conceptual Problem 7.5, show that $U(t) \sim |X|$ and $L(t) \sim -|X|$.

7.13. Let $\{X(t), t \geq 0\}$ be a BM(μ, σ) with $\mu > 0$ and $X(0) = 0$. Let $b > 0$ be given, and define

$$T(b) = \min\{t \geq 0 : X(t) = b\}.$$

Compute $E(T(b))$.

7.14. Let $Y(t) = x + \sigma B(t)$ and let $T(a, b)$ be as in Equation (7.29). Find the $x \in [a, b]$ that maximizes $E(T(a, b))$.

7.15. Redo Conceptual Problem 7.14 for $Y(t) = x + \mu t + \sigma B(t)$.

7.16. Compute the expected total cost over $[0, t]$ for a BM(μ, σ) if the cost rate is $C(t, x) = xe^{-\alpha t}$ for a given constant $\alpha > 0$.

7.17. Prove Theorem 7.17.

7.18. Derive Equation (7.37), thus showing that the maximum of a BM(μ, σ) over $[0, \infty)$ is an Exp(θ) random variable if $\mu < 0$.

7.19. Derive Equation (7.38), thus showing that the negative of the minimum of a BM(μ, σ) over $[0, \infty)$ is an Exp$(-\theta)$ random variable if $\mu > 0$.

7.20. Let $\{X(t), t \geq 0\}$ be a BM(μ, σ). Show that

$$\mathsf{E}(X(t) - \mu t | X(u) : 0 \leq u \leq s) = X(s) - \mu s, \quad 0 \leq s \leq t.$$

7.21. Let $\{X(t), t \geq 0\}$ be a BM(μ, σ), and let $\theta = -2\mu/\sigma^2$. Show that

$$\mathsf{E}(e^{\theta X(t)} | X(u) : 0 \leq u \leq s) = e^{\theta X(s)}, \quad 0 \leq s \leq t.$$

COMPUTATIONAL PROBLEMS

7.1. The amount of water in a dam is normally distributed with mean 2 million cubic meters and variance 4 million cubic meters^2. The capacity of the dam is 3 million cubic meters. What is the probability that the dam overflows?

7.2. The amount of rainfall in a year is a normal random variable with mean 22 inches and variance 16 inches squared. Federal guidelines declare the region to be drought affected if the rainfall in a year is under 12 inches. What is the probability that this region will be declared drought affected in a given year?

7.3. Consider the rainfall of Computational Problem 7.2. Suppose the rainfall in two consecutive years forms a bivariate normal random variable with marginal distribution given there and correlation coefficient .6. What is the probability that the area will be declared drought affected next year if the rainfall this year was 18 inches?

7.4. Suppose (X_1, X_2) are iid normal $(0, 1)$. Define

$$\begin{bmatrix} Y_1 \\ Y_2 \end{bmatrix} = \begin{bmatrix} 3 & 2 \\ 1 & 5 \end{bmatrix} \begin{bmatrix} X_1 \\ X_2 \end{bmatrix} + \begin{bmatrix} 2 \\ 4 \end{bmatrix}.$$

Compute the joint distribution of $(Y_1, Y_2)^{\mathsf{T}}$.

7.5. Let $(Y_1, Y_2)^{\mathsf{T}}$ be as in Computational Problem 7.4. Compute the marginal distribution of Y_1 and Y_2.

7.6. Let $(Y_1, Y_2)^\top$ be as in Computational Problem 7.4. Compute the conditional distribution of Y_1 given $Y_2 = 3$.

7.7. Let $(Y_1, Y_2)^\top$ be as in Computational Problem 7.4. Compute the conditional mean of Y_2 given $Y_1 = 0$.

7.8. Let $(Y_1, Y_2)^\top$ be as in Computational Problem 7.4. Compute the distribution of $Y_1 + Y_2$.

7.9. Suppose $\{X(t), t \geq 0\}$ is a BM(2,4). Suppose $X(0) = 5$ and $X(4) = -5$. Compute the distribution of $X(2)$.

7.10. Suppose $\{X(t), t \geq 0\}$ is a BM(2,4). Suppose $X(0) = 5$ and $X(4) = -5$. For what value of s is the variance of $X(s)$ the maximum? What is the value of the maximum variance?

7.11. Compute the probability that a BM(−2,5) starting at 3 is above its initial level at time 4.

7.12. Compute the joint distribution of $X(3) - X(2)$ and $X(10) - X(6)$ if $\{X(t), t \geq 0\}$ is a BM(−1,1) with $X(0) = 3$.

7.13. Compute the expected fraction of the time an SBM spends above zero over $[0, t]$.

7.14. Suppose $\{X(t), t \geq 0\}$ is a BM(μ, σ). Compute the covariance of $X(2)$ and $X(3) - X(1)$.

7.15. Consider the stock price process of Example 7.19. What is the probability that the stock price is above 2 at time 1?

7.16. Consider the stock price process of Example 7.19. What are the mean and variance of the stock price at time 1?

7.17. Suppose $Y(t)$, the water level in a tank at time t, is modeled as

$$Y(t) = 5 + 2B(t), \quad t \geq 0.$$

What is the probability that the tank becomes empty by time 10?

7.18. Consider the water tank of Computational Problem 7.17. Suppose the tank overflows when the water level reaches 12. What is the probability that the tank overflows before it becomes empty?

7.19. Consider the water tank of Computational Problem 7.17. What is the expected time when the tank overflows or becomes empty?

7.20. Suppose the amount of funds (in dollars) in an account at time t can be modeled as $2000 + 1000 B(t)$. What is the probability that the account is not overdrawn by time 5?

7.21. Let $\{Y(t), t \geq 0\}$ be a BM(−3,4) starting from $Y(0) = 5$. Compute the probability that $\{Y(t), t \geq 0\}$ never goes above 10.

7.22. Let $\{Y(t), t \geq 0\}$ be a BM(1,2) starting from $Y(0) = -3$. Compute the probability that $\{Y(t), t \geq 0\}$ never goes below −5.

7.23. Let $\{Y(t), t \geq 0\}$ be a BM(2,4) starting from $Y(0) = 3$. Compute the probability that $\{Y(t), t \geq 0\}$ hits level 5 before it hits zero.

7.24. Consider the stock price process $\{V(t), t \geq 0\}$ defined as

$$V(t) = V(0)e^{\sigma B(t)}, \quad t \geq 0.$$

The initial value of the stock is \$30, and the volatility parameter is $\sigma = 2$. The investor plans to sell the stock when it reaches \$40 or falls to \$25. What is the probability that the investor ends up selling at a loss?

7.25. Consider the investor of Computational Problem 7.24. What is the expected net profit per share of following his strategy?

7.26. Consider the investor of Computational Problem 7.24. What is the expected time when he liquidates his holding in this stock?

7.27. Suppose the inventory level $Y(t)$ can be modeled as

$$Y(t) = 3 - 4t + 3B(t), \quad t \geq 0.$$

How large should the inventory storage area be so that the probability that it is ever full is less than .05?

7.28. Consider the inventory model of Computational Problem 7.27. Suppose the inventory storage area has capacity 10. What is the probability that a stockout occurs before the storage area overflows?

7.29. Consider the inventory model of Computational Problem 7.27. What is the expected time when the storage area overflows?

7.30. Suppose the inventory at time t is modeled as

$$X(t) = x + \mu t + \sigma B(t), \quad t \geq 0.$$

If the inventory level is y, it costs $5y^2$ per unit time in holding and stockout costs. Compute the total expected cost over $[0, 3]$ if $x = 3$, $\mu = -2$, and $\sigma = 3$.

7.31. Consider the inventory model of Computational Problem 7.30. Find the value of the initial inventory x that minimizes this cost assuming $\mu = -2$ and $\sigma = 3$.

7.32. Consider the inventory model of Computational Problem 7.30. Find the value of the drift parameter μ that minimizes this cost assuming $x = 3$ and $\sigma = 3$.

7.33. Let $V(t)$ be the price of a stock (in dollars) at time t (in years). Suppose it is given by

$$V(t) = 20e^{3B(t)}, \quad t \geq 0.$$

Suppose an investor buys this stock continuously at rate 200 per year, so that at t years he holds $200t$ shares of this stock. What is the expected value of his portfolio at the end of the first year?

7.34. Consider the investor of Computational Problem 7.33. Suppose the stock generates 2% annual dividends continuously. What is the expected total dividend received by this investor over the first year? (We ignore the effect of dividends on the stock price.)

7.35. Consider the stock price process of Computational Problem 7.33. Suppose this stock yields a 20% dividend every six months. Suppose an investor holds 100 shares of this stock for one year (thus getting a dividend at time .5 and 1). What is the expected value of the total dividend collected by the investor over the year?

7.36. Compute the Black–Scholes price of a European call option on a stock with a current value of $10, strike price of $8, and expiration date 6 months from now. Assume that the risk-free annual rate of return is 4% and the volatility of the stock is 25% per year.

7.37. For the call option of Computational Problem 7.36, plot the price of the option as the strike price varies from $5 to $12.

7.38. For the call option of Computational Problem 7.36, plot the price of the option as the maturity period varies over 3, 6, 9, and 12 months.

7.39. For the call option of Computational Problem 7.36, compute the implied volatility if the actual market quote is $4.00.

7.40. For the call option of Computational Problem 7.36, compute the implied volatility if the actual market quote is $2.50.

7.41. Compute the Black–Scholes price of a European put option on a stock with a current value of $10, strike price of $12, and expiration date 6 months from now. Assume that the risk-free annual rate of return is 4% and the volatility of the stock is 25% per year.

7.42. For the put option of Computational Problem 7.41, plot the price of the option as the strike price varies from $8 to $15.

7.43. For the put option of Computational Problem 7.41, plot the price of the option as the maturity period varies over 3, 6, 9, and 12 months.

7.44. For the put option of Computational Problem 7.41, compute the implied volatility if the actual market quote is $2.00.

7.45. For the put option of Computational Problem 7.41, compute the implied volatility if the actual market quote is $1.50.

Appendix A
Probability

A.1 Probability Model

A probability model is a mathematical description of a random phenomenon (sometimes called a random experiment) and has three basic components:

1. sample space;
2. set of events of interest; and
3. probability of these events.

Definition A.1. (Sample Space, Ω). A sample space is the set of all possible outcomes of a random phenomenon.

A sample space is generally denoted by the Greek letter Ω (omega). Elements of Ω, generally denoted by ω, are called the *outcomes* or *sample points*.

Definition A.2. (Event). An event is a subset of the sample space.

The set of events of interest is denoted by $\mathcal{F}$. It has the following properties:

1. $\Omega \in \mathcal{F}$,
2. $E \in \mathcal{F} \Rightarrow E^c \in \mathcal{F}$,
3. $E_n \in \mathcal{F}, n \geq 1 \Rightarrow \bigcup_{n=1}^{\infty} E_n \in \mathcal{F}$.

The event Ω is called the *universal event*. The null set $\emptyset$ is in $\mathcal{F}$ and is called the *null event*. All the events encountered in this book are assumed to belong to $\mathcal{F}$, so we won't mention it every time. Table A.1 gives the correspondence between the event terminology and the set-theoretic terminology.

Events $E_1, E_2, \ldots$ are said to be *exhaustive* if at least one of the events always takes place; that is,

$$\bigcup_{n=1}^{\infty} E_n = \Omega.$$

Table A.1 Correspondence between event and set-theoretic terminology.

Event Description	Set-Theoretic Notation
E_1 or E_2	$E_1 \cup E_2$
E_1 and E_2	$E_1 \cap E_2$ or $E_1 E_2$
Not E	E^c or $\bar{E}$
At least one of $E_1, E_2, \ldots$	$\displaystyle\bigcup_{n=1}^{\infty} E_n$
All of $E_1, E_2, \ldots$	$\displaystyle\bigcap_{n=1}^{\infty} E_n$

Similarly, events $E_1, E_2, \ldots$ are said to be *mutually exclusive* or *disjoint* if at most one of the events can take place; that is,

$$E_i \cap E_j = \emptyset \text{ if } i \neq j.$$

Thus, when $E_1, E_2, \ldots$ are mutually exclusive, and exhaustive, they define a *partition* of Ω; i.e., each sample point $\omega \in \Omega$ belongs to one and only one E_n.

Definition A.3. (Probability of Events, P). The probability of an event E, written $P(E)$, is a number representing the likelihood of occurrence of the event E.

The probabilities of events have to be consistent. This is assured if they satisfy the following *axioms of probability*:

1. $0 \leq P(E) \leq 1$.
2. $P(\Omega) = 1$.
3. If $E_1, E_2, \ldots \in \mathcal{F}$ are disjoint, then

$$P\left(\bigcup_{n=1}^{\infty} E_n\right) = \sum_{n=1}^{\infty} P(E_n).$$

Axiom 3 is called the *axiom of countable additivity*. Simple consequences of these axioms are

$$P(\emptyset) = 0, \tag{A.1}$$

$$P(E^c) = 1 - P(E). \tag{A.2}$$

Let E and F be two events not necessarily disjoint. Then

$$P(E \cup F) = P(E) + P(F) - P(EF). \tag{A.3}$$

Let $E_i, 1 \leq i \leq n$, be n events not necessarily disjoint. Then

$$P\left(\bigcup_{i=1}^{n} E_i\right) = \sum_{i=1}^{n} P(E_i) - \sum_{i<j} P(E_i E_j)$$

$$+ \sum_{i<j<k} P(E_i E_j E_k) \cdots + (-1)^{n+1} P\left(\bigcap_{i=1}^{n} E_i\right). \quad (A.4)$$

The formula above is called the Inclusion–Exclusion Principle.

A.2 Conditional Probability

Definition A.4. (Conditional Probability). The conditional probability of an event E given that an event F has occurred is denoted by $P(E|F)$ and is given by

$$P(E|F) = \frac{P(EF)}{P(F)}, \quad (A.5)$$

assuming $P(F) > 0.$ ∎

If $P(F) = 0$, then $P(E|F)$ is undefined. We can write (A.5) as

$$P(EF) = P(E|F)P(F). \quad (A.6)$$

This equation is valid even if $P(F) = 0$. An immediate extension of (A.6) is

$$P(E_1 E_2 \cdots E_n) = P(E_n|E_1 E_2 \cdots E_{n-1})P(E_{n-1}|E_1 \ldots E_{n-2}) \cdots P(E_2|E_1)P(E_1). \quad (A.7)$$

The conditional probability $P(\cdot|F)$ satisfies the axioms of probability; i.e.,

1. $0 \leq P(E|F) \leq 1$ for all events $E \in \mathcal{F}$.
2. $P(\Omega|F) = 1$.
3. If $E_1, E_2, \ldots$ are disjoint, then

$$P\left(\bigcup_{n=1}^{\infty} E_n \,\middle|\, F\right) = \sum_{n=1}^{\infty} P(E_n|F).$$

A.3 Law of Total Probability

Theorem A.1. (Law of Total Probability). *Let* $E_1, E_2, E_3, \ldots$ *be a set of mutually exclusive and exhaustive events. Then, for an event* E,

$$P(E) = \sum_{n=1}^{\infty} P(E|E_n)P(E_n).$$ (A.8)

As a special case, we have

$$P(E) = P(E|F)P(F) + P(E|F^c)P(F^c).$$ (A.9)

A.4 Bayes' Rule

Let $E_1, E_2, \ldots$ be mutually exclusive and exhaustive events and E be another event. Bayes' Rule gives $P(E_i|E)$ in terms of the $P(E|E_j)$ and $P(E_j)$.

Theorem A.2. (Bayes' Theorem). *Let* $E_1, E_2, \ldots$ *be a set of mutually exclusive and exhaustive events, and let* E *be another event. Then*

$$P(E_i|E) = \frac{P(E|E_i)P(E_i)}{\sum_{n=1}^{\infty} P(E|E_n)P(E_n)}.$$ (A.10)

A.5 Independence

Definition A.5. (Independent Events). Events E and F are said to be independent of each other if

$$P(EF) = P(E)P(F).$$ (A.11)

If events E and F are independent and $P(F) > 0$, then we have

$$P(E|F) = \frac{P(EF)}{P(F)} = \frac{P(E)P(F)}{P(F)} = P(E).$$

Next we define the independence of three or more events.

Definition A.6. (Mutual Independence). Events $E_1, E_2, \ldots, E_n$ are said to be mutually independent if for any subset $S \subseteq \{1, 2, \ldots, n\}$

$$P\left(\bigcap_{i \in S} E_i\right) = \prod_{i \in S} P(E_i).$$

Appendix B
Univariate Random Variables

Definition B.1. (Random Variable). A random variable X is a function $X : \Omega \to (-\infty, \infty)$.

Definition B.2. (Probability Function of X). Let E be a subset of the real line. Then

$$P(\{X \in E\}) = P(\{\omega \in \Omega : X(\omega) \in E\}).$$

If $E = (-\infty, x]$, we write $P(X \leq x)$ instead of $P(\{X \in E\})$. Similar notation is

$$P(\{X \in (a, b)\}) = P(a < X < b),$$
$$P(\{X \in (a, b]\}) = P(a < X \leq b),$$
$$P(\{X \in (x, \infty)\}) = P(X > x),$$
$$P(\{X \in \{x\}\}) = P(X = x).$$

Definition B.3. (Cumulative Distribution Function (cdf)). The function

$$F(x) = P(X \leq x), \quad x \in (-\infty, \infty),$$

is called the cumulative distribution function of the random variable X.

We have

$$P(a < X \leq b) = F(b) - F(a),$$
$$P(X > x) = 1 - F(x),$$
$$P(X < x) = \lim_{\epsilon \downarrow 0} F(x - \epsilon) = F(x^-),$$
$$P(X = x) = F(x) - F(x^-).$$

Thus the cumulative distribution function provides all the information about the random variable. The next theorem states the four main properties of the cdf.

Theorem B.1. (Properties of the cdf).

(1) $F(\cdot)$ *is a nondecreasing function; i.e.,*

$$x \leq y \Rightarrow F(x) \leq F(y).$$

(2) $F(\cdot)$ *is right continuous; i.e.,*

$$\lim_{\epsilon \downarrow 0} F(x + \epsilon) = F(x).$$

(3) $\lim_{x \to -\infty} F(x) = F(-\infty) = 0.$
(4) $\lim_{x \to \infty} F(x) = F(\infty) = 1.$

Any function satisfying the four properties of Theorem B.1 is a cdf of some random variable.

B.1 Discrete Random Variables

Definition B.4. (Discrete Random Variable). A random variable is said to be discrete with state space $S = \{x_0, x_1, x_2, \ldots\}$ if its cdf is a step function with jumps at points in S.

Definition B.5. (Probability Mass Function (pmf)). Let X be a discrete random variable taking values in $S = \{x_0, x_1, x_2, \ldots\}$, and let $F(\cdot)$ be its cdf. The function

$$p_k = P(X = x_k) = F(x_k) - F(x_k^-), \quad k \geq 0,$$

is called the probability mass function of X.

The main properties of the probability mass function are given below.

Theorem B.2. (Properties of the pmf).

$$p_k \geq 0, \quad k \geq 0, \tag{B.1}$$

$$\sum_{k=0}^{\infty} p_k = 1. \tag{B.2}$$

Many discrete random variables take values from the set of integers. In such cases, we define $x_k = k$ and $p_k = P(X = k)$. A list of common integer-valued random variables and their pmfs is given in Table B.1.

Table B.1 Common integer-valued random variables.

Name	Parameters	Notation	State Space	pmf $P(X = k)$
Bernoulli	$p \in [0, 1]$	$B(p)$	$k = 0, 1$	$p^k (1 - p)^{1-k}$
Binomial	$n \geq 0, p \in [0, 1]$	$\text{Bin}(n, p)$	$0 \leq k \leq n$	$\binom{n}{k} p^k (1 - p)^{n-k}$
Geometric	$p \in [0, 1]$	$G(p)$	$k \geq 1$	$(1 - p)^{k-1} p$
Negative Binomial	$r \geq 1, p \in [0, 1]$	$\text{NB}(r, p)$	$k \geq r$	$\binom{k-1}{r-1}(1 - p)^{k-r} p^r$
Poisson	$\lambda \in [0, \infty)$	$P(\lambda)$	$k \geq 0$	$e^{-\lambda} \frac{\lambda^k}{k!}$

B.2 Continuous Random Variables

Definition B.6. (Continuous Random Variable). A random variable with cdf $F(\cdot)$ is said to be continuous if there exists a function $f(\cdot)$ such that

$$F(x) = \int_{-\infty}^{x} f(u)du \qquad (B.3)$$

for all $x \in (-\infty, \infty)$.

In particular, if $F(\cdot)$ is a differentiable function, then it is a cdf of a continuous random variable and the function $f(\cdot)$ is given by

$$f(x) = \frac{d}{dx} F(x) = F'(x). \qquad (B.4)$$

The function $f(\cdot)$ completely determines $F(\cdot)$ by (B.3). Hence it provides an alternate way of describing a continuous random variable and is given the special name of probability density function.

Definition B.7. (Probability Density Function (pdf)). The function $f(\cdot)$ of (B.3) is called the probability density function of X.

Theorem B.3. *A function f is a pdf of a continuous random variable if and only if it satisfies*

$$f(x) \geq 0 \text{ for all } x \in (-\infty, \infty), \qquad (B.5)$$

$$\int_{-\infty}^{\infty} f(u)du = 1. \qquad (B.6)$$

A list of common integer-valued random variables and their pdfs is given in Table B.2.

Definition B.8. (Mode of a Discrete Distribution). Let X be a discrete random variable on $S = \{0, 1, 2, \ldots\}$ with pmf $\{p_k, k \in S\}$. An integer m is said to be the mode of the pmf (or of X) if

$$p_m \geq p_k \quad \text{for all } k \in S.$$

Table B.2 Common continuous random variables.

Name	Parameters	Notation	State Space	pdf $f(x)$
Uniform	$-\infty < a < b < \infty$	$U(a,b)$	$[a,b]$	$\frac{1}{b-a}$
Exponential	$\lambda \geq 0$	$\mathrm{Exp}(\lambda)$	$[0, \infty)$	$\lambda e^{-\lambda x}$
Hyperexponential	$\lambda = [\lambda_1, \ldots, \lambda_n]$ $p = [p_1, \ldots, p_n]$	$\mathrm{Hex}(\lambda, p)$	$[0, \infty)$	$\sum_1^n p_i \lambda_i e^{-\lambda_i x}$
Erlang	$k \geq 1, \lambda \geq 0$	$\mathrm{Erl}(k, \lambda)$	$[0, \infty)$	$\lambda e^{-\lambda x} \frac{(\lambda x)^{k-1}}{(k-1)!}$
Normal	$-\infty < \mu < \infty,$ $\sigma^2 > 0$	$N(\mu, \sigma^2)$	$(-\infty, \infty)$	$\frac{1}{\sqrt{2\pi\sigma^2}} \exp\left\{ -\frac{1}{2} \left(\frac{x-\mu}{\sigma} \right)^2 \right\}$

Definition B.9. (Mode of a Continuous Distribution). Let X be a continuous random variable with pdf f. A number m is said to be the mode of the pdf (or of X) if

$$f(m) \geq f(x) \quad \text{for all } x \in S.$$

Definition B.10. (Median of a Distribution). Let X be a random variable with cdf F. A number m is said to be the median of the cdf (or of X) if

$$F(m^-) \leq .5 \quad \text{and} \quad F(m) \geq .5.$$

Definition B.11. (Hazard Rate of a Distribution). Let X be a nonnegative continuous random variable with pdf f and cdf F. The hazard rate, or failure rate, of X is defined as

$$h(x) = \frac{f(x)}{1 - F(x)} \quad \text{for all } x \text{ such that } F(x) < 1.$$

B.3 Functions of Random Variables

Let X be a random variable and g be a function. Then $Y = g(X)$ is another random variable and has its own cdf and pmf (if Y is discrete) or pdf (if it is continuous). The cdf of Y is in general difficult to compute. Here are some examples.

Example B.1. (Common Functions). Some common examples of functions of random variables are

1. linear transformation: $Y = aX + b$, where a and b are constants;
2. squared transformation: $Y = X^2$;
3. power transformation: $Y = X^m$;
4. $Y = e^{-sX}$, where s is a constant. ∎

1. (Linear Transformation). $Y = aX + b$, where a and b are constants. Then we have

$$F_Y(y) = F_X\left(\frac{y-b}{a}\right). \tag{B.7}$$

If X is continuous with pdf f_X, then so is Y, and its pdf f_Y is given by

$$f_Y(y) = \frac{1}{a} f_X\left(\frac{y-b}{a}\right). \tag{B.8}$$

2. (Square Transformation). $Y = X^2$. The cdf F_Y of Y is given by

$$F_Y(y) = F_X(\sqrt{y}) - F_X(-\sqrt{y}). \tag{B.9}$$

If X is continuous with pdf f_X, then so is Y, and its pdf f_Y is given by

$$f_Y(y) = \frac{1}{2\sqrt{y}}\left(f_X(\sqrt{y}) + f_X(-\sqrt{y})\right). \tag{B.10}$$

3. (Exponential Transformation). $Y = \exp\{\theta X\}$, where $\theta \neq 0$ is a constant. The cdf F_Y of Y is given by

$$F_Y(y) = \begin{cases} 1 - F_X\left(\frac{\ln(y)}{\theta}\right) & \text{if } \theta < 0, \ y > 0, \\ F_X\left(\frac{\ln(y)}{\theta}\right) & \text{if } \theta > 0, \ y > 0. \end{cases} \tag{B.11}$$

If X is continuous with pdf f_X, then so is Y, and its pdf f_Y is given by

$$f_Y(y) = \frac{1}{|\theta|y} f_X\left(\frac{\ln(y)}{\theta}\right), \quad y > 0. \tag{B.12}$$

B.4 Expectations

Definition B.12. (Expected Value of a Random Variable). The expected value of X is defined as

$$E(X) = \begin{cases} \sum_i x_i p_X(x_i) & \text{if } X \text{ is discrete,} \\ \int_x x f_X(x)dx & \text{if } X \text{ is continuous.} \end{cases} \tag{B.13}$$

Theorem B.4. (Expected Value of a Function of a Random Variable).

$$E(g(X)) = \begin{cases} \sum_i g(x_i) p_X(x_i) & \text{if } X \text{ is discrete,} \\ \int_x g(x) f_X(x)dx & \text{if } X \text{ is continuous.} \end{cases} \tag{B.14}$$

The linearity property of the expectation is

$$E(aX + b) = aE(X) + b, \tag{B.15}$$

where a and b are constants. In general,

$$E(g_1(X) + g_2(X)) = E(g_1(X)) + E(g_2(X)). \tag{B.16}$$

Expectations of special functions of a random variable are known by special names, as listed below.

Definition B.13. (Special Expectations).

$$E(X^n) = n\text{th moment of } X,$$
$$E((X - E(X))^n) = n\text{th central moment of } X,$$
$$E((X - E(X))^2) = \text{variance of } X. \tag{B.17}$$

From the definition of variance in (B.17), we can show that

$$\text{Var}(X) = E(X^2) - (E(X))^2 \tag{B.18}$$

and

$$\text{Var}(aX + b) = a^2\text{Var}(X). \tag{B.19}$$

Tables B.3 and B.4 give the means and variances of the common discrete and continuous random variables, respectively.

Table B.3 Means and variances of discrete random variables.

Random Variable	$E(X)$	$\text{Var}(X)$
$B(p)$	p	$p(1 - p)$
$\text{Bin}(n, p)$	np	$np(1 - p)$
$G(p)$	$1/p$	$(1 - p)/p^2$
$\text{NB}(r, p)$	r/p	$[r(1 - p)]/p^2$
$P(\lambda)$	λ	λ

Table B.4 Means and variances of continuous random variables.

Random Variable	$E(X)$	$\text{Var}(X)$
$U(a, b)$	$(a + b)/2$	$(b - a)^2/12$
$\text{Exp}(\lambda)$	$1/\lambda$	$1/\lambda^2$
$\text{Hex}(\lambda, p)$	$\sum_1^n p_i/\lambda_i$	$\sum_1^n 2p_i/\lambda_i^2 - \left(\sum_1^n p_i/\lambda_i\right)^2$
$\text{Erl}(k, \lambda)$	k/λ	k/λ^2
$N(\mu, \sigma^2)$	μ	σ^2

Appendix C
Multivariate Random Variables

Definition C.1. (Multivariate Random Variable). A mapping $\Omega : X \to S \subseteq R^n$ is called an n-dimensional multivariate random variable. It is denoted by a vector $X = (X_1, X_2, \ldots, X_n)$, and S is called its state space. $X_1, X_2, \ldots, X_n$ are called jointly distributed random variables.

When $n = 2$, $X = (X_1, X_2)$ is called a *bivariate random variable*.

Definition C.2. (Multivariate cdf). The function

$$F_X(x) = P(X_1 \le x_1, X_2 \le x_2, \ldots, X_n \le x_n), \quad x \in R^n,$$

is called the multivariate cdf or joint cdf of $(X_1, X_2, \ldots, X_n)$.

C.1 Multivariate Discrete Random Variables

Definition C.3. (Multivariate Discrete Random Variable). A multivariate random variable $X = (X_1, X_2, \ldots, X_n)$ is said to be discrete if each X_i is a discrete random variable, $1 \le i \le n$.

Definition C.4. (Multivariate pmf). Let $X = (X_1, X_2, \ldots, X_n)$ be a multivariate random variable with state space S. The function

$$p(x) = P(X_1 = x_1, X_2 = x_2, \ldots, X_n = x_n), \quad x \in S,$$

is called the multivariate pmf or joint pmf of X.

A common discrete multivariate random variable is the multinomial random variable. It has parameters n and p, where n is a nonnegative integer and $p = [p_1, \ldots, p_r]$ is such that $p_i > 0$ and $p_1 + p_2 + \cdots + p_r = 1$. The state space of $X = (X_1, X_2, \ldots, X_r)$ is

$$S = \{k = (k_1, k_2, \ldots, k_r) : k_i \ge 0 \text{ are integers and } k_1 + k_2 + \cdots + k_r = n\}$$

and the multivariate pmf is

$$p(k) = \frac{n!}{k_1! k_2! \cdots k_r!} p_1^{k_1} p_2^{k_2} \cdots p_r^{k_r}, \quad k \in S. \tag{C.1}$$

C.2 Multivariate Continuous Random Variables

Definition C.5. (Multivariate Continuous Random Variable). A multivariate random variable $X = (X_1, X_2, \ldots, X_n)$ with multivariate cdf $F(\cdot)$ is said to be continuous if there is a function $f(\cdot)$ such that

$$F_X(x) = \int_{-\infty}^{x_n} \int_{-\infty}^{x_{n-1}} \cdots \int_{-\infty}^{x_1} f(u_1, u_2, \ldots, u_n) du_1 du_2 \cdots du_n, \quad x \in R^n. \tag{C.2}$$

X is also called jointly continuous, and f is called the multivariate pdf or joint pdf of X.

A common multivariate continuous random variable is the multivariate normal random variable. Let μ be an n-dimensional vector and $\Sigma = [\sigma_{ij}]$ be an $n \times n$ positive definite matrix. $X = (X_1, X_2, \cdots, X_n)$ is called a multi-variate normal variable $N(\mu, \Sigma)$ if it has the joint pdf given by

$$f_X(x) = \frac{1}{\sqrt{(2\pi)^n \det(\Sigma)}} \exp\left(-\frac{1}{2}(x - \mu)\Sigma^{-1}(x - \mu)^\top\right), \quad x \in R^n.$$

C.3 Marginal Distributions

Definition C.6. (Marginal cdf). Let $X = (X_1, X_2, \ldots, X_n)$ be a multivariate random variable with joint cdf $F(\cdot)$. The function

$$F_{X_i}(x_i) = P(X_i \leq x_i) \tag{C.3}$$

is called the marginal cdf of X_i.

The marginal cdf can be computed from the joint cdf as follows:

$$F_{X_i}(x_i) = F(\infty, \ldots, \infty, x_i, \infty, \ldots, \infty). \tag{C.4}$$

Definition C.7. (Marginal pmf). Suppose $X = (X_1, X_2, \ldots, X_n)$ is a discrete multivariate random variable. The function

$$p_{X_i}(x_i) = P(X_i = x_i) \tag{C.5}$$

is called the marginal pmf of X_i.

The marginal pmf can be computed from the joint pmf as follows:

$$p_{X_i}(x_i) = \sum_{x_1} \cdots \sum_{x_{i-1}} \sum_{x_{i+1}} \cdots \sum_{x_n} p(x_1, \ldots, x_{i-1}, x_i, x_{i+1}, \ldots, x_n). \quad \text{(C.6)}$$

Definition C.8. (Marginal pdf). Suppose $X = (X_1, X_2, \cdots, X_n)$ is a multivariate continuous random variable with joint pdf $f(\cdot)$. The function $f_{X_i}(\cdot)$ is called the marginal pdf of X_i if

$$F_{X_i}(x_i) = \int_{-\infty}^{x_i} f_{X_i}(u)du, \quad -\infty < x_i < \infty. \quad \text{(C.7)}$$

The marginal pdf can be computed from the joint pdf as follows:

$$f_{X_i}(x_i) = \int_{x_1} \cdots \int_{x_{i-1}} \int_{x_{i+1}} \cdots \int_{x_n} f(x_1, \cdots, x_{i-1}, x_i, x_{i+1}, \cdots, x_n)$$
$$\times dx_1 \ldots dx_{i-1} dx_{i+1} \cdots dx_n. \quad \text{(C.8)}$$

Definition C.9. (Identically Distributed Random Variables). The jointly distributed random variables $(X_1, X_2, \ldots, X_n)$ are said to be identically distributed if their marginal cdfs are identical;

$$F_{X_1}(\cdot) = F_{X_2}(\cdot) = \cdots = F_{X_n}(\cdot),$$

(in the discrete case) their marginal pmfs are identical,

$$p_{X_1}(\cdot) = p_{X_2}(\cdot) = \cdots = p_{X_n}(\cdot),$$

or (in the continuous case) their marginal pdfs are identical,

$$f_{X_1}(\cdot) = f_{X_2}(\cdot) = \cdots = f_{X_n}(\cdot).$$

C.4 Independence

Definition C.10. (Independent Random Variables). The jointly distributed random variables $(X_1, X_2, \ldots, X_n)$ are said to be independent if

$$F_X(x) = F_{X_1}(x_1)F_{X_2}(x_2) \ldots F_{X_n}(x_n), \quad x \in R^n,$$

(in the discrete case)

$$p_X(x) = p_{X_1}(x_1)p_{X_2}(x_2) \ldots p_{X_n}(x_n), \quad x \in R^n,$$

or (in the continuous case)

$$f_X(x) = f_{X_1}(x_1) f_{X_2}(x_2) \dots f_{X_n}(x_n), \quad x \in R^n.$$

The jointly distributed random variables $X_1, X_2, \dots, X_n$ are called independent and identically distributed random variables, or *iid random variables* for short, if they are independent and have identical marginal distributions.

C.5 Sums of Random Variables

Let (X_1, X_2) be a bivariate random variable, and define $Z = X_1 + X_2$. If (X_1, X_2) is discrete with joint pmf $p(x_1, x_2)$, we have

$$p_Z(z) = \sum_{x_1} p(x_1; z - x_1) = \sum_{x_2} p(z - x_2; x_2). \tag{C.9}$$

If (X_1, X_2) is jointly continuous with joint pdf $f(x_1, x_2)$, we have

$$f_Z(z) = \int_{-\infty}^{\infty} f(z - x_2, x_2) dx_2 = \int_{-\infty}^{\infty} f(x_1, z - x_1) dx_1. \tag{C.10}$$

When (X_1, X_2) are discrete *independent* random variables, (C.9) reduces to

$$p_Z(z) = \sum_{x_1} p_{X_1}(x_1) p_{X_2}(z - x_1) = \sum_{x_2} p_{X_1}(z - x_2) p_{X_2}(x_2). \tag{C.11}$$

The pmf p_Z is called a discrete *convolution* of p_{X_1} and p_{X_2}. When (X_1, X_2) are continuous *independent* random variables, Equation (C.10) reduces to

$$f_Z(z) = \int_{-\infty}^{\infty} f_{X_1}(x_1) f_{X_2}(z - x_1) dx_1 = \int_{-\infty}^{\infty} f_{X_1}(z - x_2) f_{X_2}(x_2) dx_2. \tag{C.12}$$

The pdf f_Z is called a *convolution* of f_{X_1} and f_{X_2}.
  A few useful results about the sums of random variables are given below.

1. Let X_i be a Bin(n_i, p) random variable, $i = 1, 2$. Suppose X_1 and X_2 are independent. Then $X_1 + X_2$ is a Bin($n_1 + n_2, p$) random variable.
2. Let $X_i, 1 \le i \le n$, be iid B(p) random variables. Then $X_1 + X_2 + \dots + X_n$ is a Bin(n, p) random variable.
3. Let X_i be a P(λ_i) random variable, $i = 1, 2$. Suppose X_1 and X_2 are independent. Then $X_1 + X_2$ is a P($\lambda_1 + \lambda_2$) random variable.

4. Suppose $X_1, X_2, \cdots, X_n$ are iid $\text{Exp}(\lambda)$ random variables. Then $X_1 + X_2 + \cdots X_n$ is an $\text{Erl}(n, \lambda)$ random variable.
5. Suppose $X_1 \sim \text{N}(\mu_1, \sigma_1^2)$ and $X_2 \sim \text{N}(\mu_2, \sigma_2^2)$ are independent. Then $X_1 + X_2$ is an $\text{N}(\mu_1 + \mu_2, \sigma_1^2 + \sigma_2^2)$ random variable.

C.6 Expectations

Theorem C.1. (Expectation of a Function of a Multivariate Random Variable). *Let* $X = (X_1, X_2, \ldots, X_n)$ *be a multivariate random variable. Let* $g : R^n \to R$. *Then*

$$E(g(X)) = \begin{cases} \sum_x g(x) p_X(x) & \text{if } X \text{ is discrete,} \\ \int_x g(x) f_X(x) \, dx & \text{if } X \text{ is continuous.} \end{cases} \tag{C.13}$$

Theorem C.2. (Expectation of a Sum). *Let* $X = (X_1, X_2, \ldots, X_n)$ *be a multivariate random variable. Then*

$$E(X_1 + X_2 + \cdots + X_n) = E(X_1) + E(X_2) + \cdots + E(X_n). \tag{C.14}$$

Note that the theorem above holds even if the random variables are not independent!

Theorem C.3. (Expectation of a Product). *Let* $(X_1, X_2, \ldots, X_n)$ *be independent random variables. Then*

$$E(g_1(X_1)g_2(X_2)\cdots g_n(X_n)) = E(g_1(X_1))E(g_2(X_2))\cdots E(g_n(X_n)). \tag{C.15}$$

As a special case of the theorem above, we have

$$E(X_1 X_2 \cdots X_n) = E(X_1)E(X_2)\cdots E(X_n)$$

if $X_1, X_2, \cdots, X_n$ are independent.

Theorem C.4. (Variance of a Sum of Independent Random Variables). *Let* $(X_1, X_2, \ldots, X_n)$ *be independent random variables. Then*

$$\text{Var}\left(\sum_{i=1}^{n} X_i\right) = \sum_{i=1}^{n} \text{Var}(X_i). \tag{C.16}$$

Definition C.11. (Covariance). Let (X_1, X_2) be a bivariate random variable. Its covariance is defined as

$$\text{Cov}(X_1, X_2) = E(X_1 X_2) - E(X_1)E(X_2).$$

If the covariance of two random variables is zero, they are called uncorrelated. Independent random variables are uncorrelated, but uncorrelated random variables need not be independent. However, uncorrelated multivariate normal random variables are independent.

Theorem C.5. (Variance of a Sum of Dependent Random Variables). *Let $(X_1, X_2, \ldots, X_n)$ be a multivariate random variable. Then*

$$\text{Var}\left(\sum_{i=1}^{n} X_i\right) = \sum_{i=1}^{n} \text{Var}(X_i) + 2 \sum_{i=1}^{n} \sum_{j=i+1}^{n} \text{Cov}(X_i X_j). \qquad (C.17)$$

Appendix D
Conditional Distributions and Expectations

D.1 Conditional Distributions

Definition D.1. (Conditional pmf). Let (X_1, X_2) be a bivariate discrete random variable with joint pmf $p(x_1, x_2)$ and marginal pmfs $p_{X_1}(x_1)$ and $p_{X_2}(x_2)$. Suppose $p_{X_2}(x_2) > 0$. The conditional probability

$$P(X_1 = x_1 | X_2 = x_2) = \frac{p(x_1, x_2)}{p_{X_2}(x_2)}$$

is called the conditional pmf of X_1 given $X_2 = x_2$ and is denoted by $p_{X_1|X_2}(x_1|x_2)$.

Definition D.2. (Conditional pdf). Let (X_1, X_2) be a jointly distributed continuous random variable with joint pdf $f(x_1, x_2)$ and marginal pdfs $f_{X_1}(x_1)$ and $f_{X_2}(x_2)$. The conditional pdf of X_1 given $X_2 = x_2$ is denoted by $f_{X_1|X_2}(x_1|x_2)$ and is defined as

$$f_{X_1|X_2}(x_1|x_2) = \frac{f(x_1, x_2)}{f_{X_2}(x_2)}, \tag{D.1}$$

assuming $f_{X_2}(x_2) > 0$.

We can compute the probability of an event E by conditioning on a discrete random variable X by using

$$P(E) = \sum_i P(E|X = x_i)P(X = x_i) \tag{D.2}$$

or a continuous random variable X by using

$$P(E) = \int_{-\infty}^{\infty} P(E|X = x) f_X(x)\, dx. \tag{D.3}$$

D.2 Conditional Expectations

Definition D.3. (Conditional Expectation). The conditional expectation of X given $Y = y$ is given by

$$E(X|Y = y) = \sum_x x p_{X|Y}(x|y) \tag{D.4}$$

in the discrete case and

$$E(X|Y = y) = \int_{-\infty}^{\infty} x f_{X|Y}(x|y)\, dx \tag{D.5}$$

in the continuous case.

Theorem D.1. (Expectation via Conditioning). *Let (X, Y) be a bivariate random variable. Then*

$$E(X) = \sum_y E(X|Y = y) p_Y(y) \tag{D.6}$$

if Y is discrete and

$$E(X) = \int_{-\infty}^{\infty} E(X|Y = y) f_Y(y)\, dy \tag{D.7}$$

if Y is continuous.

One can treat $E(X|Y)$ as a random variable that takes value $E(X|Y = y)$ with probability $P(Y = y)$ if Y is discrete or with density $f_Y(y)$ if Y is continuous. This implies

$$E(E(X|Y)) = E(X). \tag{D.8}$$

D.3 Random Sums

Let $\{X_n, n = 1, 2, 3, \ldots\}$ be a sequence of iid random variables with common expectation $E(X)$ and variance $\mathrm{Var}(X)$, and let N be a nonnegative integer-valued random variable that is independent of $\{X_n, n = 1, 2, 3, \ldots\}$. Let

$$Z = \sum_{n=1}^{N} X_n.$$

Then one can show that

$$E(Z) = E(X)E(N), \tag{D.9}$$

$$\mathrm{Var}(Z) = E(N)\mathrm{Var}(X) + (E(X))^2 \mathrm{Var}(N). \tag{D.10}$$

Answers to Selected Problems

Chapter 2

CONCEPTUAL PROBLEMS

2.1. Proof by induction. The statement is true for $k = 1$ by the definition of transition probabilities. Suppose it is true for k. We have

$$P(X_{k+1} = i_{k+1}, \ldots, X_1 = i_1 | X_0 = i_0)$$

$$= P(X_{k+1} = i_{k+1} | X_k = i_k, \ldots, X_1 = i_1, X_0 = i_0)$$
$$\times P(X_k = i_k, \ldots, X_1 = i_1 | X_0 = i_0)$$
$$= P(X_{k+1} = i_{k+1} | X_k = i_k)$$
$$\times P(X_k = i_k, \ldots, X_1 = i_1 | X_0 = i_0)$$
$$= p_{i_k, i_{k+1}} p_{i_{k-1}, i_k} p_{i_{k-2}, i_{k-1}} \cdots p_{i_0, i_1}.$$

Here the second equality follows from the Markov property and the third one from the induction hypothesis.

2.3. Let $\bar{p} = 1 - p$ and $\bar{q} = 1 - q$. Then

$$P = \begin{bmatrix} q^3 & 3q^2\bar{q} & 3q\bar{q}^2 & \bar{q}^3 \\ \bar{p}q^2 & pq^2 + 2\bar{p}q\bar{q} & 2pq\bar{q} + \bar{p}\bar{q}^2 & \bar{q}^2 p \\ q\bar{p}^2 & 2pq\bar{p} + \bar{q}\bar{p}^2 & qp^2 + 2\bar{q}p\bar{p} & \bar{q}p^2 \\ \bar{p}^3 & 3\bar{p}^2 p & 3\bar{p}p^2 & p^3 \end{bmatrix}.$$

2.5. (1) $a_j^n = P(X_n = j) = \sum_{i=1}^{N} P(X_n = j | X_0 = i) P(X_0 = i) = \sum_{j=1}^{N} a_i [P^n]_{i,j}$. Thus, $a^n = a * P^n$.

(2) The proof is by induction on m. The statement holds for $m = 1$ due to the definition of a DTMC. Suppose it holds for $m \geq 1$. Then

$$P(X_{n+m+1} = j \mid X_n = i, X_{n-1}, \ldots, X_0)$$

$$= \sum_{k=1}^{N} P(X_{n+m+1} = j \mid X_{n+m} = k, X_n = i, \ldots, X_0)$$

$$\times P(X_{n+m} = k \mid X_n = i, \ldots, X_0)$$

$$= \sum_{k=1}^{N} p_{kj} p_{ik}^{(m)} = p_{ij}^{(m+1)}.$$

Thus the statement follows from induction.

2.7. $P(T_i = k) = P(X_0 = i, X_1 = i, \ldots, X_{k-1} = i, X_k \neq i)$
$= P(X_k \neq i \mid X_{k-1} = i) P(X_{k-1} = i \mid X_{k-2} = i) \cdots P(X_1 = i \mid X_0 = i) P(X_0 = i)$
$= p_{i,i}^{k-1} (1 - p_{i,i}).$

2.9. $P = \begin{bmatrix} 1 - p & p \\ 1 - p & p \end{bmatrix}.$

2.11. Let

$$X_n = \begin{cases} 1 \text{ if the machine is idle at the beginning of the } n\text{th minute} \\ \quad \text{and there are no items in the bin,} \\ 2 \text{ if the machine has just started production at the beginning} \\ \quad \text{of the } n\text{th minute and there are no items in the bin,} \\ 3 \text{ if the machine has been busy for 1 minute at the beginning} \\ \quad \text{of the } n\text{th minute and there are no items in the bin,} \\ 4 \text{ if the machine has been busy for 1 minute at the beginning} \\ \quad \text{of the } n\text{th minute and there is one item in the bin.} \end{cases}$$

Then $\{X_n, n \geq 0\}$ is a DTMC on $S = \{1, 2, 3, 4\}$ with the transition probability matrix

$$P = \begin{bmatrix} 1-p & p & 0 & 0 \\ 0 & 0 & 1-p & p \\ 1-p & p & 0 & 0 \\ 0 & 1 & 0 & 0 \end{bmatrix}.$$

2.13. Let

$$X_n = \begin{cases} 1 \text{ if day } n \text{ is sunny,} \\ 2 \text{ if day } n \text{ is cloudy and day } n - 1 \text{ is not,} \\ 3 \text{ if day } n \text{ is rainy,} \\ 4 \text{ if days } n \text{ and } n - 1 \text{ are both cloudy.} \end{cases}$$

Then $\{X_n, n \geq 0\}$ is a DTMC on state space $S = \{1, 2, 3, 4\}$ with the transition probability matrix

$$P = \begin{bmatrix} .5 & .3 & .2 & 0 \\ .5 & 0 & .3 & .2 \\ .4 & .5 & .1 & 0 \\ 0 & 0 & .2 & .8 \end{bmatrix}.$$

2.15. Let $A = \{X_n \text{ visits } 1 \text{ before } N\}$. Then $u_1 = 1, u_N = 0$, and for $2 \leq i \leq N-1$,

$$u_i = P(A|X_0 = i) = \sum_{j=1}^{N} P(A|X_1 = j, X_0 = i) P(X_1 = j|X_0 = i)$$

$$= \sum_{j=1}^{N} P(A|X_0 = j) P(X_1 = j|X_0 = i) = \sum_{j=1}^{N} u_j p_{i,j}.$$

2.17. $p_{i,i+1} = \mu/(\lambda + \mu) = 1 - p_{i,i-1}$ for $-4 \leq i \leq 4$ and $p_{5,5} = p_{-5,-5} = 1$.

COMPUTATIONAL PROBLEMS

2.1. 6.3613.   **2.3.** \$4.

2.5. 9.6723, 9.4307, 8.8991, 8.2678. **2.7.** [47.4028, .0114, . 679.92]

2.9. .0358.   **2.11.** 1.11 days.   **2.13.** .3528.

2.15. (a)

$$M(10) = \begin{bmatrix} 2.6439 & 2.6224 & 2.8520 & 2.8817 \\ 1.6507 & 3.7317 & 2.9361 & 2.6815 \\ 1.8069 & 2.7105 & 3.6890 & 2.7936 \\ 1.6947 & 2.6745 & 2.8920 & 3.7388 \end{bmatrix}.$$

(c)

$$M(10) = \begin{bmatrix} 3.3750 & 0 & 7.6250 & 0 \\ 0 & 3.6704 & 0 & 7.3296 \\ 2.5417 & 0 & 8.4583 & 0 \\ 0 & 2.6177 & 0 & 8.3823 \end{bmatrix}.$$

2.17. .0336 hours.   **2.19.** (a) irreducible, (c) reducible.

2.21. $\pi = \pi^* = \hat{\pi} = [.1703. , 2297, .2687, .3313]$.

2.23. π does not exist. $\pi^* = \hat{\pi} = [.50. , 10. , 15, .25]$.

2.25. (1) 0.0979. (2) 5.3686.   **2.27.** 0.1428.

2.29. .1977.   **2.31.** 0.7232.   **2.33.** 7.3152.

2.35. Both produce .95 items per minute.

2.37. $454.91 per day.   **2.39.** .0003 per day.   **2.41.** $11.02 per week.

2.43. (b) 3. (d) 3.   **2.45.** 60.7056.   **2.47.** 241.2858.

2.49. It saves $78,400 per year.

2.51. No. Annual savings under this option = $196,376.38 per year.

Chapter 3

CONCEPTUAL PROBLEMS

3.1 The mode is the largest integer less than or equal to $(n + 1)p$. If $(n + 1)p$ is an integer, there are two modes, $(n + 1)p - 1$ and $(n + 1)p$.

3.3 $\frac{k-1}{\lambda}$.   **3.5** λ.   **3.7** λ^2.

3.15 $\sum_{j=k}^{n} \binom{n}{j} p^j (1 - p)^{n-j}$.   **3.19** $e^{\lambda(s+t)} \frac{(\lambda s)^i}{i!} \frac{(\lambda t)^{j-i}}{(j-i)!}$.

3.21 $P\left(\frac{\lambda}{p}(1 - (1 - p)^n)\right)$, $n \geq 1$. **3.25** $\lambda s E(C_1^2)$.

COMPUTATIONAL PROBLEMS

3.1 Exp(1/10).   **3.3** 3.333.   **3.5** 4/9.

3.7 .3050.   **3.9** 1.667.   **3.11** .3857.

3.13 2.7534.   **3.15** .25.   **3.17** .7788.

3.19 56, 56.   **3.21** PP(.053).

3.23 .5507.   **3.25** .0194.   **3.27** 4080, 7440.

3.29 1612.5, 4957.5.

Chapter 4

CONCEPTUAL PROBLEMS

4.1. $S = \{0, 1, 2, \ldots, K\}$, $r_{i,i-1} = \lambda L, 1 \leq i \leq K$ all other $r_{i,j} = 0$.

4.3. $S = \{0, 1, 2, 3, 4, 5\}$, $r_{i,i-1} = ic\mu, 2 \leq i \leq 5$, $r_{i,0} = i(1 - c)\mu, 2 \leq i \leq 5, r_{1,0} = \mu$, all other $r_{i,j} = 0$.

4.5. $\mu_i = \mu$, $1 \le i \le K$, $\lambda_i = \lambda_1 + \lambda_2$ for $0 \le i \le M$, λ_1 for $M + 1 \le i < K$, and 0 for $i = K$.

4.7. Let $A(t) = (b, l)$ be the number of operational borers and lathes at time t. Then $\{A(t), t \ge 0\}$ is a CTMC with state space

$$S = \{(0,0), (0,1), (0,2), (1,0), (2,0), (1,1), (1,2), (2,1), (2,2)\}$$

and rate matrix

$$R = \begin{bmatrix}
0 & \lambda_l & 0 & \lambda_b & 0 & 0 & 0 & 0 & 0 \\
\mu_l & 0 & \lambda_l & 0 & 0 & \lambda_b & 0 & 0 & 0 \\
0 & 2\mu_l & 0 & 0 & 0 & 0 & \lambda_b & 0 & 0 \\
\mu_b & 0 & 0 & 0 & \lambda_b & \lambda_l & 0 & 0 & 0 \\
0 & 0 & 0 & 2\mu_b & 0 & 0 & 0 & 2\lambda_l & 0 \\
0 & \mu_b & 0 & \mu_l & 0 & 0 & \lambda_l & \lambda_b & 0 \\
0 & 0 & \mu_b & 0 & 0 & 2\mu_l & 0 & 0 & \lambda_b \\
0 & 0 & 0 & 0 & \mu_l & 2\mu_b & 0 & 0 & \lambda_l \\
0 & 0 & 0 & 0 & 0 & 0 & 2\mu_b & 2\mu_l & 0
\end{bmatrix}.$$

4.9. Let $X(t)$ be the number of customers in the system at time t if the system is up, and let $X(t) = d$ if the system is down at time t. $\{X(t), t \ge 0\}$ is a CTMC with transition rates $r_{i,d} = \theta$ for $0 \le i \le K$, $r_{d,0} = \alpha$, $r_{i,i-1} = \mu$ for $1 \le i \le K$, and $r_{i,i+1} = \lambda$ for $0 \le i \le K - 1$.

4.11. $S = \{0, 1, 2, 3, 4\}$, where the state is 0 if both components are down and one component is being repaired, the state is 1 if component 1 is up and component 2 is down, the state is 2 if component 2 is up and component 1 is down, the state is 3 if both components are up, and the state is 4 if the system is down, one component has completed repairs, and the other component is being repaired. The nonzero transition rates are $r_{0,4} = \alpha$, $r_{1,0} = \lambda p$, $r_{2,0} = \lambda q$, $r_{3,0} = \lambda r$, $r_{3,1} = \lambda q$, $r_{3,2} = \lambda p$, $r_{4,3} = \alpha$.

COMPUTATIONAL PROBLEMS

4.1. $P(0.20) = \begin{bmatrix}
.3747 & .2067 & .1964 & .1319 & .0903 \\
.2615 & .2269 & .2045 & .1841 & .1230 \\
.2424 & .1964 & .2153 & .1858 & .1601 \\
.1804 & .1947 & .2088 & .2327 & .1835 \\
.1450 & .1544 & .2103 & .2191 & .2711
\end{bmatrix}.$

4.3. $P(0.10) = \begin{bmatrix}
.1353 & .1876 & .3765 & .3006 & 0 & 0 \\
0 & .3229 & .3765 & .3006 & 0 & 0 \\
0 & .0640 & .6732 & .2627 & 0 & 0 \\
0 & .1775 & .4143 & .4082 & 0 & 0 \\
.2001 & .0801 & .1307 & .1197 & .2956 & .1738 \\
.1874 & .0696 & .1121 & .1036 & .1738 & .3535
\end{bmatrix}.$

4.5. 18.   **4.7.** .0713.   **4.9.** .0308.

$$
\textbf{4.11. } M(.20) = \begin{bmatrix}
.1167 & .0318 & .0299 & .0134 & .0083 \\
.0400 & .0834 & .0328 & .0307 & .0131 \\
.0371 & .0318 & .0760 & .0291 & .0260 \\
.0184 & .0319 & .0336 & .0864 & .0297 \\
.0132 & .0164 & .0338 & .0358 & .1007
\end{bmatrix}.
$$

$$
\textbf{4.13. } M(.10) = \begin{bmatrix}
.0432 & .0140 & .0221 & .0207 & 0 & 0 \\
0 & .0572 & .0221 & .0207 & 0 & 0 \\
0 & .0028 & .0801 & .0171 & 0 & 0 \\
0 & .0134 & .0256 & .0609 & 0 & 0 \\
.0165 & .0035 & .0051 & .0051 & .0561 & .0136 \\
.0143 & .0030 & .0042 & .0043 & .0136 & .0607
\end{bmatrix}.
$$

4.15. 3.95 minutes.   **4.17.** 9.36 hours.   **4.19.** .5289 hours.

4.21. $p = [.2528, .1981, .2064, .1858, .1569]$.

4.23. $p = [.3777, .1863, .0916, .0790, .0939, .0466, .0517, .0732]$.

4.25. .5232.   **4.27.** 4.882.   **4.29.** .4068.

4.31. 65.7519.   **4.33.** 1233.6.   **4.35.** 187.96.

4.37. $25.72.   **4.39.** $28.83.   **4.41.** Yes.

4.43. $223.73 per day.   **4.45.** .7425.   **4.47.** .2944.

4.49. 8.   **4.51.** 3.85 years.   **4.53.** $96,486.35.

4.55. $100,778.13.   **4.57.** 57.

Chapter 5

CONCEPTUAL PROBLEMS

5.7. $\lim_{t \to \infty}(Y(t)/t) = \lim_{t \to \infty}(Z(t)/t) = 1/\tau$.

5.9. No. The batch sizes depend upon the inter-arrival times.

5.11. $\dfrac{\tau_i v_i}{1 + \sum_{j=1}^{N} \tau_j v_j}$.   **5.13.** $\dfrac{1 + \sum_{j=1}^{N} \tau_j v_j}{v_i}$.   **5.15.** $\dfrac{\tau}{\tau + \sum_{i=1}^{N} \frac{1}{iv}}$.

5.17. $P = \begin{bmatrix} 0 & 1 \\ 1 & 0 \end{bmatrix}$, $w_1 = \sum_{i=1}^{N}(1/iv)$, $w_0 = \tau$.

5.19. $P = \begin{bmatrix} 1 - A_1(T) & A_1(T) \\ A_2(T) & 1 - A_2(T) \end{bmatrix}$, $w_i =$ the expected lifetime of component i, $i = 1, 2$.

5.21. $P = \begin{bmatrix} .366 & .573 & .061 \\ .3025 & .1512 & .5463 \\ .6 & .3 & .1 \end{bmatrix}$, $w = [5.5, 4.4795, 6]$.

COMPUTATIONAL PROBLEMS

5.1. $\frac{1}{3}$. **5.3.** .1818. **5.5.** .019596.

5.7. 0.2597. **5.9.** Planned: \$50.80 per year; Unplanned: \$50.00 per year.

5.11. $\geq$ \$24 per day. **5.13.** Contr. \$25.66 per year; No Contr. \$26.35 per year;

5.15. Current: \$376.67 per year; New: \$410.91 per year. **5.17.** \$5.43 per day.

5.19. .5355. **5.21.** \$16,800. **5.23.** 255.15.

5.25. .8392. **5.27.** .0629. **5.29.** .16.

5.31. \$31.20 per day. **5.33.** 0.8125. **5.35.** 0.5185.

5.37. \$533.33 per day. **5.39.** \$53.37 per day. **5.41.** 1.4274 fac./year.

5.43. 0.9136 fac./year. **5.45.** .7646. **5.47.** \$7146.

Chapter 6

CONCEPTUAL PROBLEMS

6.1. *Case* 1. $y \leq x$: $\pi_0 = \pi_0^* = \hat{\pi}_0 = 1$, $\pi_j = \pi_j^* = \hat{\pi}_j = 0$, $j \geq 1$. p_j's do not exist. Thus PASTA does not hold. *Case* 2. $y > x$: The queue is unstable.

6.11. $\lambda_i = (K - i)\lambda$, $0 \leq i \leq K$, $\mu_i = \min(i, s)\mu$, $0 \leq i \leq K$.

6.19. Yes. **6.23.** $\lambda < \min\left(K\theta, \dfrac{\alpha}{p}\right)$.

COMPUTATIONAL PROBLEMS

6.1. 7. **6.3.** 9.

6.5. L and L_q remain unchanged. W and W_q are halved.

6.7. \$46.4544 per hour. **6.9.** \$.7837 per hour.

6.11. .0311. **6.13.** 13 additional lines. Old: 1.48 minutes; New: 7.74 minutes.

6.15. 4.   **6.17.** $209.86 per hour.   **6.19.** .25.

6.21. 5.5556 minutes. 9.   **6.23.** 6.   **6.25.** $\geq$ $15.89.

6.27. 8.26 minutes.   **6.29.** .1122.   **6.31.** 5 to 5.5511.

6.33. 3.6250.   **6.35.** 14.3721.   **6.37.** 2.1540.

6.39. (1) .3 hours; (2) .1201 hours.   **6.41.** 3 per station.

6.43. < 640 per hours.   **6.45.** (1) Yes. (2) 4.4872. (3) 5.62 minutes.

6.47. P(62.6874), .2947.   **6.49.** $ 66.66 hour.

Chapter 7

CONCEPTUAL PROBLEMS

7.1. μ.

7.3. Reduce the integral to a Gamma integral.

7.5. We have

$$P(|X| \leq x) = P(-x \leq X \leq x) = P(-x/\sigma \leq X/\sigma \leq x/\sigma)$$

$$= \Phi(x/\sigma) - \Phi(-x/\sigma) = 2\Phi(x/\sigma) - 1, \quad x \geq 0.$$

We get the desired result by taking the derivative of the right-hand side with respect to x.

7.7. We have

$$E(e^{sX}) = \sqrt{\frac{1}{2\pi\sigma^2}} \int_{-\infty}^{\infty} e^{sx} \exp\left(\frac{-(x-\mu)^2}{2\sigma^2}\right) dx$$

$$= \sqrt{\frac{1}{2\pi\sigma^2}} \int_{-\infty}^{\infty} \exp\left(\frac{-(x-\mu-s\sigma^2)^2}{2\sigma^2} + s\mu + s^2\sigma^2/2\right) dx$$

$$= \exp\left\{s\mu + \frac{1}{2}s^2\sigma^2\right\} \sqrt{\frac{1}{2\pi\sigma^2}} \int_{-\infty}^{\infty} \exp\left(\frac{-(x-\mu-s\sigma^2)^2}{2\sigma^2}\right) dx.$$

The integral on the last line equals 1. Hence the result follows.

7.9. Since the variance of a random variable is nonnegative, we see from the result in Conceptual Problem 7.8 that

$$a \Sigma a^\top \geq 0 \quad \text{for any } a \in R^n.$$

This implies that Σ is positive semi-definite.

7.13. $E(T(b)) = b/\mu.$

7.15. The optimal x satisfies

$$\exp(\theta a) - \exp(\theta b) + (a + b - 2x)\theta \exp(\theta x) = 0.$$

7.17. Use the identity $\max(K - V(T), 0) - \max(V(T) - K, 0) = K - V(T)$, and the symmetry of the normal cdf Φ.

7.19. Argue that $P(L > a) = \lim_{b \to \infty} P(X(T(a,b)) = b).$

COMPUTATIONAL PROBLEMS

7.1 .3085.	**7.3** .2790.	**7.5** N(4,13).
7.7 N(−1,6.5).	**7.9** N(0,16).	**7.11** .2119.
7.13 .5.	**7.15** .3644.	**7.17** .4292.
7.19 8.75.	**7.21** .9765.	**7.23** .7395.
7.25 .82.	**7.27** 6.37.	**7.29** .7454.
7.31 1.5.	**7.33** $360,068.52.	**7.35** $39,802.
7.36 $2.219.	**7.39** $\sigma = 1.144$.	**7.41** $1.948.
7.45 $\sigma = .156$.		

Further Reading

Probability

1. W. Feller, *An Introduction to Probability Theory and Its Applications*, Vol. 1, 2nd ed., Wiley, New York, 1959.
2. M. F. Neuts, *Probability*, Allyn and Bacon, Boston, 1973.
3. E. Parzen, *Modern Probability Theory and Its Applications*, Wiley, New York, 1960.
4. S. M. Ross, *A First Course in Probability*, 3rd Ed., Macmillan, New York, 1988.

Stochastic Processes

5. E. P. C. Kao, *An Introduction to Stochastic Processes*, Duxbury Press, London 1997.
6. S. M. Ross, *Introduction to Probability Models*, Academic Press, London 1993.
7. H. M. Taylor and S. Karlin, *An Introduction to Stochastic Modeling*, Academic Press, London 1984.
8. K. S. Trivedi, *Probability and Statistics with Reliability, Queueing and Computer Science Applications*, Prentice-Hall, Englewood Ciffs, NJ, 1982.

Advanced Textbooks

9. E. Cinlar, *Introduction to Stochastic Processes*, Prentice-Hall, Englewood Ciffs, NJ, 1975.
10. D. P. Heyman and M. J. Sobel, *Stochastic Models in Operations Research*, Vol. 1, McGraw-Hill, New York, 1982.
11. S. Karlin and H. M. Taylor, *A First Course in Stochastic Processes*, Academic Press, London 1975.
12. V. G. Kulkarni, *Modeling and Analysis of Stochastic Systems*, CRC Press, London 2010.
13. S. M. Ross, *Stochastic Processes*, Wiley, New York 1983.
14. S. I. Resnick, *Adventures in Stochastic Processes*, Birkhauser, Boston, 1992.

Books on Special Topics

15. D. Bertsekas and R. Gallager, *Data Networks*, Prentice-Hall, Englewood Cliffs, NJ, 1992.
16. Bickel and Docksum, *Mathematical Statistics:Besic Idens and selected Topics;* Secord Edition,Vol 1, Upper saddic River, NJ, prentice wood.
17. J. A. Buzzacott and J. G. Shanthikumar, *Stochastic Models of Manufacturing Systems*, Prentice-Hall, Englewood Cliffs, NJ, 1993.
18. D. Gross and C. M. Harris, *Fundamentals of Queueing Theory*, Wiley, New York 1985.
19. D. Kennedy, *Stochastic Financial Models*, CRC Press, Boca Raton, FL, 2010.

Case Studies Papers

20. P. E. Pfeifer and R. L. Carraway, "Modeling Customer Relations as Markov Chains," Interactive Marketing 14(2):43–55(2000).
21. Y. Amihud and H. Mendelson, "Dealership Market: Market Making with Inventory," Journal of Financial Economics 8:31–53(1980).
22. E. P. C Kao, "Modeling the Movement of coronary Patients within a Hospital by Semi-Markov Processes," Operations Research 22(4):683–699 (1974).

Index

LaVergne, TN USA
08 January 2011
211617LV00002B/5/P